ADVANCES IN ENZYMOLOGY

AND RELATED AREAS OF MOLECULAR BIOLOGY

Volume 36

CONTRIBUTORS TO VOLUME 36

ERNEST BOREK, *Department of Microbiology, University of Colorado Medical Center, Denver, Colorado*

PEDRO CUATRECASAS, *Division of Clinical Pharmacology, The Johns Hopkins University School of Medicine, Baltimore, Maryland*

P. M. DEY, *Department of Biochemistry, Royal Holloway College (University of London), Englefield Green, Surrey, England*

VICTOR GINSBURG, *Department of Health, Education and Welfare, National Institutes of Health, Public Health Service, Bethesda, Maryland*

FREDERICK GRINNELL, *Biochemistry Research Section, Veterans Administration Hospital, Dallas, Texas*

SYLVIA J. KERR, *Department of Surgery, University of Colorado Medical Center, Denver, Colorado*

G. A. LEVVY, *Enzymology Department, Rowett Research Institute, Bucksburn, Aberdeen, Scotland*

JONATHAN S. NISHIMURA, *Department of Biochemistry, The University of Texas Medical School, San Antonio, Texas*

DAVID J. PRESCOTT, *Bryn Mawr College, Department of Biology, Bryn Mawr, Pennsylvania*

J. B. PRIDHAM, *Department of Biochemistry, Royal Holloway College (University of London), Englefield Green, Surrey, England*

SYBIL M. SNAITH, *Enzymology Department, Rowett Research Institute, Bucksburn, Aberdeen, Scotland*

CELIA WHITE TABOR, *Laboratory of Biochemical Pharmacology, National Institutes of Health, National Institute of Arthritis and Metabolic Diseases, Bethesda, Maryland*

HERBERT TABOR, *Laboratory of Biochemical Pharmacology, National Institutes of Health, National Institute of Arthritis and Metabolic Diseases, Bethesda, Maryland*

P. ROY VAGELOS, *Department of Biological Chemistry, Washington University School of Medicine, St. Louis, Missouri*

ADVANCES IN ENZYMOLOGY

AND RELATED AREAS OF MOLECULAR BIOLOGY

Founded by E. F. NORD

Edited by ALTON MEISTER

CORNELL UNIVERSITY MEDICAL COLLEGE, NEW YORK

VOLUME 36

1972
INTERSCIENCE PUBLISHERS
a Division of JOHN WILEY & SONS
New York · London · Sydney · Toronto

CONTENTS

ADVANCES IN ENZYMOLOGY

AND RELATED AREAS OF MOLECULAR BIOLOGY

Volume 36

THE tRNA METHYLTRANSFERASES

By SYLVIA J. KERR AND ERNEST BOREK, *Denver, Colorado*

CONTENTS

I. Introduction

The transfer RNA methyltransferases, a family of enzymes that methylate transfer RNA at the macromolecular level, have been adequately reviewed up to 1967 (1,2,3). Therefore, our review will concentrate on publications and information that have emerged since then.

After the discovery of methyl-deficient tRNA by the finding that synthesis of the polynucleotide chain and methylation are sequential reactions and these reactions can be separated in a "relaxed" organism of *Escherichia coli*, several groups of investigators studied the behavior of methyl-deficient transfer RNA in certain *in vitro* functional tests. Essentially, no differences were found between methyl-deficient transfer

1

RNA and the species normally endowed with the appropriate methyl groups. The reason for this is at least twofold. The first one is obvious: the methyl-deficient transfer RNA, as it was available then, consisted of an almost equal mixture of methyl-free and completely methylated RNAs. This is inherent in the method of synthesis of methyl-deficient tRNA. Prior to starvation of methionine, the organisms must be grown to a high population in presence of the amino acid; during growth they synthesize tRNA, which is normally methylated and remains stable during the starvation.

The second reason for the failure to detect any changes, especially in the transfer function to the ribosomes, is that methylation is not the only modification of transfer RNA. More recent work has resolved some of these problems. In the first place, apparently completely methyl-free tRNA is available from the very skillful large scale separations performed at Oak Ridge Laboratories. Novelli and his co-workers (4) were able to show that in methyl-free transfer RNA acceptance of amino acids is at a considerably lower rate than that in normal transfer RNA.

The modified bases involved in the recognition of the interaction between activating enzymes and tRNA have not been identified at the present time. However, in the case of the transfer function to ribosomes, there is a well-defined, absolute requirement. In all transfer RNAs except $tRNA^{fMet}$, the base adjacent to the anticodon at the $3'$ position is modified. It has been shown in three different cases (5,6,7) that if the modification of the base next to the anticodon is lacking, then the tRNA, while it still accepts its appropriate amino acid, will no longer be able to transfer it to the ribosomes. Gefter and Russell (5) studied the suppressor $tRNA^{Tyr}$ produced in *E. coli* by the transducing phage derived from phage $\phi80$. This tRNA has an adenine next to its anticodon that is triply modified. It contains an isopentenyl group in the N6 position and a methylthio group in the 2 position (at least three different enzymes must achieve these modifications.) After infection, the extent of these modifications can be controlled by cultural conditions. If all of the modifications are lacking, the inability to attach to ribosomes is absolute. If it is partially modified, the ribosomal attachment is only partially suppressed. Thiebe and Zachau (6) have studied $tRNA^{Phe}$ yeast which has the modified base Y next to its anticodon. If this base is excised by exposure to mild acid, the ability of the tRNA for ribosomal attachment disappears.

The effect of the removal of the modification of still another anti-codon-neighboring base is one studied by Fittler and Hall (7). They removed the isopentenyl moiety from tRNA by oxidation with permanganate. The tRNA was still functional in the amino acid charging reaction, but the ribosomal attachment function was eliminated.

It is obvious from these examples that any *in vitro* study of methyl-deficient tRNA in its ribosomal transfer function would be meaningless if the modification of the anticodon-neighboring base were other than methylation.

It has also been shown in Peterkofsky's laboratory (8) that with methyl-deficient tRNA the codon response is different from the codon response of the normal species. Therefore, at least three functions of the modifications of tRNA have emerged in the recent past. The answer to what other functions they may have awaits the availability of transfer RNAs with specific modifications missing from known positions. However, even with such products, great care will have to be taken in the interpretation of findings. Should any of these modifications have a role in regulation, then *in vitro* assays may fail to reveal such a mechanism. The above observations provide partial confirmation for the hypothesis proposed some years ago that the modifications of tRNA are required for protein synthesis (9).

Two model tRNAs may be available for studying the role of specific modifications on the structure and function of tRNA. The first of these would be tRNA^{Ala} transcribed from the gene synthesized by Khorana and his co-workers (10).

Another such model, a biological one, turned up unexpectedly in Strominger's hands during his monumental studies on the mechanism of cell wall synthesis. Among the populations of tRNAs extracted from *Staphyloccoccus epidermis*, Strominger and his group have isolated a tRNA^{Gly} that can be charged by the mixture of charging enzymes in this organism and participates in *in vitro* peptidoglycan synthesis but is totally inert in *in vitro* protein synthesis (11,12). Analysis of this purified tRNA^{Gly} revealed that it contains but one modified base, 4-thiouridine (12). The whole coterie of the other modified bases, dihydrouridine, pseudouridine, and all of the methylated bases are absent from its structure. Physicochemical studies of this substrate should yield highly valuable information on the role of modifications in determining the secondary and tertiary structure of tRNA.

Another intriguing question that may be answered by the structure of this unique tRNA is how it resists modification by the variety of enzymes. Strominger suggests that perhaps it is an adventitious placing of the thio group in the chain which prevents further modification. This is a plausible hypothesis, for some of the modifications of tRNAs are known to be incorporated into tRNA before its methylation (13). Another example of sequential modification is provided from work in Brenner's laboratory. The su tRNA^{Tyr} discussed above has three modifications on the adenine residue adjacent to the anticodon. Three different enzymes are needed for the modification of this base, and they apparently react in the following order. One enzyme introduces the isopentenyl group; another enzyme introduces the sulfur group; and finally still another enzyme methylates the thio group (14). All other modifications of this tRNA molecule apparently occur prior to the modifications of the base adjacent to the anticodon.

A structural modification that methylation is known to confer on tRNA is a hypochromic effect, which implies an increment in secondary and/or tertiary structuring. This has been shown both by enzymatic methylation of methyl-deficient tRNA from $E.$ $coli$ by its homologous enzymes (15), as well as by chemical methylation (16).*

At any rate, one might safely retract now what was said a few years ago, that the "tRNA methyltransferases are enzymes in search of a function."

II. Genetics

Very interesting studies on the genetic origins of the tRNA methyltransferases have come from the groups studying these enzymes at Uppsala. Kjellin-Straby and Boman (18) observed that a methionine auxotroph of $Saccharomyces$ $cerevisiae$ accumulates partially methyldeficient RNA during methionine deprivation. In an extension of these observations Kjellin-Straby and Phillips studied a number of other methionine auxotrophs of yeast. Among these they found a mutant which lacks in its tRNA N^2-dimethyl guanine (19). That this deficiency is due to the lack of the appropriate enzyme, rather than to some change in the tRNA sequence, was unequivocally demonstrated

---

* For an outstanding review of the tertiary structure of tRNA read Cramer(17). Structural interrelationships that are baffling on the basis of the two-dimensional cloverleaf model become obvious when considered on the basis of Cramer's model of the intricate tertiary structure of tRNA.

by these authors by introducing *in vitro* into the tRNA from these mutants a dimethyl group at the N^2 position of guanine by enzymes extracted from the wild-type organism. These observations are very important, for they highlight the great genetic specificity in the formation of the methylated bases in tRNA. There are two separate genes for the synthesis of two enzymes that monomethylate guanine and dimethylate it.

For a study of *in vivo* function of tRNA methylation, Bjork and Isaksson (20) have undertaken the formidable task of isolating mutants of *E. coli* defective in some methyltransferases. Their method of screening was based on the assumption that if a tRNA methyltransferase disappears from a clone, the tRNA extracted from those organisms should be methyl-deficient with respect to the battery of enzymes present in wild-type *E. coli*. They exposed *E. coli* to extensive mutagenesis by ethylmethane sulfonate, and isolated survivors and analyzed the clones as indicated above. From a screening of 3000 clones, they were able to isolate six mutants that could accept methyl groups *in vitro* into uracil in tRNA. Therefore, these mutants presumably lack uracil methyltransferase. The fact that no other base specific methyltransferase mutants could be detected is significant: one must assume that those mutations are lethal.

Bjork and Neidhardt (21) have submitted these mutants and their isogenic pairs to very intensive study; they have observed that the absence of 5-methyluridine (ribothymidine) from tRNA can decrease the growth rate of *E. coli* cells reproducibly and to a significant extent. These conclusions came from studies of growth rates at different temperatures, 43 and 37°, and in rich and minimal media. That the absence of the uracil methylase is a disadvantage to the mutant was also demonstrated by competitive growth experiments in the same medium. The authors mixed a population of *E. coli* with the normal enzyme capacity, designated as Trm+, with a population of the organisms lacking the enzyme, designated as Trm−. The initial ratio of the two species of *E. coli* was 13 and 87%, respectively. After approximately 22 doublings, the culture was cloned and the individual clones were assayed for the presence of undermethylated tRNA. This assay revealed that only 6% of the population was Trm−; therefore, the Trm− were at a distinct disadvantage during growth in the presence of the wild type of organism. These very interesting studies will, of course, have to be extended. For example, the authors point out that

E. coli is known to have the ability to adjust several parts of its protein-synthesizing machinery, for example, the number of ribosomes, to compensate for the lack of some optimal conditions. Nor is it known whether the Trm⁻ mutants that have been isolated may not have had some other mutations in the rest of the tRNA molecules to compensate for the loss of the methyl group in the uracil. We must await sequence analysis of pure tRNAs isolated from such mutants. The outcome of these experiments will be awaited with particular interest because so far every tRNA examined, except some from mycoplasma, contains the 5-methyluridine in position 23 from the 3'-OH end. The availability of tRNA lacking this modification may be a useful tool in elucidating the function of this base in the tRNAs of most organisms. It should be noted that preliminary studies *in vitro* by Bjork and Neidhardt revealed no anomalies of the tRNA extracted from the Trm⁻ organism.

As we mentioned above, the tRNA of mycoplasma is very low in ribothymidine (22). The level of this minor component is less than 1/80 nucleotides. Therefore, one could expect some tRNAs to be devoid of this modification. This prediction has been born out in Söll's laboratory, where tRNA^{Ile} from a strain of mycoplasma was analyzed and no ribothymidine was found (23). This rules out the need for this base as a site of recognition for charging or for ribosomal binding. However, as the authors point out, "the presence of ribothymidine instead of uridine in tRNA may possibly alter the kinetic parameters of aminoacylation or polypeptide formation. Much more subtle experiments would be required to test such possibilities." Of course, it is also possible that the function of ribothymidine, if any, in the tRNA of most other species, is dispensed with or not yet acquired in tRNA^{Ile} of that strain of mycoplasma.

III. Interaction of S-Adenosylmethionine and Methyltransferases

Several groups have studied the specificity of S-adenosylmethionine and its derivatives in the methyltransferase reactions for which this compound serves as the methyl donor. The methylation of tRNA is in this category. Various manipulations in the structure of the molecule have been made in order to probe the exact three-dimensional requirements for interaction with the methyltransferases.

If the sulfur atom is replaced by selenium (Se-adenosyl-seleno-methionine), enzymatic methylation can still take place (24).

The sulfonium ion is also a center of stereoisomerism and, as in most biological reactions, only one of the stereoisomers is preferred by the enzymes (25).

It has been observed that L-ethionine is activated by the methionine-activating enzymes of rat liver (26) and yeast (27) to form S-adenosyl-ethionine. In intact animals, upon administration of ethionine, ethyl groups are incorporated into tRNA and possibly to a very slight extent into DNA (26). It was presumed that S-adenosylethionine was the active intermediate. However, the work of Ortwerth and Novelli (28) casts serious doubt on this conclusion. After injection of rats with L-ethionine-ethyl-1-^{14}C, they isolated the DNA, ribosomal RNA, and transfer RNA. They found negligible labeling in the purified DNA and ribosomal RNA; essentially all the labeled ethyl groups were in the transfer RNA. They also concluded that S-adenosylethionine was not the active intermediate in the ethylation of tRNA on the basis of the following considerations: the rate of tRNA ethylation was not related to the metabolic pool size of S-adenosylethionine; that is, while the S-adenosylethionine content of the liver remained constant for 3 to 24 hr after injection, tRNA ethylation reached a maximum at 3 hr and had ceased by 24 hr. It appeared that ethylation of tRNA *in vivo* was dependent on the presence of free ethionine in the liver rather than S-adenosylethionine. Also, coinjection of levels of methionine that inhibit the formation of S-adenosylethionine actually stimulated the transfer of labeled ethionine into tRNA. Even when the levels of methionine were raised so as to completely block the formation of S-adenosylethionine, the ethylation of tRNA was not completely inhibited. Finally, an *in vitro* assay of the tRNA methylating enzymes using S-adenosylethionine did not result in any ethylation of tRNA. Thus, while some methyltransferases may utilize S-adenosylethionine (26), it is not a functional substrate for the transfer RNA methylating enzymes.

By far the most extensive and precise dissection of the S-adenosyl-methionine molecule has been carried out in the laboratory of Schlenk (29). This group removed the two amino groups and the carboxyl group of the molecule singly and in combination, and tested the resulting compounds in *in vitro* enzyme assays, both as methyl donors and as inhibitors. Figure 1 shows the compounds they synthesized and tested on three different purified methyltransferase enzymes: histamine methyltransferase, acetylserotonin methyltransferase, and homocysteine methyltransferase. These three enzymes involve methyl transfer to nitrogen, oxygen, and sulfur atoms, respectively. Compound I in Figure 1 is S-adenosylmethionine.

The 2-amino group of the methionine proved to be an absolute requirement. The deamination product, S-adenosyl-L(2-hydroxy-4-methylthio)butyric acid (II), was inactive either as a methyl donor or inhibitor in all three enzyme systems. The doubly deaminated compound, S-inosyl-L-(2-hydroxy-4-methylthio)butyric acid (III), was, of course, also inactive. When only the adenosine moiety was deaminated, the product, S-inosylmethionine (IV), was ineffective as a methyl donor in the histamine methyltransferase and acetylserotonin methyltransferase reactions but worked perfectly well in the homocysteine methyltransferase system. The latter result is not too surprising, as this particular enzyme system will also utilize S-methyl-L-methionine which has no adenosine moiety, as a methyl donor.

Compound V, S-adenosyl-(5′)-3-methylthiopropylamine, the decarboxylated form that occurs naturally as an intermediate in the synthesis of polyamines, was also inactive in the histamine and acetylserotonin methyltransferase systems but active in homocysteine methyltransferase reaction.

On the basis of these results, Schlenk and co-workers proposed a general model for the binding of S-adenosylmethionine to methyltransferase enzymes that involves three to four recognition sites as shown in Figure 2. The presence of the amino group on the methionine moiety appears to be obligatory. The modified compounds have not

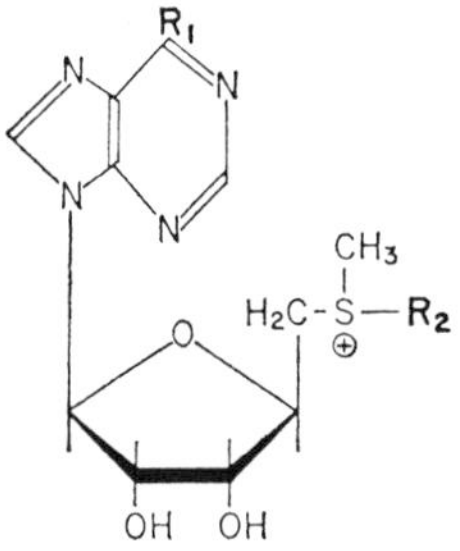

R_1	R_2	
$-NH_2$	$-CH_2CH_2CH(NH_2)COOH$	I
$-NH_2$	$-CH_2CH_2CH(OH)COOH$	II
$-OH$	$-CH_2CH_2CH(OH)COOH$	III
$-OH$	$-CH_2CH_2CH(NH_2)COOH$	IV
$-NH_2$	$-CH_2CH_2CH_2NH_2$	V

Fig. 1. S-Adenosylmethionine (I) and some of its derivatives. From Zappia et al. (29).

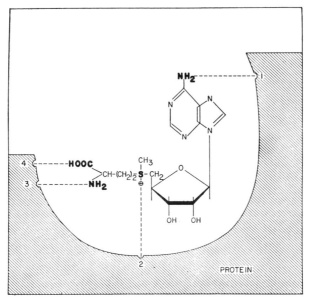

Fig. 2. Proposed binding sites for S-adenosylmethionine in methyl transfer enzymes. From Zappia et al. (29).

as yet been tested in any transfer RNA methylating systems; it will be interesting to study whether the tRNA methyltransferases conform to the general principles delineated above.

IV. Recognition Sites of the tRNA Methyltransferases

The tRNA methyltransferases and their substrates offer a unique opportunity for the study of the interaction of macromolecules with their modifying enzymes. This family of enzymes is base specific, species specific, and organ specific (30,31); their synthesis as shown above is under rigid genetic control. The determination of the primary sequence of several tRNAs revealed an extreme site specificity of the enzymes. A uridine in the 23rd position from the CCA terminal is always methylated. The question of the recognition signals for these remarkably specific interactions was therefore posed.

Baguley and Staehelin brought an ingenious approach to these studies (32). They noted that yeast tRNA contains but 0.9% 1-methyladenine; therefore, only 70% of the individual tRNAs can contain 1-methyladenine. Thus, as expected from the species specificity of the enzymes, and more particularly from the above analytical data the tRNA of yeast should serve as a substrate for methylation by the mammalian enzyme.

1-Adenine methyltransferase from the 100,000-g supernatant of rat-liver homogenate was purified by sequential chromatography on DEAE-cellulose and on G-200 Sephadex, followed by precipitation with $(NH_4)_2SO_4$. The enzyme was allowed to methylate purified yeast tRNA with ^{14}C-methyl-S-adenosylmethionine.

Oligonucleotide fragments of the tRNAs were prepared by digestion with pancreatic RNase, followed by chromatography. Sequence analysis of the oligonucleotides was performed by subjecting the tRNA to the usual battery of nucleases—T_1 RNase, micrococcal nuclease, E. coli alkaline phosphatase, and snake venom phosphodiesterase—and then separating the products by two-dimensional, thin layer chromatography and thin layer electrophoresis. As can be seen from the data in Table I, the trinucleotide and tetranucleotide sequences containing 1-methyladenine were identical in rat-liver tRNA and in yeast tRNA which had been methylated *in vitro* with enzyme from liver. *Therefore, the enzyme from the liver methylates sites in vitro that the yeast enzyme omitted in vivo.* The recognition system, consequently, must be other than the short

TABLE I

Approximate Contents of Methyladenine-containing Sequences in Liver and Yeast tRNA[a]

Sequence	Sequences/molecule		Percent radioactivity *in vitro* methylation
	Rat-liver tRNA	Yeast tRNA	
Total trinucleotide sequences	0.35	0.30	47
GpMeApUp	0.20	0.20	35
ApMeApUp	0.05	0.05	4
GpMeApCp	0.10	0.05	8
Total tetranucleotide sequences	0.55	0.25	53
GpMeApApUp	0.30	0.10	23
ApMeApApUp	0.20	0.10	20
GpMeApApCp	0.05	0.05	10

[a] From Baguley and Staehelin (32).

sequence of bases around the target adenine. What this system of recognition is, is conjectural at present. It may be a longer sequence or it may be the tertiary structure of the tRNA. It would be informative to repeat these interesting studies on purified specific tRNAs from the two sources.

The need for studying the methyltransferases with purified, specific tRNA substrates is apparent from the beautiful work in Nishimura's laboratory. He and his co-workers studied the guanine tRNA methyltransferase from rat liver and its interaction with several, pure, specific tRNAs (33). They fractionated the enzymes on a column of hydroxylapatite and precipitated various enzyme fractions with 60% saturated ammonium sulphate. This purification procedure apparently stabilizes the tRNA methyltransferases from mammalian sources, which in itself is a significant advance in the technology of these enzymes. They obtained three different fractions of methyltransferases, designated I, II, and III. All of the fractions methylated guanine residues. It is evident from the data in Table II that these different enzymes

TABLE II

Methyl Acceptor Capacities of Individual *E. coli* tRNAs with Liver Methylases *in vitro*[a]

tRNA	Extent of methylation		
	Methylase I	Methylase II	Methylase III
tRNA^{Asp}	210	290	150
tRNA$_2^{\text{Glu}}$	440	530	140
tRNA$_2^{\text{Leu}}$	1790	2340	170
tRNA^{fMet}	4580	4110	210
tRNA$_1^{\text{Met}}$	74	110	510
tRNA^{Phe}	33	240	770
tRNA$_1^{\text{Ser}}$	32	220	520
tRNA$_3^{\text{Ser}}$	2380	3270	110
tRNA$_2^{\text{Tyr}}$	260	480	210
tRNA$_1^{\text{Val}}$	64	350	710
E. coli tRNA	1440	1280	460
Methyl-deficient *E. coli* tRNA	1730	1600	540
Rat-liver tRNA	45	84	34
Yeast tRNA	230	240	70

[a] From Kuchino and Nishimura (33).

react selectively with the various tRNAs offered as substrates. Thus methyltransferase I is essentially inert toward $tRNA_1^{met}$, $tRNA^{Ser}$, $tRNA^{Phe}$, and $tRNA_1^{Val}$. To characterize the bases methylated, large-scale methylations were carried out on the specific tRNAs and the methylated bases were isolated. Only guanine was methylated, both with methyltransferase II and with methyltransferase III. Surprisingly, methyltransferase II produced both N^2-methylguanine and N^2,N^2-dimethylguanine. This finding is in direct contrast with the earlier mentioned genetic findings on these two enzymes in yeast. This discrepancy may be due, of course, to some species variation, but perhaps more likely, it is due to the incomplete separation of the methyltransferases under fraction II.

Nishimura and his colleague also determined the position of the methylguanine introduced by methyltransferase II into $tRNA^{fMet}$ and introduced into $tRNA_1^{Val}$ by methyltransferase III. It is apparent from Figure 3 that the two different guanine methyltransferases introduced their methyl groups into different positions in the two tRNAs. Therefore, we must conclude that the methyltransferases are

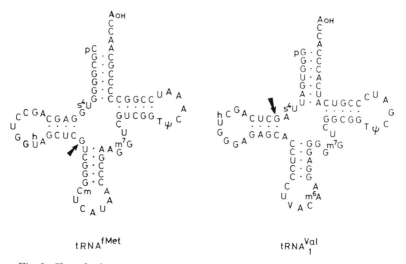

Fig. 3. Cloverleaf structure of *E. coli* tRNAs, indicating the site of methylation in tRNA^{fMet} by methyltransferase II and in tRNA$_1^{Val}$ by methyltransferase III. Enzyme II did not accept valyl tRNA as a substrate, while enzyme III would not act on formylmethionyl tRNA. V stands for uridin-5-oxyacetic acid. From Kuchino and Nishimura (33).

both tRNA specific and base and position specific. This is a highly welcome demonstration, for it must clarify in the minds of investigators that simply because guanine is methylated by a complex of enzymes in a complex of tRNAs, the enzyme involved is not necessarily the same. Many faulty conclusions have been drawn on the identity of the enzymes from tissues from different sources because of lack of understanding of the above (34).

Another line of evidence on the great specificity of the enzymes comes from studies of Svensson, Bjork, and Lundahl (35,36). These investigators fractionated enzymes from yeast and obtained three different fractions, all of which methylated uracil exclusively. The substrate was the mixture of methyl-deficient tRNAs obtained from *E. coli* K12W6 after methionine deprivation. With limiting amounts of substrate tRNA they obtained entirely different extents of methylation with the three separate enzymes. They obtained 3- to 10-fold differences in total capacity of incorporation, again emphasizing large differences in substrate specificity of a base-specific enzyme. Chromatography of T_1 RNase digests of the enzymatically methylated tRNA also revealed differences in sequences methylated by the three enzyme fractions.

V. Ontogenetic and Organ Specificity of the tRNA Methyltransferases

Hancock et al. observed that crude extracts of tRNA methyltransferases in embryonic liver have a capacity to introduce methyl groups into a heterologous substrate several fold higher than extracts of adult organs (37). Similar observations were made on extracts of embryonic versus adult brain by Simon et al. (38). These observations have been confirmed in our laboratory (39). That these differences in enzyme capacity may have relevance in the intact organs as well, finds some support in the following observations. As a general rule, the tRNA methyltransferases from a given source are inert toward the tRNA extracted from the homologous organ in the same phase of growth. (The background level of such interactions varies somewhat with the method of extraction of the tRNA.) If, indeed, the tRNA methylases in the embryonic organ have a capacity *in vivo* higher than their counterparts in the adult organs, then the tRNA isolated from the latter should be under-methylated with respect to the enzymes from the embryo and should serve as a receptive substrate for methylation. It was shown in our laboratory that this is, indeed, the case (39).

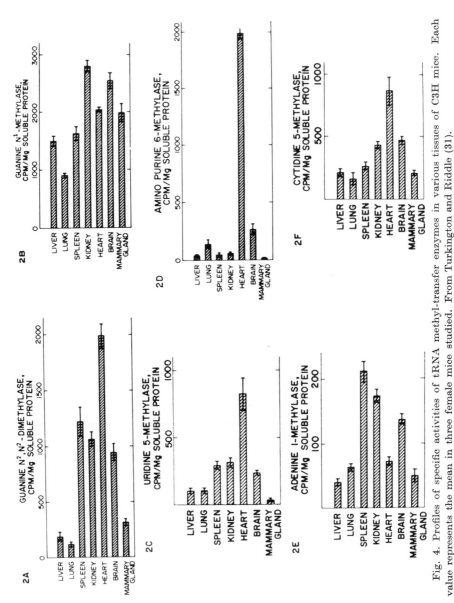

Fig. 4. Profiles of specific activities of tRNA methyl-transfer enzymes in various tissues of C3H mice. Each value represents the mean in three female mice studied. From Turkington and Riddle (31).

14

It might also be added that examination of the population of tRNAs from adult and embryonic tissue by MAK column chromatography reveals differences in elution profiles (40). However, whether these differences stem from variations in the primary sequence or in modifications of the macromolecules remains to be analyzed.

Clear evidence for organ specificity of the tRNA methyltransferases has been presented by Turkington and Riddle (31), who studied the relative distribution of six different enzymes in seven different organs of mice. The large variations in the extracts from the different organs are readily apparent from the data in Figure 4. Even greater organ specific variation was found with respect to the enzyme 7-methylguanine methyltransferase. (See Table III.) This enzyme activity, which is high in liver, is absent from the heart, brain, and mammary gland of normal mice.

TABLE III

Specific Activity of Guanine 7-Methylase in Various Tissues of the C3H Mouse[a]

Tissue	Guanine 7-methylase activity (cpm/mg soluble protein)
Liver	102
Lung	39
Spleen	45
Kidney	51
Heart	0
Brain	0
Mammary gland	0

These results are representative of three such experiments.

[a] From Turkington and Riddle (31).

The biological significance of the organ-specific distribution of this enzyme and its product, 7-methylguanine, is obscure at present. The presence of 7-methylguanine in the tRNA of normal rat (41) and pig liver (42) has been demonstrated earlier.

The distribution of the tRNA methyltransferases and their products, the methylated tRNAs, in tumor tissues is the subject of another review (34).

VI. Modulation of the tRNA Methyltransferases in Various Biological Systems

A. BACTERIOPHAGE INDUCTION AND INFECTION

Variations in the tRNA methyltransferases in biological systems undergoing changes in regulatory controls were first observed by Wainfan (43). She studied the tRNA methyltransferases in lysogenic *E. coli* at various intervals after ultraviolet induction and observed a profound reduction in total enzyme capacity from 10 to 15 min after the induction. Thirty minutes after the induction the enzyme capacity returned to the level of extracts from noninduced organisms. The changes among the various base-specific enzymes were not uniform; the reduction in the capacity of the uracil methyltransferases accounted for most of the diminution of the total methylase capacity. Wainfan traced the mechanism of the diminution of enzyme activity to the appearance of an inhibitor of the enzymes in the induced cells (44). We shall discuss this more appropriately in a later section. Large changes in the tRNA methyltransferases were also observed by Wainfan and her collaborators (43) after infection by T2 bacteriophage. That the changes in the methyltransferases are also manifest in their products, the tRNAs themselves, was shown by Boezi and his co-workers (45).

B. NONONCOGENIC AND ONCOGENIC VIRUS INFECTION

Inhibition of the tRNA methyltransferases in mammalian cells in tissue culture infected by polio virus (46) and by the virus of hoof and mouth disease (47) has been observed by two different groups of investigators. These observations are quite in contrast to the effect of oncogenic viruses on cells in tissue culture.

Kit et al. (48) have observed no change in the methylating enzymes after productive or abortive infection of cells by SV 40. However, in SV-40 transformed cell lines there was an increase of 2- to 4-fold in tRNA methyltransferase capacity.

Gallagher et al. (49) observed a 3- to 6-fold increase in the rate of tRNA methyltransferase activity and a 7-fold increment in the capacity to methylate in a polyoma-transformed rat-embryo cell line.

C. HORMONE-INDUCED MODIFICATION OF THE tRNA METHYL-TRANSFERASES

Baliga et al. (50) have observed large changes in the total capacity of tRNA methyltransferases in an insect during its metamorphosis. This is a very complex sequential biological process, stages of which are known to be under hormonal control. We have explored the possibility of hormonal control over the tRNA methyltransferases in other biological systems as well. In preliminary experiments, it was observed by Christman and Borek (51) that the methyltransferase capacity in extracts of uteri of ovariectomized rats was much lower than that of normal uteri. These preliminary observations were extended and confirmed. Pooled uteri of ovariectomized rats were extracted and the methyltransferase capacity determined at saturation levels; the base-specific enzymes were determined individually in the methylated tRNA product (52). Comparisons were made of two of the enzymes that are most abundant in rat uterus. The N^2-monomethylguanine methyltransferase and the N^2,N^2-dimethylguanine methyltransferase. In the normal uterus, these two enzyme activities are expressed in nearly equal ratios. However, in the ovariectomized animal the ratio of the two enzyme activities is reduced to about 0.5. Administration of physiological levels of estradiol restored both the total enzyme capacity and the ratio of these two base-specific enzymes to normal. The effect of ovariectomy and the administration of estradiol on the tRNA methyltransferase was confirmed in the ovariectomized pig uterus by Sharma and Borek (53). These workers were able to determine that one of the sources of diminution of methyltransferase capacity may be the appearance of an inhibitor of these enzymes. Natural inhibitors of the methyltransferases will be discussed in a later section.

The availability of large amounts of tissue from the pig permitted analysis of the population of tRNAs in the ovariectomized and normal uterus by MAK column chromatography. It was observed that there is a novel tRNA^{Ser} among the population of the tRNAs from ovariectomized uterus. Administration of physiological levels of estradiol for 3 days caused the elimination of this tRNA^{Ser}. Whether this novel tRNA^{Ser} is different in primary sequence or in some alteration of the macromolecular structure remains to be determined.

Very interesting studies were reported by Turkington (54) on the effect of hormones on the tRNA methylase in mammary epithelial cells in tissue culture. The administration of either prolactin or insulin produced a high elevation in the capacity of the methyltransferases extracted from the stimulated cells.

Sheid and his co-workers (55) studied the effects of massive doses of estrogen on a rat. They observed a pronounced elevation in the tRNA methyltransferase capacity in the liver of the treated animals. These findings are contrary to the findings with physiological levels of estradiol reported above. When hormone is administered at physiological levels, its effect on the methyltransferases can be observed only in the target organ, the uterus, not in the liver.

On the other hand, the possibility that the administration of massive doses of estradiol may produce an effect on a nontarget organ is known from the administration of estrogens to roosters. The administration of massive doses of estrogen elicits the synthesis of phosvitin—a vitelline protein—in the liver of the rooster (56). The tRNA methyltransferases of roosters administered large doses of estrogen were investigated in our laboratory. We observed changes both in capacity and base-specific activity in extracts of such rooster liver (57). That the tRNA methyltransferases can be under hormonal influence has been shown in another system by Pillinger et al. (58), who studied the enzymes in the giant bull frog, *Rana catesbeiana*, during thyroxine-induced metamorphosis. Three days after the administration of physiological levels of thyroxine into the environment of the bull frog tadpole, there was a diminution in methyltransferase capacity of extracts of both the liver and tail to one-half of that in the untreated animals. Four days after the administration, there was a beginning of return to normal in the capacity of the enzymes. Eight days after the administration of the hormone the enzyme capacity was almost that of the untreated animals.

It should be fruitful to pursue these preliminary observations with hormones upon the methyltransferase capacity of affected organs in depth. It should be emphasized that effects of hormones upon the tRNA methyltransferases represents a *qualitative* intercession of the hormones upon the protein-synthesizing system in the appropriate organs. The changes in the tRNA methyltransferases are not simply quantitative; that is, there is not just more enzyme per cell or per unit amount of tissue, but rather the specificities of the enzyme are changed. They can recognize more or fewer sites for methylation in the same target substrate. In turn, the tRNAs modified by these altered enzymes are also qualitatively different. There is not just more of a specific tRNA but rather a novel tRNA with a novel structure.

VII. Inhibitors of the tRNA Methyltransferases

A. NATURALLY OCCURRING INHIBITORS

As stated earlier, Wainfan et al. discovered an inhibitor of the tRNA methyltransferases in induced lysogenic *E. coli* (44). The inhibitor appears during a well-defined period after the induction. It is not the result of the irradiation per se, for no such inhibitor could be detected in a nonlysogenic organism after ultraviolet irradiation. The inhibitor was a dialyzable product. Analysis of the tRNAs methylated, in the presence of the inhibitor, revealed that the latter essentially inhibits the uracil tRNA methyltransferases. Unfortunately, the preparation of large amounts of induced lysogenic organisms under well controlled conditions is a very difficult task; therefore, attempts at the isolation of this inhibitor and studies of its nature have not progressed.

More widely distributed inhibitors of the tRNA methyltransferases, however, became available for study. It was observed by Hancock et al. that in neonatal and adult tissues, the tRNA methyltransferases have a much lower capacity than the enzymes extracted from embryonic tissue (37). Kerr and Dische (59) also observed the disappearance of tRNA methyltransferase activity in extracts of developing bovine lens tissue. A hypothesis was proposed by Kerr that the diminution or disappearance of the methyltransferase activity in these systems may be due to the presence of an inhibitor (39). Upon admixture of extracts of embryonic and adult tissues, she observed a diminution of activity expected from the embryonic extracts alone. Even more convincing a demonstration for the presence of an inhibitor in adult extracts was made possible by a fortuitous circumstance. If the pH of a crude $100,000 \times g$ supernatant extract of adult organs is adjusted to 5, there is a precipitate that contains methyltransferase activity. It turned out to be the N^2-methylguanine methyltransferase. When the precipitated enzyme is redissolved to its original volume, it has an approximately 5-fold higher methyltransferase capacity than the original extract, indicating that some inhibitor of the enzyme had been removed by the differential precipitation. If the supernatant fluid from the pH 5 precipitation is added to the redissolved enzyme, the activity of the latter is inhibited.

Purification of the inhibitory activity revealed that it has two obligatory components (60). If the extract is fractionated, on Sephadex G-25, two components are eluted. One of these is a large molecular weight component which is a protein. The second component is of much lower molecular weight. Both components are needed for inhibition of the tRNA methyltransferases. Examination of extracts of foetal tissue revealed that the large molecular weight component of the inhibitory system is absent from them. This component was also absent, or its effects eliminated, from three different tumor tissues examined: Morris hepatoma, Novikoff hepatoma, and Ehrlich ascites cells. The low molecular weight component, however, is present both in embryonic tissue and in the tumor tissues examined. It will be interesting to purify these inhibitors and to study their interaction with purified base-specific tRNA methyltransferases.

Halpern and Smith and their collaborators (61) have also noted the absence of an inhibitor of the tRNA methyltransferases from a Walker 256 carcinoma. This group identified this inhibitor, which is found in normal tissues as nicotinamide (62).

Like many other enzyme systems, the tRNA methyltransferases are subject to competitive product inhibition, in their case S-adenosyl-homocysteine (S-adenosylmethionine + tRNA → methyl-tRNA + S-adenosylhomocysteine). Hurwitz et al. (63) were the first to note this for the tRNA methyltransferases in *E. coli*. Its effectiveness as an inhibitor was varied for the different base-specific methyltransferases. Zappia et al. (29) have extended this observation to some methyltransferases from mammalian sources. They also point out that this could be a significant control mechanism, since the levels of S-adenosyl-homocysteine found in animal tissues approximate that of S-adenosylmethionine (64).

Another observation by Schlenk and his co-workers was that the decarboxylated form of S-adenosylmethionine (Compound V, Fig. 1), which occurs naturally in the pathway of polyamine synthesis, can act as an inhibitor of reactions when S-adenosylmethionine is the methyl donor. Thus the decarboxylating enzyme not only competes with the methyltransferases for S-adenosylmethionine but also abets this competition by inhibiting those enzymes with its product (29).

Other naturally occurring compounds that can inhibit the tRNA methyltransferases *in vitro* include adenosine (63,65), adenine (65,66), and some cytokinins, including kinetin riboside, N^6-($\triangle^2$-isopentenyl)-adenosine, zeatin riboside, and 6-benzylaminopurine riboside (67). Whether these compounds have a similar function *in vivo* remains to be demonstrated.

B. SYNTHETIC INHIBITORS

Certain pharmaceutical products used in cancer chemotherapy have been demonstrated to interfere with tRNA methyltransferases.

Wainfan (65) has shown that tubercidin (7-deazaadenosine) inhibited the enzymes from both bacterial and mammalian sources *in vitro*, while isopentenyladenosine, also an antitumor agent, was active only against enzymes from mammalian sources.

Chloramphenicol (68) and 5-fluorouracil (69) have been shown to cause *in vivo* inhibition of tRNA methylation in *E. coli*.

An interesting observation along the opposite line comes from the work of Moore and Smith (66). Ethionine is a well-known hepatic carcinogen in rodents (70). In the liver, ethionine can be activated to S-adenosylethionine (26). Moore and Smith were able to demonstrate that *in vitro* S-adenosylethionine could inhibit competitively the tRNA methyltransferases of *E. coli* and rat liver, as could ethylthioadenosine and adenine. However, as in the experiments of Ortwerth and Novelli (28) mentioned earlier, they could not demonstrate any *in vitro* ethylation of tRNA using S-adenosylethionine.

When Moore and Smith sought to confirm the inhibitory effect of S-adenosylethionine in *in vivo* experiments, they found instead that in ethionine-fed rats, the methyltransferase activity was twice that of the controls. This hyperactivity appears before evidence of any tumor. A similar observation had been made in mice by Hancock (71).

VIII. Stimulation of the tRNA Methyltransferases

Littauer and his co-workers have found that very high concentrations of ammonium ions, 0.36 M, enhance the total tRNA methyltransferase capacity of extracts of rat liver tissue (72). Kaye and Leboy (73) have suggested that altered ion concentrations in tissues from various sources, that is, normal tissue and tumor tissue, may affect the activity of the tRNA methyltransferases *in vivo*. However, the high concentration of ammonium ion required for maximal stimulation of the enzymes makes it improbable that such conditions would be encountered *in vivo*. Also, the effect of ammonium ions does not appear to be a general one for the tRNA methyltransferases. Several workers have reported tRNA methyltransferase systems that do not respond to ammonium ion (39,49,74).

Leboy has also studied the effect of polyamines upon the tRNA methyltransferases *in vitro*, and these observations may have greater biological validity (75). Spermine and spermidine at concentrations not exceeding those present in liver increased by several fold both specific methyl-transferase activity and capacity. Whether these effects are exerted *in vivo* depends upon the geographical localization of the enzymes and the polyamines in the cell.

IX. Other Modifications of tRNA

The number of modified nucleotides found in tRNA is increasing. By last count they were past 40.

Some of these modifications are startling in their complexity, as can be seen from three examples diagrammed in Figure 5.

As discussed in the introduction, Compound I requires at least three enzymes for its synthesis. A projection of the synthesis of compound Y—as well as its nomenclature—is beyond our ability.

The biological cost of the synthesis of these compounds is enormous, and for the rationale of their existence one can only conjecture that during the evolution of tRNA, there was an increasing need for the expansion of variability.

There are evidences of varying degrees of firmness indicating that all of these modifications are achieved at the macromolecular level by highly specific enzymes, but the discussion of these enzymes is beyond the scope of this review. However, a cautionary note, to obviate some needless work, and publications, may be in order. The requirement for the exact positioning of, say, thymine, is so compelling that to forestall the accidental insertion of ribothymidine, no kinase for the synthesis of its triphosphate exists. Indeed, the absence of this enzyme was an early confirmation of the enzymatic synthesis of ribothymidine at the macromolecular level—a welcome confirmation indeed, for the existence of a novel pathway of synthesis of thymine in RNA was baffling enough to elicit incredulity.

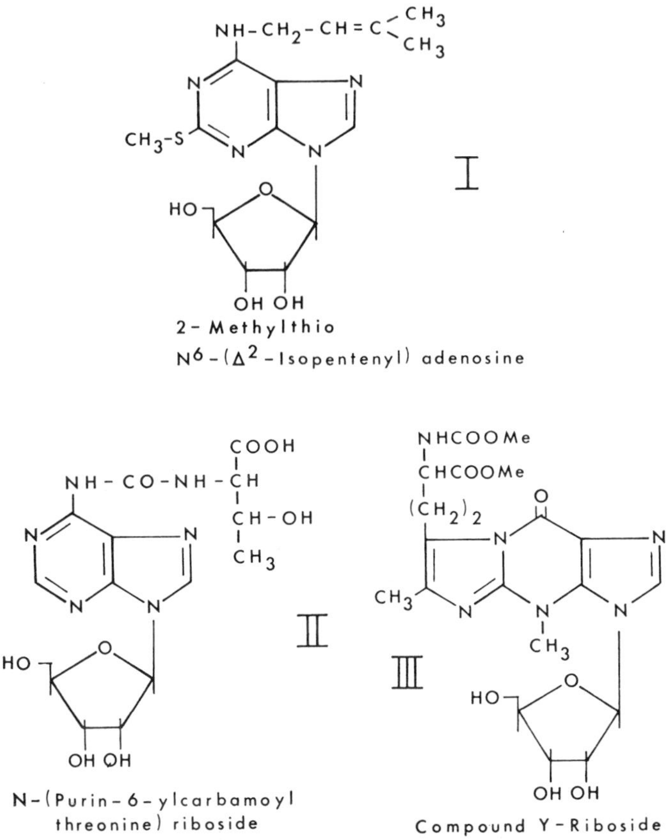

Fig. 5. Three of the modified nucleosides found in transfer RNA. I, 2-methylthio, N^6-(Δ^2-isopentenyl) adenosine, from *E. coli* tRNA (79). II, N-purin-6-ylcarbamoylthreonine riboside, *E. coli*, yeast, tobacco, and mammalian tRNAs (80). Base Y, from phenylalanyl tRNAs of yeast, wheat germ, and rat liver (81).

The synthesis of isopentenyl adenosine at the macromolecular level is well documented by the work of Fittler et al. (76). The evidence rests on the reconstitution of isopentenyl adenosine in tRNA, which is deficient in this modification. The deficiency was created by the elimination of the modifying moiety by oxidation with permanganate.

It was surprising, therefore, to read a recent publication on the *in vitro* phosphorylation of isopentenyladenosine (77). However, inspection revealed that the material was tentatively identified as the monophosphate. Thus, until a kinase for the synthesis of the triphosphate is demonstrated, the generalization about the exclusion of the modified nucleotides at transcription still stands.

We should also point out that variability of the other modifying enzymes in biological systems undergoing changes in regulatory process may also be expected. Indeed, Weiss has shown such alterations in uracil-thiolating enzymes of *E. coli* after phage infection (78).

X. Summary

It is increasingly evident that the structure of tRNA is much more complex than had been previously visualized, the macromolecule being subject to modifications of its primary bases and having extensive secondary and tertiary structure. For precise interaction with proteins and other nucleic acids, its three-dimensional conformation must, apparently, be intact. (17).

Methylation and other modifications of tRNA are indispensable for the attainment of its native conformation. The extreme specificity required for recognition between the tRNA methyltransferases and tRNA is just emerging: Primary sequence alone apparently is insufficient; some specific conformations may be required. It is apparent that we are only at the threshold of the complete elucidation of the mechanism of these interactions. Future investigation on the structural requirements for the many functional roles of this most complex of macromolecules should be rewarding and will be awaited with interest.

Note added in proof: Both the macromolecular and the small compound of the inhibitory system described by Kerr have now been identified. The small component is glycine and the protein is glycine *N*-methyltransferase. This enzyme regulates the tRNA methyltransferases by competing for the SAM pool and by producing *S*-adenosylhomocysteine which is a powerful inhibitor of the tRNA methyltransferases but a weaker inhibitor of glycine methyltransferase. For details see Reference 82.

References

1. Borek, E., and Srinivasan, P. R., *Ann. Rev. Biochem.*, *35* (1966).
2. Srinivasan, P. R., and Borek, E., *Progr. Nucleic Acid Res. and Mol. Biol.*, *5*, 157 (1966).

3. Starr, J. L., and Sells, B. H., *Physiol. Rev.*, *49*, 623 (1969).
4. Shugart, L., Novelli, G. D., and Stulberg, M. P., *Biochim. Biophys. Acta*, *157*, 83 (1968).
5. Gefter, M. L., and Russell, R. L., *J. Mol. Biol.*, *39*, 145 (1969).
6. Thiebe, R., and Zachau, H. G., *Eur. J. Biochem.*, *5*, 546 (1968).
7. Fittler, F., and Hall, R. H., *Biochem. Biophys. Res. Commun.*, *25*, 441 (1966).
8. Capra, J. D., and Peterkofsky, A., *J. Mol. Biol.*, *33*, 591 (1968).
9. Borek, E., *Cold Spring Harbor Symp. Quant. Biol.*, *28*, 139 (1963).
10. Agarwal, K. L., Büchi, H., Caruthers, M. H., Gupta, N., Khorana, H. G., Kleppe, K., Kumar, A., Ohtsuka, E., Rajbhandary, U. L., Van de Sande, J. H., Sgaramella, V., Weber, H., and Yamada, T., *Nature*, *227*, 27 (1970).
11. Bumstead, R. M., Dahl, J. L., Söll, D., and Strominger, J. L., *J. Biol. Chem.*, *243*, 770 (1968).
12. Stewart, T. S., Roberts, R. J., and Strominger, J. L., *Nature*, *230*, 36 (1971).
13. Mandel, L. R., and Borek, E., *Biochemistry*, *2*, 560 (1963).
14. Gefter, M. L., *Biochem. Biophys. Res. Commun.*, *36*, 435 (1969).
15. Borek, E., and Christman, J., *Fed. Proc.*, *24*, 292 (1965).
16. Pillinger, D. J., Hay, J., and Borek, E., *Biochem. J.*, *114*, 429 (1969).
17. Cramer, F., *Progr. in Nucleic Acid Res. and Mol. Biol.*, *11*, 391 (1971).
18. Kjellin-Straby, K., and Boman, H. G., *Proc. Nat. Acad. Sci. U.S.*, *53*, 134 (1965).
19. Phillips, J. H., and Kjellin-Straby, K., *J. Mol. Biol.*, *26*, 509 (1967).
20. Bjork, G. R., and Isaksson, L. A., *J. Mol. Biol.*, *51*, 83 (1970).
21. Bjork, G. R., and Neidhardt, F. C., *Cancer Res.*, *31*, May (1971).
22. Hayashi, H., Fisher, H., and Söll, D., *Biochemistry*, *8*, 3680 (1969).
23. Johnson, L., Hayashi, H., and Söll, D., *Biochemistry*, *14*, 2823 (1970).
24. Mudd, S. H., and Cantoni, G. L., *Nature*, *180*, 1052 (1957).
25. DeLa Haba, G., Jamieson, G. A., Mudd, S. H., and Richards, H. H., *J. Amer. Chem. Soc.*, *81*, 3975 (1959).
26. Stekol, J. A., in S. K. Shapiro and F. Schlenk, Eds., *Transmethylation and Methionine Biosynthesis*, University of Chicago Press, Chicago, 1965, p. 231.
27. Mudd, S. H., and Cantoni, G. L., *J. Biol. Chem.*, *231*, 481 (1958).
28. Ortwerth, B. J., and Novelli, G. D., *Cancer Res.*, *29*, 380 (1969).
29. Zappia, V., Zydek-Cwick, C. R., and Schlenk, F., *J. Biol. Chem.*, *244*, 4499 (1969).
30. Srinivasan, P. R., and Borek, E., *Proc. Nat. Acad. Sci. U.S.*, *49*, 529 (1963).
31. Turkington, R. W., and Riddle, M., *Cancer Res.*, *30*, 650 (1970).
32. Baguley, B. C., and Staehelin, M., *Biochemistry*, *9*, 1645 (1970).
33. Kuchino, Y., and Nishimura, S., *Biochem. Biophys. Res. Commun.*, *40*, 306 (1970).
34. Borek, E., and Kerr, S., *Advan. Cancer Res.*, *15*, in press.
35. Bjork, G. R., and Svensson, I., *Eur. J. Biochem.*, *9*, 207 (1969).
36. Svensson, I., Bjork, G. R., and Lundahl, P., *Eur. J. Biochem.*, *9*, 216 (1969).
37. Hancock, R. L., McFarland, P., and Fox, R. R., *Experientia*, *23*, 806 (1967).
38. Simon, L. N., Glasky, A. J., and Rejal, T. H., *Biochim. Biophys. Acta*, *142*, 99 (1967).
39. Kerr, S. J., *Biochemistry*, *9*, 690 (1970).

40. Kerr, S. J., *Fed. Proc.*, *29*, 893 Abs. (1970).
41. Craddock, V. M., Villa-Trevino, S., and Magee, P. N., *Biochem. J.*, *107*, 709 (1968).
42. Dunn, T. B., *Biochem. J.*, *86*, 14P (1963).
43. Wainfan, E., Srinivasan, P. R., and Borek, E., *Biochemistry*, *4*, 2845 (1965).
44. Wainfan, E., Srinivasan, P. R., and Borek, E., *J. Mol. Biol.*, *22*, 349 (1966).
45. Boezi, J. A., Armstrong, R. L., and DeBacker, M., *Biochem. Biophys. Res. Commun.*, *29*, 281 (1967).
46. Grado, C., Fridlender, B., Ihl, M., and Contreras, G., *Virology*, *35*, 339 (1968).
47. Vande Woude, G. F., Polatnick, J., and Ascione, R., *J. Virology*, *5*, 458 (1970).
48. Kit, S., Nakajima, K., and Dubbs, D. R., *Cancer Res.*, *30*, 528 (1970).
49. Gallagher, R. E., Ting, R. C. Y., and Gallo, R. C., *Proc. Soc. Exp. Biol. Med.*, *136*, 819 (1971).
50. Baliga, B. S., Srinivasan, P. R., and Borek, E., *Nature*, *208*, 555 (1965).
51. Borek, E., and Christman, J., unpublished observations (1966).
52. Sharma, O. K., Kerr, S. J., Wiesner, R. L., and Borek, E., *Fed. Proc.*, *30*, 167 (1971).
53. Sharma, O. K., and Borek, E., *Biochemistry*, *9*, 2507 (1970).
54. Turkington, R. W., *J. Biol. Chem.*, *244*, 5140 (1969).
55. Sheid, B., Bilik, E., and Biempica, L., *Arch. Biochem. Biophys.*, *140*, 437 (1970).
56. Greengard, O., Gordon, M., Smith, M. A., and Acs, G., *J. Biol. Chem.*, *239*, 2079 (1964).
57. Mays, L. L., Abs. Amer. Soc. Biol. Chem. (1971).
58. Pillinger, D. J., Borek, E., and Paik, W., *J. Endocrinol.*, *49*, 547 (1971).
59. Kerr, S. J., and Dische, Z., Invest. Ophth., *9*, 286 (1970).
60. Kerr, S. J., *Proc. Nat. Acad. Sci. U.S.*, *68*, 406 (1971).
61. Chaney, S. Q., Halpern, B. C., Halpern, R. M., and Smith, R. A., *Biochem. Biophys. Commun.*, *40*, 1209 (1970).
62. Halpern, R. M., Chaney, S. W., Halpern, B. C., and Smith, R. A., *Biochem. Biophys. Res. Commun.*, *42*, 602 (1971).
63. Hurwitz, J., Gold, M., and Anders, M., *J. Biol. Chem.*, *239*, 3474 (1964).
64. Salvatore, F., Zappia, V., and Shapiro, S. K., *Biochim. Biophys. Acta*, *158*, 461 (1968).
65. Wainfan, E., and Borek, E., *Mol. Pharmacol.*, *3*, 595 (1967).
66. Moore, B. G., and Smith, R. C., *Can. J. Biochem.*, *47*, 561 (1969).
67. Wainfan, E., and Landsberg, B., *Abstr. Amer. Soc. Biol. Chem.* (1971).
68. Gordon, J., Boman, H. G., and Isaksson, L., *J. Mol. Biol.*, *9*, 831 (1964).
69. Lowrie, R. J., and Bergquist, P. L., *Biochemistry*, *7*, 1761 (1968).
70. Farber, E., and Ichinose, H., *Cancer Res.*, *18*, 1209 (1958).
71. Hancock, R. L., *Biochem. Biophys. Res. Commun.*, *31*, 77 (1968).
72. Rodeh, R., Feldman, M., and Littauer, U. Z., *Biochemistry*, *6*, 451 (1967).
73. Kaye, A. M., and Leboy, P. S., *Biochem. Biophys. Acta*, *157*, 289 (1968).
74. McFarlane, E. S., *Can. J. Microbiol.*, *15*, 189 (1969).
75. Leboy, P. S., *Biochemistry*, *9*, 1577 (1970).

76. Fittler, F., Kline, L. K., and Hall, R. H., *Biochem. Biophys. Res. Commun.* *31*, 571 (1968).
77. Hacker, B., *Biochim. Biophys. Acta,* *224*, 635 (1970).
78. Weiss, S. B., Hsu, W. T., Foft, J. W., and Scherberg, N. H., *Proc. Nat. Acad. Sci. U.S.,* *61*, 114 (1968).
79. Burrows, J., Armstrong, D. J., Skoog, F., Hecht, S. M., Boyle, J. T. A., Leonard, N. J., and Occolowitz, J., *Science,* *161*, 691 (1968).
80. Chedda, G. B., Hall, R. H., Magrath, D. I., Mozejko, J., Schweizer, M. P., Stasiuk, L., and Taylor, P. R., *Biochemistry,* *8*, 3278 (1969).
81. Nakanishi, K., et al., *J. Amer. Chem. Soc.,* *92*, 7617 (1970).
82. Kerr S. J., *J. Biol. Chem.,* July 1972 (in press).

AFFINITY CHROMATOGRAPHY OF MACROMOLECULES

By PEDRO CUATRECASAS, *Baltimore, Maryland*

CONTENTS

I. Introduction

The use of selective adsorbents having biological specificity for the purification of macromolecules has been referred to as affinity chromatography (1–7). Conventional procedures for the purification of enzymes and other proteins depend on differences in the physicochemical properties of the various proteins present in the mixture. These differences are generally not unique, and most of the available separation procedures of a preparative nature are not sufficiently discriminative to permit facile separation of molecules whose physicochemical differences are subtle. Therefore, purification is generally laborious and incomplete.

In contrast, affinity chromatography is a "functional" purification approach that exploits the most unique property of macromolecules, their biological function. Most biological or natural macromolecules possess localized regions (active sites) highly specialized to perform unique functions. The tertiary structure of enzymes is generally believed to be highly selective for maintaining the integrity of these unique regions. The function of enzymatic active sites can be fundamentally recognized as involving two separate chemical processes, those of recognition and of catalysis of selected small molecules or of restricted regions of other macromolecules (Scheme I).

$$E + S \underset{k_{-1}}{\overset{k_1}{\rightleftharpoons}} ES \overset{k_2}{\longrightarrow} E + P$$

It is the former property, that of recognition, which forms the theoretical basis upon which the principles of affinity chromatography have been developed. Although the principles and procedures have generally been based primarily on enzyme systems, virtually any interacting system composed of two or more unique species can be approached by these procedures. The fundamental requirement is that the comparative rate constants (k_1 and k_{-1}) reflect reasonable affinity, and that the qualitative nature of E and S reflect reasonable stereochemical specificity. It is highly desirable that the interacting system be chosen such that the complex (ES) does not proceed to the right, that is, that the interacting species not be chemically altered as a result of the interaction. For this reason virtually all enzyme systems successfully approached by affinity chromatography have utilized competitive inhibitors rather than substrates. The potential use of substrates in these procedures will be developed in a later section.

It is immediately clear that a number of nonenzymatic interacting systems which do not normally result in chemical alterations of the structure of the interactants are ideally suited to affinity chromatography. Such procedures can therefore be applied, at least in principle, to such molecules as antigens, antibodies, vitamin- and drug-binding proteins, biological receptors, and transport proteins.

Basically, the methodology of affinity chromatography involves preparation of a selective adsorbent by covalent immobilization of one of the interacting species, usually a small ligand, to a suitable insoluble matrix or solid support. As will be discussed in a later section, the ligand must be coupled in a way that does not interfere with its ability to be recognized by the enzyme. The sample containing the enzyme to be purified is applied to a column containing the selective adsorbent, and the column is washed with the buffer used to equilibrate the column. Several column chromatographic elution patterns are possible, depending on the effectiveness of the adsorbent under the experimental conditions utilized (Fig. 1). If the solid support is unsubstituted, or if the adsorbent is totally ineffective, the specific enzyme to be purified will emerge in coincidence with the breakthrough, contaminating protein [Fig. 1(a)]. The specific enzyme may be retarded relative to the major breakthrough protein in its downward migration through the column, resulting in one of several patterns [Fig. 1(b), (c), (d)]. In successful affinity chromatography the enzyme will adsorb very strongly to the column [Fig. 1(e)]. Although the adsorbed enzyme would be expected to emerge eventually from the column without altering the conditions of the buffer, the time necessary for this to occur may be prohibitively long. The case illustrated by Figure 1(e) therefore, is, simply an exaggeration of case (d). There is a very important practical difference, however. A truly effective adsorbent, which would be defined operationally as one in which the buffer conditions must be modified (change in pH, ionic strength, etc.) to affect elution of the adsorbed enzyme, will yield a purified enzyme in a very concentrated form as a sharp elution peak.

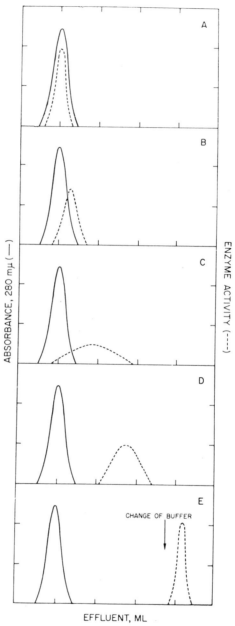

Fig. 1. Theoretical affinity chromographic patterns obtained by passing a crude protein mixture containing a specific enzyme to the purified (– – –) on a column containing an adsorbent specific for that enzyme. Successful application is depicted in (e). Various degrees of ineffectiveness are demonstrated in the other column emergence patterns.

For successful application of affinity chromatography, the experimental conditions used must resemble as much as possible those that exist when the components are free in solution. Careful consideration must be given to the nature of the solid carrier; stereochemical alteration of the ligand resulting from its covalent linkage to the matrix; the distance separating the ligand from the backbone of the solid matrix; the steric properties and flexibility of the group interposed between the ligand and the backbone; the stability of the chemical bonds; and the experimental conditions selected for adsorption, washing, and elution.

One of the disadvantages of affinity chromatography compared to conventional purification procedures is that the detailed and specific procedures utilized by the former are relatively more individualized and peculiar to the specific enzyme in question. Considerable thought and planning must precede specific laboratory procedures; some knowledge regarding substrate and inhibitor specificity, as well as other properties, such as the role of ions, metals, pH, and temperature, must be known. A specific and selective adsorbent must be deliberately prepared for every enzyme, and this will in many cases require relatively complex synthetic procedures.

Affinity chromatography, can, in principle, be applied to a wide variety of macromolecular-ligand systems. For example, specific adsorbents may be prepared to purify enzymes, antibodies, antigens, nucleic acids, vitamin-binding proteins, transport proteins, repressor proteins, drug or hormone receptors, sulfhydryl-containing proteins, peptides formed by organic synthesis, intact cell populations, specific polyribosomal complexes, and multienzyme complexes.

Also, affinity chromatography may be useful in concentrating dilute solutions of protein, in removing denatured forms of a purified enzyme, and in the resolution and separation of various components that result from specific chemical modifications of purified proteins.

Some of the advantages of this form of purification include the rapidity and ease of the procedure, the rapid separation of the protein to be purified from inhibitors and destructive contaminants, for example, proteases, protection from denaturation during purification by stabilizing the tertiary structure of the protein by ligand binding to the active site, the possibility of rapidly separating the soluble and solid components by centrifugation or filtration, and the frequent reutilization of the same adsorbent.

The basic principles described above for affinity chromatography are similar to those utilized for many years in the preparation and use of immunoadsorbents for the purification of antigens and antibodies (9,10). The problems involved in enzyme purification are considerably greater than those encountered in immune complex systems, as has recently been discussed in detail (2). A few scattered reports of enzyme purification by utilization of ligand–polymer systems have appeared during the past 20 years (11–13), but these had been regarded more as interesting curiosities than as examples of a potentially powerful tool with broad and exciting possible applications. During the past few years, the principles of this type of technique have been outlined clearly, new methodology has been developed, and many examples have been presented to illustrate this technique (1–7). As a result, much attention has recently been focused on this area, in which the basic precepts are so simple and rational. The general feasibility of affinity chromatography has been demonstrated to depend very critically on proper experimental execution. Now it is possible to approach intelligently the purification of any given enzyme, or other macromolecule, systematically and with a reasonable chance of success by affinity chromatography.

II. The Concept of Progressively Perpetuating Effectiveness

One of the most basic and important, but until now unrecognized, features of affinity chromatography is illustrated in Figure 2. The concentration of the component of the interacting system (S, ligand, in Scheme I) which is linked to the solid matrix is known and is fixed. The concentration of the other component (E, Scheme I), however, is known only in the starting sample. Once the sample enters the column, the enzyme molecules will have successful collisions with the ligand and will be slightly retarded. Enzyme molecules subsequently entering the column will begin to add to the medium which contains those molecules which had been retarded. The concentration of the specific enzyme on the column, therefore, will be continually changing (increasing), and there will be a dynamic concentration gradient over a certain region of the column. A fundamental property of dissociably interacting processes (Scheme I) is that the effective concentration of the complex (ES) increases as the concentration of the reactants is increased. As the downward movement of enzyme molecules begins to

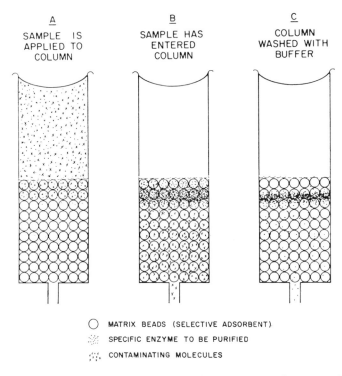

Fig. 2. Steps involved in the adsorption of enzyme to a column containing a specific adsorbent during affinity chromatography. The formation of a sharp, highly concentrated enzyme band on the column is demonstrated. The enzyme is progressively concentrated over a narrow band during application of the sample to the column. The progressively increasing concentration of enzyme during the progress of the procedure greatly increases the strength with which the enzyme adsorbs to the column.

decrease, and as the enzyme concentration thus begins to rise, the strength with which the band of enzyme will adsorb to the column will increase. The effectiveness of enzyme adsorption in affinity chromatography, therefore, must be considered in terms of a dynamic process which perpetuates itself once an initial effective interaction (retardation) has been established. This is one of the most unique precepts of column affinity chromatography, which is readily understood on theoretical grounds by examination of the laws that govern interactions of the type described in Scheme I.

The foregoing concepts have other important practical and theoretical implications. The importance of using column procedures, rather than batch-type procedures, can be readily understood. Only rarely (very high affinity complexes) will batch-type adsorption be effective in affinity chromatography. Another important implication is that if an adsorbent is obtained which results in some retardation of the specific enzyme, then this should be taken seriously as an encouraging sign. Very slight changes in pH, ionic strength, temperature, degree of ligand substitution, and so forth, may rapidly and dramatically transform a nearly ineffective adsorbent into a very effective one. These concepts probably explain the rarity with which chromatographic patterns of the type described in Figure 1(b), (c) and (d) are encountered. Most frequently, the extremes will be encountered: the adsorbent is totally ineffective or very effective. In fact, patterns such as (b), (c), and (d) (Fig. 1) should raise suspicions that adsorption is occurring through nonspecific ionic or hydrophobic, rather than specific, active-site-directed interactions.

The notion of self-perpetuating, increasing effectiveness during addition of the sample to an affinity column also explains why certain low affinity (K_i, 10^{-3} M) systems, such as that of β-galactosidase (8), can be successfully attacked. Without such considerations, it is most difficult to understand how samples containing low enzyme concentrations can be adsorbed, even to columns that contain very high (10 μmoles/ml) ligand concentrations.

Definition of adsorbent "capacity" for enzyme must also be carefully viewed in this light. Because of the dynamic character of the processes occurring during the chromatographic procedure, it may be very difficult to precisely define "capacity," even in an operational sense. Caution must also be exercised in deciding when to cease addition of more sample to the column. Very frequently, no more sample is added if enzymatic activity is found to emerge from the column. This, of course, may be misleading, since it may be possible to add many times more enzyme while losing only insignificant quantities in the breakthrough buffer. It is not uncommon for small quantities of enzyme activity to leak out in the column breakthrough, especially when the sample volume is many times greater than that of the column.

The concepts detailed above are also important in certain considerations relating to elution of the enzyme adsorbed to the column. Buffers of composition (pH, ionic strength) that do not permit sufficiently effective enzyme–ligand interactions for initial adsorption of the enzyme to the column may be totally inadequate for elution of the enzyme that is already adsorbed to the column. The latter, adsorbed enzyme may be present in the column in very concentrated form as a sharp band (Fig. 2), and the conditions required for elution (to disrupt effective interaction) may be much more extreme under these circumstances. For example, if sufficiently high concentrations of E and S (Scheme I) are present, a change in pH leading to a 5-fold decrease in effective affinity of the complex may not disrupt or cause the rapid downward migration of the enzyme. On the other hand, such a fall in affinity may be critical if the concentration of one of the species (enzyme) is low, as is usually the case when a protein sample is first applied to a column. The sharp localization of the enzyme band on an affinity column has been documented with staphylococcal nuclease (7). A purified sample of this enzyme was modified by reaction with dinitrodifluorobenzene. This reagent reacts randomly with ϵ-amino groups of the enzyme without affecting the catalytic function. The resulting yellow protein was applied to an affinity column; the enzyme adsorbed as a very sharp, intense yellow band at the top of the column. Elution with 0.1 M acetic acid caused a rapid downward movement of the sharply demarcated yellow band.

The above considerations also explain why one of the most effective means of eluting a strongly adsorbed enzyme is to remove the uppermost portion of the adsorbent and to dilute this in a beaker with a relatively large volume of an appropriate buffer.

III. Selection of Insoluble Carriers

An ideal insoluble carrier for affinity chromatography of enzymes should possess the following properties: it must interact very weakly with proteins in general to minimize the nonspecific adsorption of proteins; it should exhibit good flow properties that are retained after coupling; it must possess chemical groups that can be activated or modified under conditions innocuous to the structure of the matrix to allow the chemical linkage of a variety of ligands (these chemical groups should be abundant in order to allow attainment of a high effective concentration of coupled inhibitor, that is, capacity, so that satisfactory adsorption can be obtained even with protein-inhibitor systems of low affinity); it must be mechanically and chemically stable to the conditions of coupling and to the various conditions of pH, ionic strength, temperature, and presence of denaturants, such as urea, guanidine-hydrochloride, and detergents which may be needed for adsorption or elution because such properties also permit repeated use of the specific adsorbent; and it should form a very loose, porous network that permits uniform and unimpaired entry and exit of large macromolecules throughout the entire matrix.

In addition, the gel particles should preferably be uniform, spherical, and rigid. A high degree of porosity is an important consideration of ligand–protein systems of relatively weak affinity (dissociation constant of $10^{-4}\,M$ or greater), since the concentration of the ligand freely available to the protein must be quite high to permit interactions strong enough to physically retard the downward migration of the protein through the column. Figure 3 depicts the relationship between the porosity of a solid matrix support (bead) and the distribution of enzyme in the bead and surrounding medium. It is possible to have derivatives with extremely high ligand substitution, the bulk of which is not readily accessible to the enzyme (Fig. 3, bottom). This will be illustrated shortly with studies on the purification of β-galactosidase.

Considerable experience has been accumulated on the use of various derivatives of cellulose and polystyrene as insoluble carriers for enzymes and antibodies (10,14–20). As discussed recently in detail (2), these derivatives have generally been found to be inferior to the agarose supports in affinity chromatography. Beaded derivatives of agarose (21), a polysaccharide polymer, have nearly all the properties of an ideal adsorbent (1–4) and they are available commercially.

Since the beaded agarose derivatives are very loose structured, molecules with molecular weights in the millions diffuse readily through the matrix. The cross-linked polysaccharides readily undergo substitution reactions when activated with cyanogen halides (22,23), are very stable, and have a moderately high capacity for substitution. Numerous procedures are available for covalently linking a variety of ligands to agarose gels (4,24). Nearly all recent reports of functional enzyme purification have been performed with agarose, and most of the discussion in the remainder of the chapter will therefore be concerned with agarose derivatives.

Synthetic polyacrylamide gels also possess many desirable features and are available commercially as spherical beads in pregraded sizes and porosities. Derivatization procedures have been described (4,25)

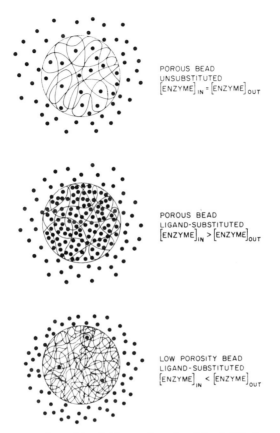

Fig. 3. Representation of insoluble carriers (beads) of differing porosity to illustrate importance of free accessibility of the enzyme to be purified to the ligand inside the bead meshwork.

that permit the attachment of a variety of ligands or proteins to polyacrylamide beads. The beads have uniform physical properties and porosity, and the polyethylene backbone gives them physical and chemical stability. Preformed beads are available that permit penetration of proteins with molecular weights of about one-half million, but their porosity is diminished during the chemical modifications required for the attachment of ligands (4). In this respect the acrylamide beads are inferior to agarose. The principal advantage of polyacrylamide is its very large number of modifiable groups (carboxamide).

Highly substituted derivatives may thus be prepared for purifying enzymes for which only weakly binding ligands are available.

The importance of bead porosity, discussed earlier (Fig. 3), is illustrated by studies of the purification of *E. coli* β-galactosidase (8). *p*-Aminophenyl-β-D-thiogalactopyranoside was attached in identical manner to various derivatives of agarose and polyacrylamide. Although the degree of ligand substitution of the acrylamide derivatives was at least 10 times greater than that of the agarose derivatives, none of the former were effective in adsorbing the enzyme, whereas the agarose derivatives adsorbed the enzyme strongly. Most of the ligand on the acrylamide beads must be located inside the bead, inaccessible to enzyme. It might be argued that the relatively large size of this enzyme (mol. wt. greater than 200,000) precludes facile penetration of the bead pores. It has also been found that a much smaller enzyme, staphylococcal nuclease (mol. wt. = 17,000) also has some difficulty entering the acrylamide pores, since affinity adsorbents of agarose are more effective than similar acrylamide derivatives despite the much higher degree of ligand substitution of the latter (4).

It is important to recognize that for certain kinds of purifications, the porosity of the matrix or bead will not be an important consideration. This may be the case in the purification of very high affinity systems, where protein–ligand binding may approach stoichiometry, in systems where the capacity of the matrix is not a limiting factor, or in systems involving extremely large complexes (polysomes, intact cells, and membrane fractions) which cannot reasonably be expected to penetrate the pores of the bead. This can be illustrated by the successful utilization of hapten–polyacrylamide derivatives to separate by selective adsorption lymphocytes specifically capable of recognizing the haptene (26). Similarly, insulin–agarose derivatives can retain fat cell ghosts that contain surface receptors for insulin (27), and glucagon–agarose derivatives can selectively adsorb liver membranes that contain specific receptors for this hormone (28). In all these examples, interactions of very high affinity and specificity occur between ligands on the *surface* of the bead and the macromolecular species or cell in the surrounding medium. In such studies the porosity of the beads is not important. It is necessary, however, that the matrix not act as a mesh to physically trap the large particles in solution. Large spherical beads, with large spaces *between* adjacent beads, are ideal for this purpose.

A technique for the immobilization of organic substances on glass surfaces has been described (29.) Trypsin and pepsin have been covalently attached to porous glass beads with a silane coupling agent (20). Inorganic carriers, such as glass, have the inherent advantages of resistance to microbial attack, dissolution by organic solvents, and configurational alterations resulting from changes in pH or buffer composition. Little experience, however, has been obtained in the use of glass beads for selective functional purifications. Glass beads coated with antigen have been moderately successful in the separation of immune lymphoid cells (31,32). However, nonspecific adsorption of proteins to glass surfaces is a serious problem, and the porosity of the available beads does not readily permit easy entry or exit of large macromolecules. Glass beads may be particularly useful in special cases that deal with high affinity systems, in which the ligand to be immobilized is highly insoluble in water. In such cases coating of the bead *surface* with the ligand may provide sufficient substitution, and the beads can be conveniently washed extensively with organic solvents to remove adsorbed ligand before use. For example, estradiol has been coupled to amino-alkyl-substituted glass beads (Bio-Glass 2500, 200–325 mesh) by procedures similar to those described for agarose substitutions (4) to extract estradiol-binding proteins of calf uterus (33). The flow rates of columns packed with such glass beads are prohibitively slow, and the tight packing of the beads results in mechanical trapping of particulate materials that causes significant problems on passing large samples through the column. Although column chromatographic procedures are therefore difficult, batchwise methods can be used successfully, provided the system under study displays a high degree of affinity, such as that of the estradiol-receptor interactions.

IV. Preparation of Selective Adsorbents

A. SELECTION OF THE LIGAND

In preparing selective enzyme adsorbents, the small molecule to be covalently linked to the solid support must display special affinity for the macromolecule to be purified. It can be a substrate analogue, an effector, a cofactor, a vitamin, and, in special cases, a substrate. Enzymes requiring two substrates for reaction may be approached by immobilizing one of the substrates, provided that one substrate is sufficiently well bound to the enzyme in the absence of the other. Also, a substrate may be used if it binds to the enzyme under some conditions that do not favor catalysis, that is, in the absence of metal ion, at low temperatures, or if the pH dependence of K_m and k_{cat} is different.

The small molecule to be rendered insoluble must possess chemical groups that can be modified for linkage to the solid support without abolishing or seriously impairing interaction with the complementary protein. If the strength of interaction of the free complex in solution is very strong, for example, a K_i of about 10^{-9} M, a decrease in affinity of 3 orders of magnitude upon preparation of the insoluble derivative may still leave an effective and selective adsorbent. The important parameter is the effective experimental affinity, that displayed between the protein in solution and the insolubilized ligand under specific experimental conditions. In practice it has been very difficult to prepare insoluble adsorbents for systems having dissociation constants greater than 5 mM unless the ligand is attached to the carrier backbone by a long arm (4). It is possible in theory to prepare adequate adsorbents for such systems if enough of the inhibitor can be coupled to the solid support, but the porosity of the insoluble support is an important factor because it is possible to have a very large amount of ligand bound to a carrier in such a way that most of it is not freely accessible to the protein.

B. SEPARATION OF THE LIGAND FROM THE MATRIX BACKBONE

For successful purification by affinity chromatography, the part of the inhibitor critical for interaction with the macromolecule to be purified must be sufficiently distant from the solid matrix to minimize steric interference with binding (2,4,8). Steric considerations seem most important in proteins of high molecular weight. The problem may be approached by preparing an inhibitor with a long hydrocarbon chain or "arm" attached to it. The arm can be attached in turn to the insoluble support (1). Alternatively, such a hydrocarbon extension arm can be attached to the solid support first (4,24).

The importance of these considerations will be illustrated in a later section dealing with the purification of α-chymotrypsin (1), β-galactosidase (4,8), tyrosine aminotransferase (34), and acetylcholinesterase (35).

Although steric factors appear to be important in explaining the dramatic improvement that can frequently result from introducing a very long hydrocarbon extension between the ligand and the matrix backbone, it is not clear whether there are other contributory processes. For example, the added flexibility and mobility of the ligand attendant to its attachment to a long "arm" may be an important factor.

C. COVALENT LINKING REACTIONS

1. Cellulose and Polystyrene

Perhaps the most common covalent linkage of enzymes and antigens to cellulose and polystyrene derivatives is by the reaction of histidyl and tyrosyl residues of these proteins with diazotized derivatives of the matrix. Pressman et al. (36) were the first to utilize this technique in binding ovalbumin to sheep erythrocytes using bisdiazobenzidine as a divalent coupling agent. Campbell et al. (9,37) subsequently characterized immunoadsorbents prepared by coupling ovalbumin to diazotized p-aminobenzyl cellulose in detail. Antigens are now commonly coupled to diazonium cellulose (16,38–45) and diazotized poly-p-aminostyrene (15,46–50). Antigens have been coupled by their amino groups to carboxymethyl cellulose (amide linkage) with dicyclohexylcarbodiimide in organic solvents (46–48). Solid supports containing isothiocyanate groups have been used to link proteins by means of their amino groups (51). Other proteins have been attached through their amino groups to insoluble polystyrene and cellulose supports by reactions involving acyl halides (52–54), acid anhydrides (55), acyl azides (56–57), and sulfonyl chlorides (52). Nucleotides having monoesterified phosphate groups have been linked to cellulose in organic (58,59) and aqueous (60) solvents by carbodiimide reagents.

2. Agarose

a. General Considerations. As discussed earlier, beaded agarose has been the most generally useful solid support for affinity chromatography. Axén, Porath, and Ernback described a gentle method for coupling compounds containing primary aliphatic amino groups to cross-linked polysacchardies, such as dextran and agarose (22,23). The method has been modified and adapted for the preparation of a number of specific adsorbents for affinity chromatography (1,4,5).

This method is based on the activation of the gel with cyanogen bromide at pH 11, followed by washing and coupling the activated matrix with ligands or proteins under appropriate buffer conditions. The exact chemical nature of the intermediate formed by cyanogen halide treatment of polysaccharide derivatives is unknown, but the principal products formed upon coupling with amino groups appear to be mixed derivatives of iminocarbonate and isourea (61). Because of the absence of adjacent hydroxylic groups in agarose, it is likely that compounds are coupled to this support primarily if not exclusively by isourea linkage. It is important that both linkage forms result in the retention of positive charge on the amino nitrogen. This may be important in those cases where loss of the charge of the amino group of the ligand results in loss of specific binding to the enzyme to be purified.

Although the chemical nature of the cyanogen bromide-activated agarose intermediate is unknown, it has been established that covalent reaction with primary amino groups occurs with the unprotonated form of the amino group (4). Compounds containing primary aliphatic amino groups, such as the ϵ-amino group of lysine or aminoalkyl compounds, couple optimally at pH 10. Higher pH values result in lower coupling efficiency because the stability of the activated agarose is greatly reduced; this is particularly striking above pH 10.5 (4). Glycine, ammonium acetate, ammonium bicarbonate, and Tris buffers should not be used, since they interact with and thus interfere with the coupling reaction. Sodium borate or sodium bicarbonate buffers are recommended for coupling reactions at the higher pH values. The most facile coupling occurs with compounds bearing aromatic amines, due to the low pK of the amino group (4). Very high coupling efficiencies are obtained at pH values between 8 and 10 (1,4). Compounds containing an α-amino group, such as alanine, will couple optimally at a pH of 9.5 to 10 (Table I).

The cyanogen bromide-activated agarose intermediate is rather unstable, having a half-life of about 15–20 min at 4°. Washing of this activated derivative to remove excess cyanogen bromide prior to the coupling reaction must therefore be performed quickly and with buffers at low temperature (4°). Although very little of the activated group would be expected to remain on the agarose after the usual time periods (12–16 hr) used in the coupling step, it is sometimes wise to wash the final coupled derivative at room temperature with 0.1 M

TABLE I

Effect of pH on the Coupling of ^{14}C-Alanine to Activated Agarose

Conditions for coupling reaction	pH	Alanine coupled (μM/ml agarose)
0.1 M sodium citrate	6.0	4.2
0.1 M sodium phosphate	7.5	8.0
0.1 M sodium borate	8.5	11.0
0.1 M sodium borate	9.5	12.5
0.1 M sodium carbonate	10.5	10.5
0.1 M sodium carbonate	11.5	0.2

Packed sepharose-4B, 60 ml, was mixed with 60 ml water and treated with 15 g cyanogen bromide. To the cold, washed activated agarose were added 2.2 mM of ^{14}C-L-alanine (0.1 μCi/μM) in 60 ml of cold distilled water; 20-ml samples of the mixed suspension were added rapidly to beakers containing 5 ml cold 0.5 M buffer 8f composition described in the table. The final concentration of alanine was 0.015 M. After 24 hr the suspensions were thoroughly washed; aliquots were hydrolyzed by heating at 110° in 6 N HCl for 24 hr; and the amount of ^{14}C-alanine released was determined. Data from Cuatrecasas (4).

glycine buffer, pH 9.0. This will neutralize any unreacted agarose groups that could potentially react covalently with protein solutions subsequently added to the derivative. This washing step may be especially useful when the derivative is to be used soon after preparation or when the derivative is to be used for the adsorption of minute quantities of a protein that can be detected by sensitive procedures. Although the occurrence of such protracted reaction of residual agarose groups has not been unequivocally demonstrated, the washing step is recommended as a precautionary step in certain cases. Because of the still unresolved nature of the chemistry of the activation step, it is possible that cyanogen bromide reaction results in quantitatively minor but active products of greater stability.

The quantity of ligand coupled to agarose can in part be controlled by the amount of ligand added to the activated agarose (1,4). When highly substituted derivatives are desired, the amount of ligand added should, if possible, be 20–30 times greater than that which is desired in the final product. For ordinary procedures, 100–150 mg of cyanogen bromide are used per milliliter of packed agarose, but much higher coupling yields can be obtained if this amount is increased to 250–300 mg (4). The pH at which the coupling stage is performed will also determine the degree of coupling, since, as described earlier, it is the unprotonated form of the amino group that is reactive.

The procedures for estimating the amount of ligand coupled to agarose and determining the operational capacity for specific protein adsorption have been described (1,5). The degree of ligand substitution should be represented as micromoles of ligand bound per milliliter of packed gel. This means of expression is more useful than those represented on the basis of dry gel weight.

b. Specific Procedures. A number of chemical derivatives of agarose have recently been prepared, innovating a variety of methods for attaching ligands and proteins to agarose (4,24). These derivatives, which increase the general versatility of affinity chromatography, are proving especially useful in cases where hydrocarbon arms of varying length are to be interposed between the matrix and the ligand, where amino groups are not present on the ligand, and where it is desirable to remove the intact protein–ligand complex by specific chemical cleavage of the ligand–matrix bond. These derivatives facilitate the preparation of specific adsorbents, since unlike the cyanogen bromide-activated derivate, they are quite stable in storage. Furthermore, the unpleasant use of cyanogen bromide is avoided.

Perhaps the most versatile derivative is ω-aminoalkyl-agarose, which can either be utilized directly or can be derivatized further, as will be described shortly. Aliphatic diamines of the general formula $NH_2(CH_2)_xNH_2$ (e.g., ethylenediamine and hexamethylenediamine) can be attached by one of the amino groups (with little cross-linking) by the cyanogen bromide procedure (4). In this way it is easy to insert extensions of considerable length, depending on the nature of $(CH_2)_x$. One of the most useful diamines that has been used extensively in this laboratory is 3,3′-diaminodipropylamine (4). The aminoagarose derivatives thus obtained can be used for ligand attachment in a variety of ways by procedures performed easily and quickly in aqueous media (Figure 4).

Virtually any ligand that contains a carboxylic group can be coupled at pH 4.7 to aminoagarose with a water-soluble carbodiimide [Fig. 4(a)]. The carbodiimide used most extensively has been 1-ethyl-3-(3-dimethylaminopropyl) carbodiimide (4). Bromoacetyl agarose can be readily prepared by treating aminoagarose with O-bromoacetyl-N-hydroxysuccinimide (62). The agarose derivative can then be used to couple ligands that contain sulfhydryl, amino, phenolic, or imidazole groups [Fig. 4(b)]. Carboxylic agarose derivatives can be prepared from aminoalkyl agarose by reacting with succinic anhydride. Virtually any ligand containing a primary aliphatic or aromatic amino group can be coupled to this derivative with a water-soluble carbodiimide (Fig. 4, C). In this way it is possible to immobilize through a long hydrocarbon extension any ligand that would otherwise be coupled to agarose by the cyanogen bromide procedure. It is thus seldom, if ever, necessary to couple the ligand directly by the cyanogen bromide procedure.

Ligands that contain imidazole or phenolic groups react rapidly and in higher yield under mild conditions with diazonium derivatives of agarose [Fig. 4(d)]. The latter can be prepared easily by reaction with p-nitrobenzoyl azide (4). The derivative can be cleaved rapidly by reduction with sodium dithionite. Sulfhydryl derivatives of agarose can be prepared very easily by treating aminoalkyl agarose with N-acetylhomocysteine thiolactone [Fig. 4(e)]. The latter can be used to couple ligands via thiol ether or ester linkage. The ester derivative can be cleaved readily with neutral hydroxylamine or under mildly basic conditions.

Ligands that contain aromatic amino groups, and thus can be converted to corresponding diazonium by mild reaction in nitrous acid, can be coupled readily to agarose derivatives containing imidazole or phenolic groups. The latter are conveniently prepared by coupling with the cyanogen bromide procedure peptides having carboxy-terminal histidine or tyrosine (4,24). The ligand can thus be placed at some distance from the matrix backbone. Such derivatives have proved useful for the purification of various neuraminidases (63), staphylococcal nuclease (4), and β-galactosidase from $E.$ $coli$ (8).

Derivations of agarose can also be performed in certain organic solvents. This may be potentially very useful, even though it has not generally been recognized and has not yet been exploited for the preparation of selective adsorbents. For example, fatty acids activated by reaction with dicyclohexylcarbodiimide (in dioxane) have been coupled in very high yield to aminoalkyl agarose in anhydrous dioxane (27). Furthermore, this solvent, as well as dimethylformamide (50%, v/v) and ethylene glycol (50%, v/v), can be used in the above coupling

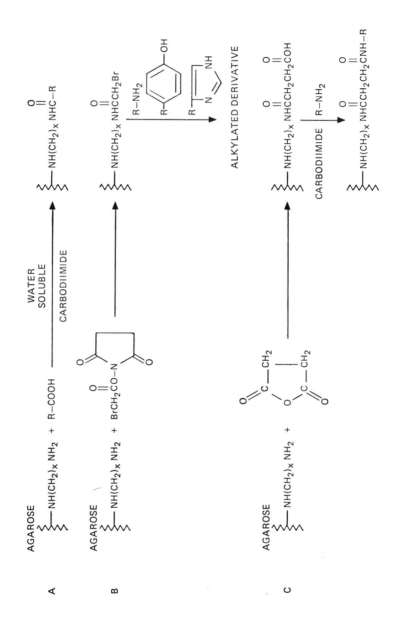

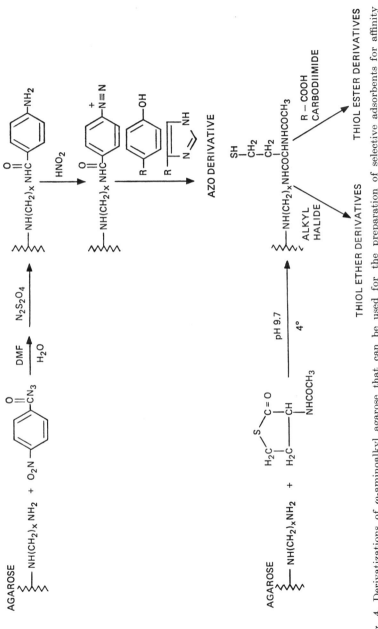

Fig. 4. Derivatizations of ω-aminoalkyl agarose that can be used for the preparation of selective adsorbents for affinity chromatography. The detailed procedures for the preparation of these derivatives are described in ref. 4.

49

reactions in cases of limited solubility of the ligand in aqueous medium and in the washing of coupled agarose derivatives prepared from such ligands (4).

c. Stability of Derivatives. Beaded agarose gels, unlike the cross-linked dextrans (Sephadex), cannot be dried or frozen, since they will shrink severely and essentially irreversibly (4). Similarly, they will not tolerate many organic solvents. Dimethylformamide (50%, v/v) and ethylene glycol (50%, v/v) do not adversely affect the structure of these beads. These solvents are quite useful in situations in which the compound to be coupled is relatively insoluble in water (e.g., steroids, thyroxine, and tryptophan derivatives), since the coupling step can be carried out in these solvents. Similarly, the final, coupled derivative can be washed with these solvents to remove strongly adsorbed or relatively water-insoluble material. The coupled, substituted adsorbents of agarose can be stored at 4° in aqueous suspensions, with antibacterial preservatives, for periods of time limited only by the stability of the bound ligand. Agarose beads tolerate 0.1 M NaOH and 1 M HCl for at least 2–3 hr at room temperature without adverse alteration of their physical properties and without cleavage of the linked ligand (4). These beads, therefore, can be used repeatedly even after exposure to relatively extreme conditions. Agarose beads also tolerate quite well exposure to 6 M guanidine · HCl or 7 M urea solutions for prolonged periods; slight shrinkage of the gel is observed under these conditions. These protein denaturants may therefore be used to aid in the elution of specifically bound proteins, to thoroughly wash columns in preparation for reuse, and to wash off protein that may be tightly adsorbed during the coupling procedure.

3. Acrylamide

Procedures described by Inman and Dintzis for polyacrylamide derivatization have been used to prepare selective adsorbents for affinity chromatography (4,5,25). The carboxamide side groups are converted to the hydrazide groups by heating with hydrazine, and the hydrazide is converted into acyl azide by treatment with nitrous acid (Fig. 5). The acyl azide groups react rapidly with primary aliphatic or aromatic amino groups without intermediate washings or transfers.

Almost all of the agarose derivatives described in Figure 4 can be prepared from polyacrylamide (4). The principal advantage of polyacrylamide, its high capacity for ligand substitution, is severely offset

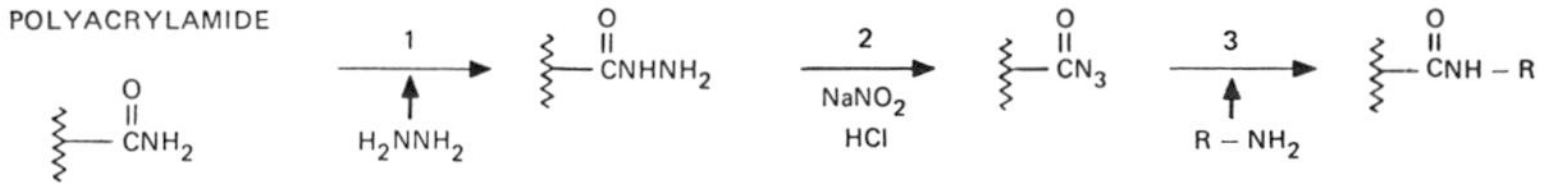

POLYACRYLAMIDE

$$\text{—CNH}_2 \xrightarrow[\text{H}_2\text{NNH}_2]{1} \text{—CNHNH}_2 \xrightarrow[\substack{\text{NaNO}_2 \\ \text{HCl}}]{2} \text{—CN}_3 \xrightarrow[\text{R—NH}_2]{3} \text{—CNH—R}$$

Fig. 5. Various polyacrylamine derivatizations useful in the preparation of selective adsorbents for affinity chromatography (4,25).

by the low porosity of these derivatives. This porosity is decreased further upon formation and reaction of the acyl azide step (4).

D. LINKING OF PROTEINS TO AGAROSE

The cyanogen bromide coupling procedures have been used to covalently coupled various proteins to agarose (4,22,23,64–82). Some specific and unique biochemical uses of such derivatives will be described in a later section.

In addition to attaching proteins or peptides directly through their amino groups to agarose by the cyanogen bromide procedure, it is possible to use the bromoacetyl, diazonium, or sulfhydryl agarose derivatives (4,5,24) described earlier in this report. Proteins attached in the latter fashion will extend some distance from the matrix backbone by an arm, which may be very useful in overcoming steric difficulties when interactions with other macromolecules are being studied. The study of the interaction of certain insolubilized proteins or polypeptides with intact cells or cell structures may best be achieved with such derivatives. For example, glucagon has been attached by various linkage groups to agarose (Fig. 6) to explore the interaction of this peptide hormone with specific receptors in liver cell membranes (83). In all cases vigorous attempts were made to obtain derivatives having a single bridge linking the peptide to the gel. Derivatives B, C, and D, which contain the peptide hormone at some distance from the matrix backbone, were effective in activating liver membrane adenyl cyclase. Derivative B has been used in column experiments to selectively adsorb membrane fragments containing receptors for the hormone. Similarly, insulin–agarose derivatives in which the hormone is separated from the gel backbone by long hydrocarbon extensions are the most useful in adsorbing insulin receptor structures solubilized from fat cell and liver cell membranes (27).

An important consideration in the covalent attachment of a biologically active protein to an insoluble support is that the protein should

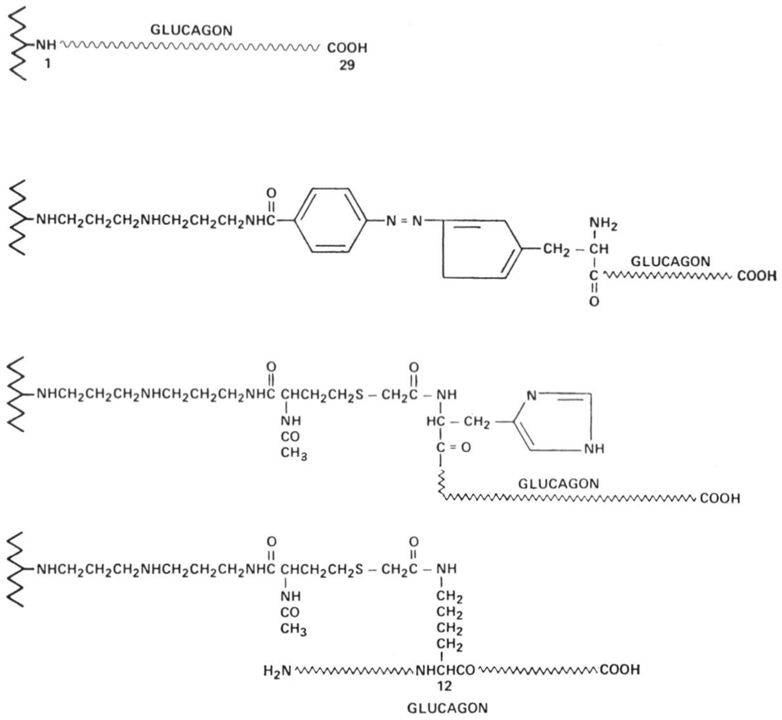

Fig. 6. Glucagon–agarose derivatives used to study the interaction of this hormone with receptor-containing liver membrane fragments (83). The derivatives, starting from the top, are glucagon coupled at pH 5.5 to cyanogen bromide–activated agarose; glucagon coupled by its single, N-terminal histidyl residue to a diazotized agarose derivative [Fig. 4(d)]; glucagon bearing an N-terminal bromoacetyl group coupled to a sulfhydryl agarose derivative [Fig. 4(e)]; and glucagon bearing a bromoacetyl group on the —amino group of the single lysyl residue coupled to sulfhydryl agarose [Fig. 4(e)].

be attached to the matrix by the fewest possible bonds (2,4,5,64). This will increase the probability that the attached macromolecule will retain its native tertiary structure, and its properties may more nearly resemble those of the native protein in solution. Proteins react with cyanogen bromide-activated agarose through the unprotonated form of their free amino groups. Since most proteins are richly endowed with lysyl residues, most of which are exposed to solvent, it is likely that such molecules will have multiple points of attachment to the resin when the coupling reaction is done at pH 9.5 or higher, as is usually the case. The problem may be circumvented by carrying out the coupling procedure at a less favorable pH. For example, it has been demonstrated that if antibodies are coupled to agarose at pH 6.0 to 6.5, the resultant immunoadsorbent has a much greater capacity for antigen than that which is prepared by performing the coupling procedure at pH 9.5. For example, (4,5,27) sheep antiporcine insulin antibodies, purified with insulin-agarose (66), have been linked in nearly quantitative yield to Sepharose-4B at pH 6.5 (0.2 M sodium citrate); 40 mg of protein were added to 5 ml of packed agarose that had been activated with 1.5 g of cyanogen bromide, in a total volume of 10 ml. This derivative was capable of binding approximately 80 % of the theoretical capacity for insulin. In contrast, an immunoglobulin–agarose derivative, prepared in an identical manner except that the pH of the coupling step was 9.5, could bind only 7 % of the theoretical insulin capacity. Since the total protein content of both derivatives is the same, the latter derivative must contain immunoglobulin that is incapable of effectively binding antigen. Adsorbents containing a high concentration of protein can be prepared at the lower pH values if a large amount of cyanogen bromide (250 mg/ml of packed gel) is used for activation, and if a relatively high concentration of protein (10 mg/ml) is used in the coupling step.

V. Adsorption and Elution of Affinity Columns

As discussed earlier, selective adsorbents for enzymes are best used in column, rather than batchwise, procedures. Further comments here will therefore be restricted to column chromatographic experiments. The specific buffer conditions used for equilibrating the affinity column in preparation for sample application should reflect the specific properties of the interacting system under study. The nature of the buffer used, and its pH and ionic strength, should be those that are optimal for the ligand–enzyme interaction in solution. The enzyme sample that is to be applied to the column should be present in the same buffer used to equilibrate the column. If possible, the sample should be dialyzed against this buffer before application on the column. It should be remembered that the pH which may be optimal for the catalytic function of the enzyme may be quite different from the pH which is optimal for the binding of substrates or inhibitors.

After application of the sample and washing with the starting buffer, it is sometimes desirable to wash the column containing the specifically adsorbed enzyme with buffers different from that used for equilibration and sample application. In this way it may be possible to remove nonspecifically adsorbed proteins or proteins bound to gel by non-specific ionic forces.

Protein specifically and strongly adsorbed to a column of a selective adsorbent [Fig. 1(a)] can be eluted by several possible maneuvers. The conditions of the buffer can be changed such that the affinity of the ligand-enzyme complex falls sufficiently to destroy effective binding on the column. This can be done by selectively altering the charge state of active site or ligand ionization groups to a state that does not support binding; this can be done by changing the pH and/or ionic strength of the buffer. Because of the highly concentrated nature of the enzyme on the column (Fig. 2) it should be remembered that more extreme buffer conditions may be required to achieve elution than are necessary to alter the binding interaction significantly, as measured by most enzymic assays in solution. For example, a 100-fold loss of affinity may not be sufficient to remove an enzyme that has been adsorbed in very high quantity to a small portion of the gel by application of large sample volumes to the column (see earlier discussion). For this reason, it is unlikely that in most cases elution will follow a simple or partial change in the charge state of a specific functional group involved in the binding interaction. In most cases, therefore, elution by application of a buffer of changed pH or ionic strength will involve a partial or major structural transition of the tertiary structure of the protein. Ideal elution in such cases involves utilization of a buffer system that causes sufficient alteration in the conformation of the protein to appreciably decrease the affinity of the protein for the ligand, but is not sufficient to completely unfold the protein. This form of elution, which may involve, for example, the use of 0.1 M acetic acid or 0.1 M ammonium hydroxide, generally results in elution of the adsorbed protein in a small volume. The eluted protein should be neutralized, diluted, or dialyzed immediately to permit prompt reconstitution of the native structure. The effectiveness of such a reconstitution can be tested by rechromatographing the purified protein to determine changes in adsorption to the affinity column.

To elute certain tightly bound enzymes, it is sometimes necessary to use buffers that are very strongly basic or acidic, or contain potent protein denaturants, such as urea or guanidine·HCl. Since these conditions frequently result in irreversible denaturation and loss of enzymic function, alternative methods of elution must be available. Elution of a specifically adsorbed protein can frequently be achieved by using a solution containing a high concentration of a specific inhibitor or substrate. The inhibitor can either be the same as the one that is covalently linked to the matrix (and must be used at higher concentrations), or preferably another, stronger, competitive inhibitor. In the past, examples in which the protein has been eluted with a buffer containing substrates or inhibitors show that the protein has emerged in larger effluent volumes than those obtained when elution is effected by changes in the pH or ionic strength of the buffer. This is particularly striking in those cases that involve high affinity enzyme–ligand systems. In such cases, removal of specifically bound enzyme may be very slow, despite utilization of very high concentrations of inhibitor in the eluting buffer.

Although not generally appreciated, the reasons for this are readily apparent by consideration of the basic principles underlying reversible interactions of the type exploited in affinity chromatography (Scheme I). The only factor that will govern the pattern of enzyme removal with inhibitor-containing solutions is the *rate of dissociation* (k_{-1}) of the complex. If the affinity (k_1/k_{-1}) of the complex is very high $(Ki < 10^{-7}\ M)$, the *time* required for dissociation may be quite appreciable. For example, a rate constant with a half-life of 15 min will require 75 min to dissociate to an extent greater than 95%. This *time requirement* will *not* be affected by the concentration of free inhibitor in the medium, since the rate of dissociation is a first order process that is function-dependent only on the concentration of the complex. Free substrate or inhibitor cannot affect this rate; their only role is to prevent *reassociation* of the enzyme with the specific ligand or inhibitor attached to the gel. It is important to consider that these effects will be considerably exaggerated at lower temperatures, conditions that are customary for performing enzymic purifications, such as affinity chromatography. Dramatically improved inhibitor elutions can therefore be performed by (*a*) performing the elution step at room temperature (or higher if tolerated), and (*b*) passing the inhibitor-containing buffer into the column and then *stopping* the flow through the column for an appropriate time interval that must be determined by the nature of k_{-1}.

Another useful maneuver in eluting enzymes very tightly bound to an affinity column is to unpack or remove the gel containing the adsorbed enzyme and to dilute this in a tube or beaker with an appropriate buffer. Frequently, only the upper $\frac{1}{4}$ to $\frac{1}{3}$ of the column material must be removed, since the bulk of the adsorbed material is in this portion of the column. In this procedure, there will be simultaneous dilution of both the inhibitor and the enzyme. Thus the effectiveness of complex formation will be drastically reduced. Furthermore, the suspension can be stirred, heated, or supplemented with free substrates, inhibitors, or salts, and the pH of the suspension can be accurately adjusted as desired. This has been a particularly useful procedure for elution of immunoglobulins tightly bound to antigen columns.

In some cases, it may be of value to remove the intact protein–ligand complex from the solid support. Three of the derivatization products described earlier are readily adapted to specific chemical cleavage of the ligand–gel bond under relatively mild conditions. The azo-linked derivatives can be cleaved by reduction with sodium dithionite at pH 8.5 (4). This procedure has been of considerable value in studies of the purification of serum estradiol-binding protein with derivatives of agarose to which estradiol is attached by the azo linkage (33). The serum estradiol-binding protein is denatured irreversibly by exposure to pH 3 or 11.5, and by low concentrations of guanidine·H3l (3 M) or urea (4 M). The protein, which binds estradiol very tightly (K_i about 10^{-9} M), can be removed in active form from the agarose–estradiol gel by reductive cleavage of the azo link with dithionite. Carboxylic acid ester derivatives of various ligands have also been linked to agarose by the procedures described here. Such bonds can be cleaved readily by short periods of exposure to pH 11.5 at 4°. Ligands attached to agarose by thiol esters can be similarly cleaved by short exposure to alkaline pH or by treatment with neutral hydroxylamine. These procedures present alternative ways of removing intact ligand–protein complexes from insoluble supports.

VI. Selected Applications

A. PURIFICATION OF ENZYMES

1. Staphylococcal Nuclease

The inhibitor of staphylococcal nuclease, 3′-(4-aminophenylphosphoryl)deoxythymidine 5′-phosphate (pdTp-aminophenyl) is an ideal

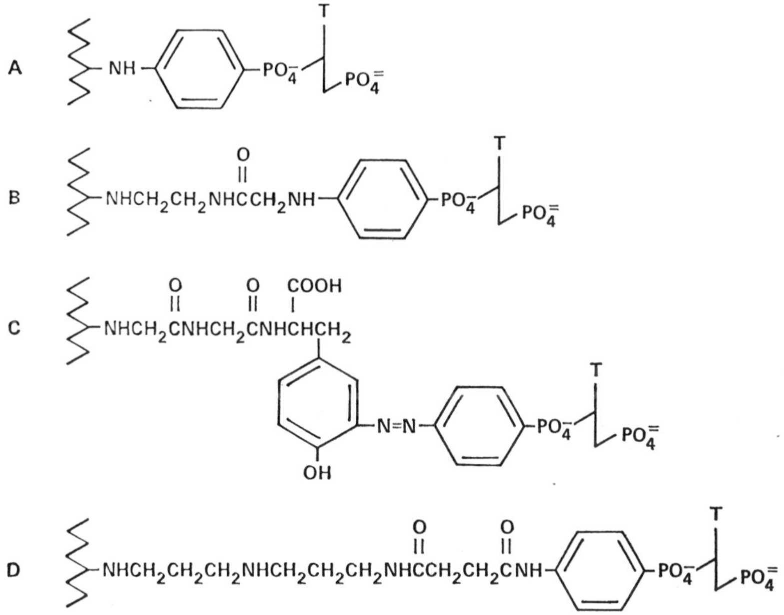

Fig. 7. Specific staphylococcal, nuclease affinity chromatographic adsorbents prepared by attaching the competitive inhibitor, pdTp-aminophenyl (84–86), to various derivatives of Sepharose-4B or Bio-Gel P-300 by the procedures described in ref. 4. In (a) the inhibitor was attached *directly* to agarose, after activation of the gel with cyanogen bromide, or to polyacrylamide via the acyl azide step. In (b), ethylenediamine was reacted with cyanogen bromide-activated agarose, or with the acyl azide polyacrylamide derivative. This amino gel derivative was then reacted with N-hydroxysuccinimide ester of bromoacetic acid to form the bromoacetyl derivative; the latter was then treated with the inhibitor. In (c), the tripeptide, Gly-Glu-Tyr, was attached by the α-amino group to agarose or polyacrylamide by the cyanogen bromide or acyl azide procedure, respectively; this gel was then reacted with the diazonium derivative of the inhibitor. In (d), 3,3′-diaminodipropylamine was attached to the gel matrix by the cyanogen bromide or acyl azide step. The succinyl derivative, obtained after treating the gel with succinic anhydride in aqueous media, was then coupled with the inhibitor with a water-soluble carbodiimide. The *jagged vertical lines* represent the agarose or polyacrylamide backbone.

ligand for the preparation of specific adsorbents because it is a strong competitive inhibitor (K_i, 10^{-6} M), its 3'-phosphodiester bond is not cleaved by the enzyme, it is stable at pH values 5 to 10, the pK of the aromatic amino group is low, and the amino group is relatively distant from the basic structural unit (pTp-X) recognized by the enzymatic active site (84–86). Figure 7 describes the various ways in which this ligand has been attached to agarose and to polyacrylamide beads (1,4). Derivative A, although it has the ligand closer to the matrix than other derivatives, is a very effective adsorbent for the enzyme (1). The capacity of derivatives B, C, and D for the enzyme, however, is higher than that of A. The less dramatic effect of "arm" length here, compared to some other enzymes, may be in part explained by the small size (mol. wt. = 17,000) of this enzyme, by the rather shallow binding or active site of the enzyme (87), and the effectiveness (low K_i) of the inhibitor used in the purification.

Crude solutions of the enzyme (culture medium) can be passed on columns containing any of the adsorbents depicted in Figure 7. After adsorption of enzyme, the column can be washed extensively with the equilibrating buffer. Quantitative elution of the enzyme is achieved in a very small volume by using 0.1 M acetic acid or ammonium hydroxide. The specific activity of the purified enzyme is as high as that obtained by the more tedious conventional procedures. The polyacrylamide derivatives comparable to the agarose adsorbents (Fig. 7) are much less effective than the latter, despite the fact that they contain much larger quantities of the adsorbent.

The agarose adsorbents can be used repeatedly, and they are effective in the separation of active and inactive nuclease derivatives from samples subjected to chemical modification (64,88,89).

2. β-Galactosidase

Several adsorbents have been prepared (8) by linking a substrate-analogue competitive inhibitor of *E. coli* β-galactosidase, *p*-aminophenyl-β-D-thiogalacto-pyranoside, to agarose (Fig. 8). Derivative *a*, which contains the inhibitor linked directly to agarose, was totally ineffective, since the enzymic activity was present entirely in the breakthrough protein peak. This occurred, despite having as much as 10 μM of the inhibitor per milliliter of packed agarose. In contrast, derivative *b* caused a slight retardation of the migration of the enzyme [as in Fig. 1(*b*)], and derivatives *c* and *d* caused very strong adsorption

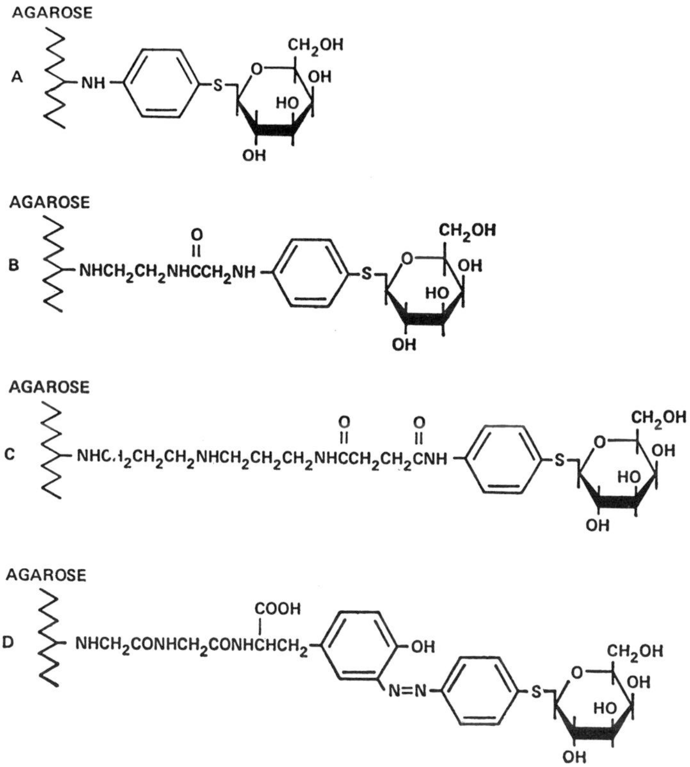

Fig. 8. Agarose adsorbents prepared for the selective purification of β-galactosidase from various sources by affinity chromatography. Adapted from Steers, Cuatrecasas, and Pollard (8).

of the enzyme. This is perhaps the best example of the dramatic effects that can result upon placing the ligand at an increasingly longer distance from the matrix backbone. Derivative c can be diluted 5-fold with unsubstituted agarose without losing the capacity of this adsorbent to extract β-galactosidase from $E.$ $coli$ extracts.

The enzyme adsorbed tightly at neutral pH to columns containing the resin c, and elution was achieved with buffers having a pH of 10 (Fig. 9). Despite the fact that the pH of the fractions containing the leading edge of the activity peak (fractions 91–93) is between 8.5 and 9.0, it is not possible to elute the enzyme directly with buffers of pH 7.5–9.5. The enzyme could not be readily eluted with buffers

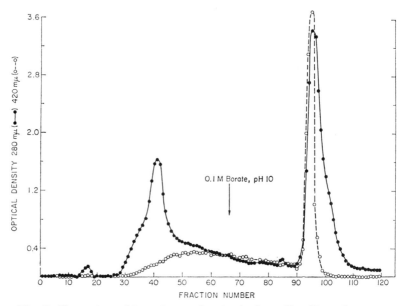

Fig. 9. Chromatographic patterns of extracts from *E. coli* on the agarose derivative substituted with *p*-aminophenyl-β-D-galactopyranoside, as depicted in Fig. 8(*c*). The column was equilibrated and chromatographed with 0.05 *M* Tris · HCl buffer, pH 7.5. Elution of adsorbed protein was carried out with 0.1 *M* sodium borate, pH 10.05. The extract (20 ml) was applied to a column, 1.5 × 22 cm. The column was run at room temperature (23°) with a flow rate of 80 ml/ hr. Fractions containing 0.8 ml were collected. Polyacrylamide disk gels of fractions 40 and 96 showed a single band. Protein was measured spectrophotometrically at 280 mμ (●———●); enzymatic activity at 420 mμ (○———○). Data from Steers, Cuatrecasas, and Pollard (8).

containing lactose, *o*-nitrophenyl-β-D-galactopyranoside, or isopropyl-β-D-galactopyranoside. The enzyme retains enzymatic activity despite being tightly bound to the agarose column. This was visually shown by the appearance of bright yellow color (*o*-nitrophenol) in the column and effluent during attempts to elute the enzyme with buffers containing the substrate, *o*-nitrophenyl-β-D-galactopyranoside. The enzyme could, however, be eluted with substrate-containing buffers if the column contained agarose having a low degree of ligand substitution. Elution from such columns was gradual, however, and the enzyme recovered was very dilute. An important advantage of elution with buffers of alkaline pH is in the quantitative recovery of the enzyme in a sharp boundary at the interface in a concentrated form. Elution with alkaline pH also makes it possible to use highly substituted resins, which have a large capacity for the enzyme, for large scale preparative purification.

The enzyme obtained by these procedures was pure by disk gel electrophoresis, and it possessed the physical and chemical characteristics of the protein purified by conventional procedures (8). The specific activity was between 300,000 and 320,000 units/mg. The procedure can be carried out on a preparative scale, and the adsorbent may be used several times without appreciable impairment of the binding capacity. The enzyme remains active while bound to the column; it can be readsorbed and eluted without apparent deleterious effects.

Of interest is the finding that the monomer form of the enzyme, previously identified in purified preparations of β-galactosidase (90) binds to the thiogalactoside agarose to the same extent as the tetramer form under the conditions studied and is coeluted with the tetramer at pH 10. It is apparent from these observations that while the monomer form of β-galactosidase is devoid of hydrolytic activity, it is nevertheless capable of binding the substrate analogue. This may prove to be especially useful in the purification of certain biologically inactive mutant forms of the enzyme. Several mutant forms of β-galactosidase have been shown to exist in either the nomoner or dimer form (91–93). In one such case, the mutant 3310, it was found that the molecule existed as a dimer, which retained full immunologic activity (91). Although enzymatically inactive, the catalytic site of this mutant is essentially unimpaired, since it is capable of forming fully active hybrid tetramer molecules when renatured from urea solutions containing wild-type enzyme (94). The basic defect, therefore, appears to be the loss of ability to associate into the active tetramer form. The purification of such molecular forms by conventional means is complicated by the absence of enzymatic activity, requiring complementation or immunological methods of assay. Furthermore, many mutant forms are considerably more labile to conventional purification procedures, and they are present initially in greatly reduced amounts in comparison with the wild type.

It is significant that a highly effective agarose adsorbent for β-galactosidase was prepared with a relatively weak (K_i about 5 mM) inhibitor. This strongly supports the view that affinity chromatography must not be restricted to those enzymes for which very potent competitive inhibitors are available or can be developed.

Polyacrylamide adsorbents of identical makeup to the agarose derivatives shown in Figure 8 were totally ineffective in adsorbing or retarding the migration of purified or crude samples of β-galactosidase from *E. Coli* (8).

The purification of β-galactosidase from *E. coli* has also been performed with β-thiogalactoside ligands attached to extensively cross-linked, insoluble, bovine β-globulin (95). Although polymerized bovine γ-globulin served as an adequate adsorbent in this case, its granular, irregular, and charged nature make it far from ideal as a solid support for column chromatography. Specifically, the use of this supporting gel has the following limitations: the lack of porosity impairs free communication between the aqueous and solid phases and restricts the protein-adsorbing capacity, nonspecific protein binding by electrostatic interactions is a potential hazard, and there are severe limitations in column flow rates.

3. Proteolytic Enzymes

A number of proteolytic enzymes, of which α-chymotrypsin and carboxypeptidase A are examples, are capable of binding, but not hydrolyzing, significantly, the enantiomorphic substrate analogue. This has been used to advantage to purify these two proteolytic enzymes (1). D-Tryptophan methyl ester, attached directly to agarose by the α-amino group, results in an adsorbent that can weakly interact with chymotrypsin. Increasing the length of this inhibitor, by the insertion of an ϵ-aminocaproyl group to the α-amino group, increased dramatically the effectiveness of the adsorbent obtained by attaching this ligand to agarose by the ϵ-amino group. Chymotrypsinogen, pancreatic ribonuclease, subtilisin, and DFP-trypsin did not bind to columns prepared with these derivatives.

Carboxypeptidase A adsorbs quite strongly to agarose derivatives containing the enantiomorphic inhibitor, L-tryosyl-D-tryptophan (Fig. 10).

Trypsin has also been purified on ovomucoid–agarose (75). The purification of a single, highly specific protease is not easily achieved with adsorbents of the latter type, which have complex macromolecular substitutes. Chymotryptic-like enzymes have been purified with agarose derivatives containing 4-phenylbutylamine (96).

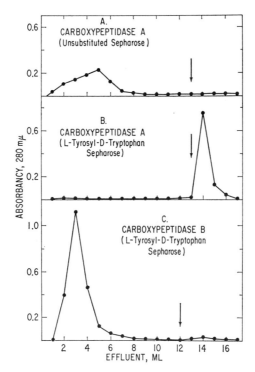

Fig. 10. Affinity chromatography of carboxypeptidase A on a column (0.5 × 6 cm) of agarose coupled with L-tyrosine-D-tryptophan. The buffer used was 0.05 M Tris·HCl, pH 8.0, containing 0.3 N NaCl. About 1 mg of pure carboxypeptidase A (a, b) and 1.8 mg of carboxypeptidase B (c), in 1 ml of the same buffer, were applied to the columns. Elution was accomplished with 0.1 M acetic acid (arrow). Data from Cuatrecasas, Wilchek, and Anfinsen (1).

Papain has been purified with an agarose resin containing the peptide, gly-gly-tyr(Bzl)-Arg (97). Approximately 50% of the protein of a purified papain preparation did not adsorb to a column containing this derivative. This material was catalytically inactive. In contrast, the adsorbed enzyme, which could be eluted with distilled water, contained 1 mole of SH and twice the specific activity of the starting pure enzyme. Another proteolytic enzyme, plasminogen, has been purified from human plasma by affinity chromatography on L-lysine-agarose columns (98). A 200-fold purification was achieved of material which on disk gel electrophoresis demonstrated seven enzymatically active bands.

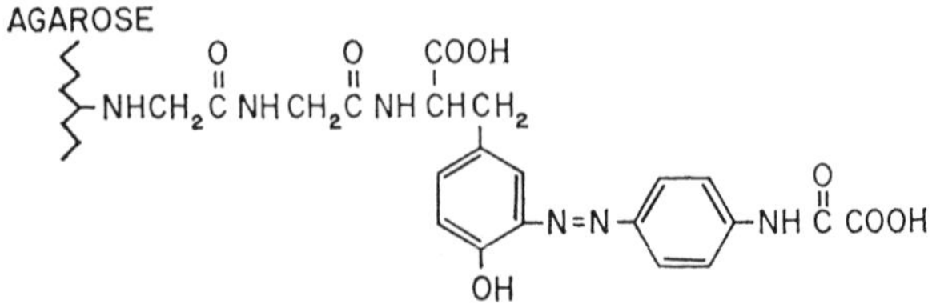

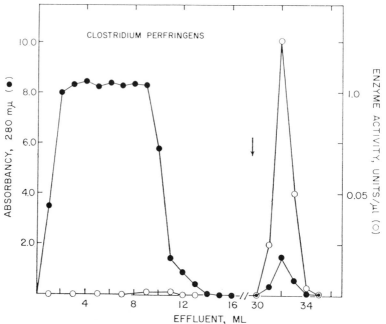

Fig. 11. Adsorbent used in the selective purification of neuraminidases by affinity chromatography (63). The diazonium derivative of N-(p-aminophenyl) oxamic acid was coupled to agarose-gly-gly-tyr.

Fig. 12. Affinity chromatography of neuraminidase from *Clostridium perfringens* on the agarose derivative shown in Scheme I. Seven milliliters of a commercially (Worthington) "purified" enzyme containing 73 mg of protein were dialyzed for 16 hr at 4° against 4 liters of 0.05 M sodium acetate buffer, pH 5.5, containing 2 mM $CaCl_2$ and 0.2 mM EDTA. This material was applied to a 0.4 × 4 cm column containing the absorbent, which had been equilibrated with the same buffer. Elution was achieved with 0.1 M $NaHCO_3$ buffer, pH 9.1 (arrow: the pH of the eluted sample was immediately lowered to 6.0 with 1 N NaOH.) The progress of the eluting buffer through the column can be followed conveniently by the intensification of the color of the adsorbent, which results upon ionization of the azo moiety by the higher pH of the buffer. The purification data of this experiment is summarized in Table II. Data from Cuatrecasas and Illiano (63).

4. Neuraminidases

A selective adsorbent for neuraminidase was prepared (63) by attaching through azo linkage an inhibitor of this enzyme, N-(p-aminophenyl)oxamic acid, to agarose beads containing the tripeptide, glycyl-glycyl-tyrosine (Fig. 11). Columns containing this adsorbent can completely extract the enzymatic activity present in extracts of *Clostridium perfringens* (Fig. 12) and *Vibrio cholerae*, and quantitative elution is readily achieved by modifying the pH and ionic strength of the buffer (Table II). Intact influenza virus particles are also reversibly

TABLE II

Affinity Chromatography of Neuraminidases from Various Sources

Purification step	Source of enzyme		
	C. perfringens	*V. cholerae*	Influenza virus[a]
Sample applied on column			
Protein (mg)[b]	73	92	1.2[c]
Volume (ml)	7	12	5
Specific activity[d]	3.6	0.071	0.86
Fraction eluted from column			
Protein (mg)[b]	1.1	0.2	0.5[c]
Specific activity[d]	164	30	1.9
Yield of activity (%)	105	97	91
Purification	45	420	2.2

These data summarize the chromatographic experiment for purification of the *C. perfringens* neuraminidase, illustrated in Figure 12. Influenza virus and the enzyme from *V. cholerae* were purified under identical conditions. Data from Cuatrecasas and Illiano (65).

[a] Purified monovalent inactivated influenza virus vaccine strain A2/Aichi/2/68, containing 400 CCA units/ml, courtesy of Dr. Allen F. Woodhour, Merck Institute for Therapeutic Research.

[b] Determined by absorbancy at 210 mμ.

[c] For convenience, this was determined by ultraviolet spectroscopy using $E_{280}^{0.1\%} = 3.0$.

[d] Micromoles of N-acetylneuraminic acid formed per minute per milligram of protein.

adsorbed by these columns, indicating the superficial location of neuraminidase in this virus and emphasizing the feasibility of using functional purification procedures for resolving complex and particulate biological structures.

5. Acetylcholinesterase

An elegant one-step purification procedure has been developed to extract acetylcholinesterase from the electric tissue of *Electrophorus electricus* and bovine erythrocyte membranes (35). The latter enzyme had not been successfully purified previously by conventional procedures. The following inhibitors were synthesized and coupled in various ways to agarose: trimethyl(p-aminophenyl)ammonium chloride hydrochloride, trimethyl(p-acetamidophenyl)ammonium iodide, trimethyl(m-aminophenyl)ammonium chloride hydrochloride, and the N-methyl-N-(p-aminophenyl)carbamate ester of m-(trimethylamino)-phenol. The derivatives prepared by coupling the inhibitors directly to the matrix backbone were totally ineffective, whereas excellent selective adsorbents resulted when the inhibitors were coupled by long extensions or "arms."

The eel enzyme was quantitatively adsorbed to the affinity column, and about 80% of the enzymatic activity could be eluted with 0.01 M tensilon, a strong inhibitor (K_i, 10^{-6} M) of the enzyme (35). The specific activity of the purified enzyme was 16,000 units/mg, which is at least as good as the best preparations obtained by conventional procedures, and on disk gel electrophoresis the purified protein migrated as a single electrophoretic component. Purification of the erythrocyte enzyme (35) was equally dramatic; an electrophoretically homogeneous preparation was obtained after an approximately 2500-fold purification.

A less effective affinity adsorbent for acetylcholinesterase from the electric eel has been described by Kalderon et al. (99). A relatively strong competitive inhibitor (K_i, 6×10^{-6} M), ϵ-aminocaproyl-p-aminophenyl-trimethyl-ammonium bromide was linked to agarose by the cyanogen bromide procedure. Small amounts of enzyme were adsorbed to relatively large columns; a substantial (more than one-third) amount of the total enzyme activity was unretarded, for some reason. The material adsorbed was not removed after washing with a volume of starting buffer equal to five times the column volume. It was eluted, however, by increasing the ionic strength (1 M NaCl) of the buffer; the enzyme was eluted in rather dilute form. A 17-fold purification was achieved by these procedures, which have not yet been applied in a preparative scale.

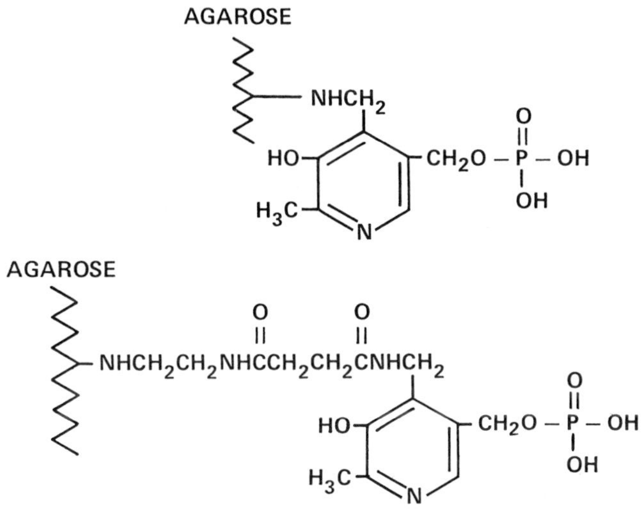

Fig. 13. Agarose derivatives of pyridoxamine phosphate used for the purification of tyrosine aminotransferase from hepatoma-tissue culture cells (34).

6. Tyrosine Aminotransferase

The behavior of tyrosine aminotransferase on affinity adsorbents has been studied by Miller, Cuatrecasas, and Thompson (34). Crude preparations of the enzyme from mouse hepatoma cells in tissue culture do not bind to adsorbents prepared by attaching pyridoxamine phosphate directly to agarose. If the ligand is attached to succinylated aminoethyl agarose (4), strong binding occurs (Fig. 13); the enzyme can be eluted with buffers containing pyridoxal phosphate. If the adsorbent contains the ligand attached by an even longer arm, succinylated aminodipropylamino agarose (4), binding of the enzyme is so strong that elution becomes difficult. As will be described later, these adsorbents have been used to enrich ribosomes capable of selectively synthesizing this enzyme.

7. Avidin

Biocytin–agarose adsorbents have been used by Cuatrecasas and Wilchek (100) to obtain 1.7 mg of pure avidin (70% yield) from 8 g of crude egg white in a single batch-type procedure. The binding of

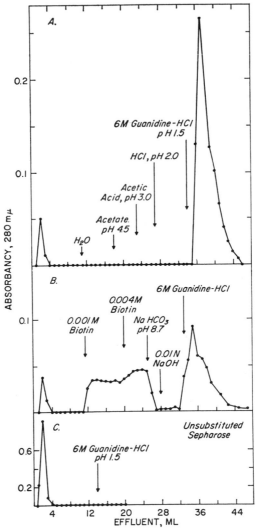

Fig. 14. Affinity chromatography of purified avidin on bicytin-agarose (a, b) and unsubstituted agarose (c) columns. The columns $(0.5 \times 5 \text{ cm})$ were equilibrated with 0.2 M NaHCO$_3$, pH 8.7, and 0.75 mg of avidin (in 0.5 ml of the same buffer) was applied to each column. One-milliliter fractions were collected; the flow rate was about 30 ml/hr; the experiments were performed at room temperature. Elution was attempted by varying the conditions as indicated (arrows). The small protein peak that emerges early in (a) and (b) represents an impurity that does not bind biotin [14]C. Data from Cuatrecasas and Wilchek (100).

this protein to the adsorbent is extremely tight in this case, reflecting the low dissociation constant of the complex in solution, 10^{-15} M (101). Elution requires the combined use of 6 M guanidine·HCl and pH of 1.5 (Fig. 14). A major portion of the binding energy of the complex is therefore retained in this insoluble derivative. This contrasts with the biotin–cellulose derivative prepared by McCormick (102), which, although capable of binding avidin, was susceptible to elution merely by slight changes in the ionic strength of the buffer.

8. cAMP-dependent Protein Kinases

A casein–agarose column has been used by Reimann et al. (103) to separate the catalytic and regulatory subunits of a cAMP-dependent protein kinase from rabbit skeletal muscle. The column was chromatographed in the presence of cAMP in the buffer; cAMP binds to the regulatory subunit, which presumably results in the release of the catalytic unit in active form. The latter binds to the casein on the column. The catalytic subunit is eluted from the column with an NaCl gradient. Upon addition of this subunit to the breakthrough protein, the enzyme is reconstituted.

Figure 15 depicts some agarose derivatives of cAMP that have been prepared in this laboratory (104) to purify the cAMP-dependent protein kinase from adipose tissue. When a crude homogenate from isolated adipose tissue cells is applied on a column containing these derivatives, the breakthrough protein contains all the original kinase activity (using histone as substrate), which is maximally stimulated in the absence of added cAMP. No free cAMP is present in this peak, and no regulatory or binding subunit can be detected. The latter subunit, which is adsorbed strongly to the selective adsorbent, can be eluted with a buffer containing 5 mM cAMP.

9. Chorismate Mutase

A very useful affinity adsorbent for chorismate mutase from *Claviceps paspali* was prepared by Sprossler and Lingens (105) by attaching the allosteric activator, L-tryptophan, to agarose by the cyanogen bromide procedure. Virtually all the activity from a 9-mg sample of DEAE–cellulose purified sample adsorbed strongly to a small column (1.5 × 9 cm). Elution of more than 80% of the activity resulted with phosphate buffers supplemented with 1 mM L-tryptophan. Agarose containing an inhibitor of this enzyme, L-phenylalanine, did not significantly

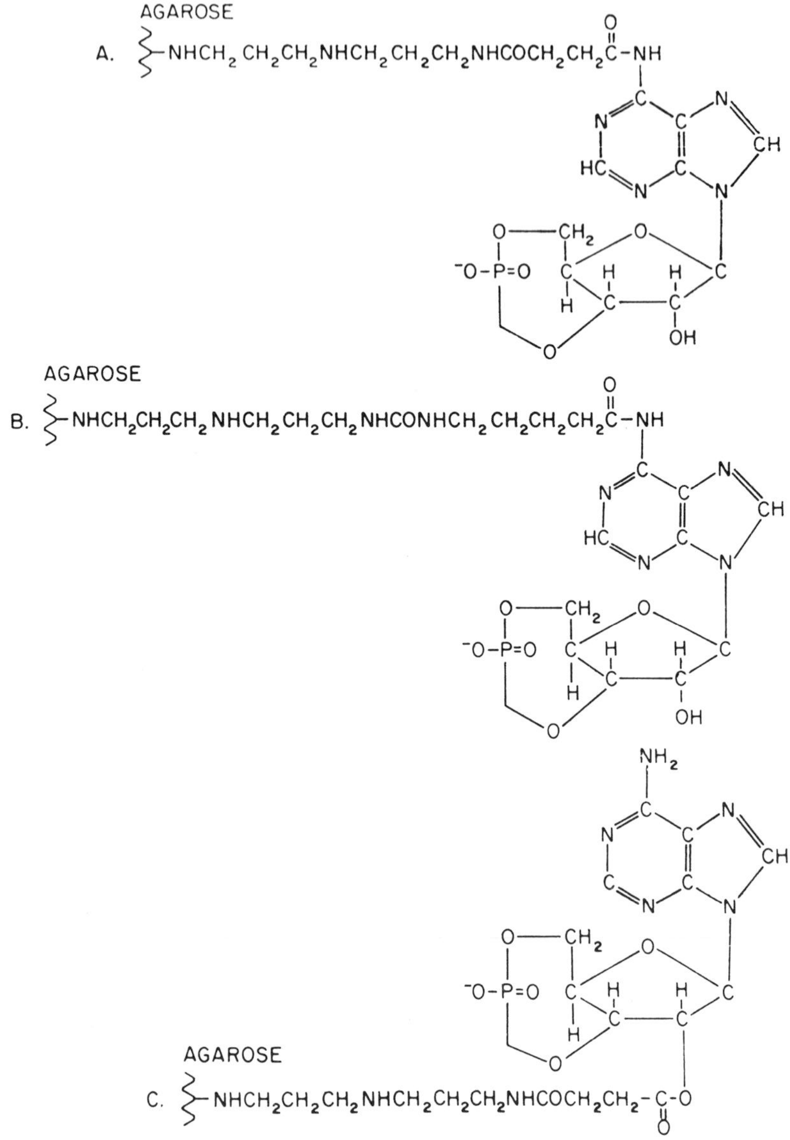

AGAROSE

A. ξ—NHCH$_2$CH$_2$CH$_2$NHCH$_2$CH$_2$CH$_2$NHCOCH$_2$CH$_2$C(=O)—NH

(purine–ribose–phosphate structure)

AGAROSE

B. ξ—NHCH$_2$CH$_2$CH$_2$NHCH$_2$CH$_2$CH$_2$NHCONHCH$_2$CH$_2$CH$_2$CH$_2$C(=O)—NH

(purine–ribose–phosphate structure)

(adenine–ribose–phosphate structure with NH$_2$)

AGAROSE

C. ξ—NHCH$_2$CH$_2$CH$_2$NHCH$_2$CH$_2$CH$_2$NHCOCH$_2$CH$_2$—C(=O)—O

Fig. 15. Selective adsorbents used for the purification of the regulatory subunit of a cAMP-dependent protein kinase from adipose tissue (104). N^6-Succinyl-cAMP is attached by carbodiimide reaction (4) to agarose containing 3,3'-diaminodipropylamine (upper). N^6-aminocaproyl-cAMP is coupled with a water-soluble carbodiimide to a succinylated diaminodipropylamine derivative of agarose (middle). 3'-OH-succinyl-cAMP is linked through a carbodiimide step to diaminodipropylamino agarose (lower).

70

retard the enzyme. Schmit, Artz, and Zalkin (106) used phenylalanine agarose to study the nature of the catalytic and regulatory sites of chorismate mutase-prephenate dehydratase. A mutant form of this enzyme, insensitive to inhibition by phenylalanine, binds to the specific adsorbent. Along with other chemical evidence, the authors interpreted these findings to suggest topographically distinct regulatory and catalytic sites. Thus affinity columns may provide rapid qualitative tests for binding of an end product inhibitor to a feedback-insensitive enzyme; this may be especially useful, since crude enzyme preparations may be used for such tests.

10. T4 Bacteriophage-specific Dihydrofolate Reductase

Dihydrofolate reductase specified by bacteriophage T4 adsorbs very strongly to columns containing polyacrylamide beads to which N^{10}-formylaminopterin(4-amino-10-formylpteroylglutamate) is attached covalently (107). An enzyme preparation partially purified on DEAE-cellulose was applied to such an adsorbent. About 80% of the activity could be eluted with 0.2 mM dihydrofolate. The enzyme, which was purified approximately 3000-fold by this step, was homogeneous on SDS disk gel electrophoresis.

11. Miscellaneous Enzymes

The principles of enzyme–substrate interactions have been used sporadically for a great many years for the purification of certain proteins (108). The binding of amylase to insoluble starch has been known for at least 60 years (109), and elution of the enzyme can be achieved with soluble starch (110). Similarly, glycogen synthetase binds tightly to particulate glycogen (111), and starch synthetase to bean starch granules (112). Some cases have been reported of specific substrate elution of enzyme that has been adsorbed to a carrier by strictly physicochemical forces. For example, dilute pyrophosphate solutions have been used to elute yeast pyrophosphatase from C alumina gel (113); fructose 1,6-diphosphate solutions can remove fructose 1,6-diphosphatase and aldolase differentially from CM-cellulose (114); and a chicken pancreas nuclease is eluted from phosphorylated cellulose with RNA solutions (115). Insoluble substrates have also been used to trap, and thus partially purify, chymotrypsin (116) and trypsin (117).

Perhaps the first example of purification by what is now called affinity chromatography was Lerman's demonstration of adsorption of tyrosinase on an adsorbent prepared by reacting diazotized aminophenol to cellulose that contained resorcinol residues in ether linkage (11). Elution of the enzyme could be achieved with buffers having a pH of about 9.5. Arsenis and McCormick described adsorption and elution of liver flavokinase on columns containing flavin covalently linked to cellulose and CM–cellulose (12,13).

Chan and Takahasi reported a 100-fold purification of 3-deoxy-D-arabino-heptulosonate-7-phosphate synthetase on a column containing an effector of this enzyme, L-tyrosine, attached to agarose (118). The phenylalanine-sensitive enzyme was not retarded. Some purification was achieved, although the protein did not bind tightly to the column. The downward migration of the enzyme was merely retarded with respect to the bulk of the protein, similar to the behavior reported earlier for α-chymotrypsin when the inhibitor was attached *directly* to the matrix backbone (1). Although such chromatographic patterns can be useful in certain circumstances, the procedures become very cumbersome, and the purification too incomplete, when dealing with large samples of crude starting material. It is tempting to speculate that strong adsorption of this enzyme would occur if the tyrosyl group were placed farther from the matrix backbone, as occurs in the several cases already discussed. This is especially likely, since the dissociation constant (5×10^{-5} M) for L-tyrosine is quite small (119).

A hemoglobin-agarose column prepared by the cyanogen bromide procedure, was used by Chua and Bushuk to purify proteases from crude extracts of malted wheat flour (120). High recoveries (90%) of the proteolytic activity applied to the column were achieved on elution with 0.1 N acetic acid, and virtually all the nonproteolytic components were separated by this procedure. Three major and one minor protein of similar electrophoretic mobility were detected in the eluted protein.

In a manner quite analogous to that used in the purification of staphylococcal nuclease [ref. 1; Fig. 7(a)], Wilchek and Gorecki (121) coupled an inhibitor of pancreatic ribonuclease, aminophenyl-p-U-cyclic-p, to agarose. The pancreatic enzyme adsorbed well to this adsorbent; reduced and oxidized ribonuclease did not bind at all.

Holshan, Mahazan, and Fondy (122) have made some interesting observations on the behavior of glycerol-3-P dehydrogenase on affinity adsorbents. Although the cytoplasmic form of this enzyme, which requires NAD (NAD-G3PDH), has been obtained in pure form in the past, the mitochondrial flavin-linked enzyme (F-G3PDH) has resisted purification by classical procedures. The latter enzyme binds to an agarose adsorbent prepared by reacting hexamethylenediamine-treated (4) agarose with 1-Cl-glycerol-3-P, or to an adsorbent prepared by coupling (with cyanogen bromide) the compound obtained by reacting hexamethylenediamine with 1-Cl-glycerol-3-P. The K_i of the latter is rather large, 2 mM. Elution of a 30-fold purified enzyme can be affected with 0.5 M DL-glycerol-3-P. The cytoplasmic NAD-G3PDH also binds to this column, although not as tightly as the flavin enzyme, and is independent of the presence of NAD. If these enzymes can be satisfactorily purified in good yield, it will be possible to compare the properties of the two forms from several tissue sources.

Kristiansen et al. (123) have described the purification of L-asparaginase from *E. coli* by affinity chromatographic procedures. As demonstrated for other enzymes (1,4,8,34,35), *direct* attachment of D-asparagine to agarose resulted in an ineffective adsorbent, and it was necessary to insert an aliphatic extension or arm to achieve a satisfactory adsorbent. Elution was achieved with 1 mM D-asparagine. Although insufficient amounts of the enzyme were purified to determine the specific activity, it is likely that the degree of purification was considerable. The authors expressed some concern that large-scale purifications may be complicated by the fact that D-asparagine can be hydrolyzed by the enzyme. It has been pointed out (2) that in some cases substrate-agarose columns can be used quite effectively if the experimental conditions (such as temperature, buffer, pH, or salt concentration) are altered so that the binding function is favored and the catalytic rate decreased. Particularly valuable in such cases would be the performance of the chromatographic steps at depressed temperatures, which would suppress the catalytic rates by large factors.

Shaper, Barker, and Hill (124) have recently described the coupling of uridine, adenine, and guanosine nucleotide diphosphate hexitolamine derivatives to carboxylic acid agarose. The uridine diphosphate agarose derivative adsorbs lactose synthetase from whey. Glycogen synthetase also adsorbs to this derivative, and it can be eluted with glycogen solutions. These derivatives hold considerable promise in the purification of a variety of glycosyltransferases and other enzymes that can bind these nucleotide cofactors.

An affinity adsorbent prepared by coupling p-aminobenzyl-1-thio-β-D-xylopyranoside to agarose has been used to purify β-D-xylosidase from *Bacillus pumilus* (125). 3-Iodotyrosine-agarose has been used to assist purification of brain tyrosine hydroxylase (126). Erythrocyte carbonic anhydrase has been purified on agarose derivatives containing the inhibitor, p-aminomethylbenzene sulfonamide (127).

B. PURIFICATION OF SULFHYDRYL PROTEINS

The preparation of an agarose organomercurial derivative by very simple procedures was described by Cuatrecasas (4). p-Chloromercuro-benzoic acid is coupled to aminoalkyl agarose (Fig. 16) with a water-soluble carbodiimide. The resultant agarose has a very high capacity for sulfhydryl proteins and may be useful in a variety of separative procedures. For example, this material can separate, from purified thyroblobulin preparations, molecules that have no free sulfhydryl groups (do not bind) from molecules that have a single, free sulfhydryl (128). Elution can be achieved with buffers containing chelating or reducing reagents. Sluyterman and Wijdenes (129) have attached p-aminophenylmercuric acetate to agarose by the cyanogen bromide procedure (4). The adsorbent is very useful in the separation of active papain (binds very strongly) from the enzymatically inactive enzyme. Elution could be affected with solutions containing 0.5 mM $HgCl_2$, or 0.5 mM mercaptoethanol. The procedure was readily adapted to preparative separations (2.6 g of papain). Insoluble sulfhydryl derivatives of cellulose (130) and of cross-linked dextran (131) have been prepared; however, the procedures are much more complicated and the products much less effective than with the agarose derivatives described above.

C. RESOLUTION OF CHEMICALLY MODIFIED ENZYMES

Studies of the functional effects of chemical modification of purified enzymes frequently reveal incomplete loss of enzymic activity. It is often difficult to determine if this activity represents residual native

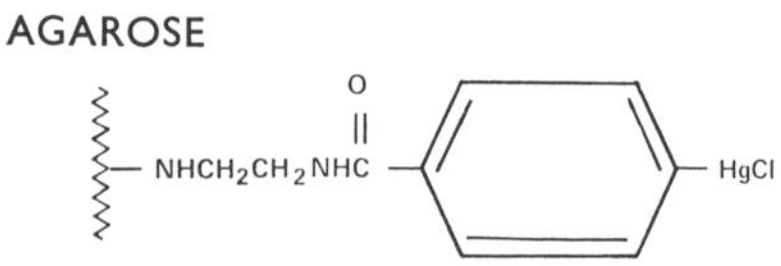

Fig. 16. Organomercurial agarose derivative used for the purification of sulfhydryl-group containing proteins (4).

enzyme or an altered protein with diminished catalytic power. Separation of the active native and the catalytically inert protein is often difficult, but in certain cases the difficulty may be resolved by using affinity adsorbents. For example, modification of staphylococcal nuclease by attachment of a single molecule of an affinity-labeling reagent through an azo linkage to an active site tyrosyl residue resulted in loss of 83% of the enzyme activity (88). Chromatography of this enzyme solution on a nuclease-specific agarose column (1) revealed that the residual activity was caused entirely by a 20% contamination of native enzyme; complete resolution of the two components was possible by affinity chromatography. Similar separations were made with partially active preparations of staphylococcal nuclease modified with bromoacetamidophenyl affinity-labeling reagents (62), and residual native nuclease could be separated from an inactive peptide fragment obtained by specific tryptic cleavage (144). Whitney (127) utilized an affinity column to separate native erythrocyte carbonic anhydrase from enzyme that was modified at its active site by substitution of a carboxamidomethyl group.

D. PURIFICATION OF AFFINITY-LABELED ACTIVE-SITE PEPTIDES

Givol et al. (132) described a clever application of affinity chromatography that can be used to purify uniquely labeled peptides obtained during affinity-labeling studies of enzymes. The native enzyme, linked to agarose by the cyanogen bromide procedure, is used to adsorb selectively the ligand-bound peptide present in the peptide mixture obtained by proteolytic digestion of the affinity-labeled enzyme The success of this technique, of course, will depend on how well the native enzyme will recognize the ligand after it has been altered by covalent linkage to a peptide. The investigators (132) demonstrated the usefulness of the procedures with studies on affinity labeling of pancreatic RNase and of goat anti-DNP antibodies. Wilchek (133) used the same methods to purify labeled peptides derived from affinity labeling of staphylococcal nuclease with bromoacetyl-substrate analogues (62).

E. PURIFICATION OF PROTEIN INHIBITORS, HORMONE TRANSPORT, AND VITAMIN-BINDING PROTEINS

The single-step purification of a biotin-binding protein (avidin) was described earlier. No other reports have yet appeared on the purification of vitamin-binding proteins, but we can anticipate that these will occur in the near future, since these are macromolecules that should be well suited to functional purification. Pensky and Marshall (134) have described the purification of human serum thyroxine-binding globulin with L-thyroxine-agarose columns. Pages, Cahnmann, and Robbins have also successfully prepared and used columns containing agarose with N-(6-2-minocaproyl) thyroxine, or thyroxine itself (135). The serum thyroxine-binding globulin, which has a dissociation constant of about 10^{-10} M for thyroxine and which is present at very low concentrations in serum (10–20 $\mu g/ml$), remains strongly adsorbed to the agarose after washing the column with 0.1 M $NaHCO_3$ buffer, pH 9.0, and is eluted with 4 mM NaOH, pH 11.4. Although not entirely pure, the eluted protein is enriched considerably by this single-step procedure.

Marchesi (136) has purified wheat germ agglutinin to homogeneity (on disk gel electrophoresis) by passing a crude solution of material on an affinity column containing ovomucoid coupled to agarose by the cyanogen bromide procedure (4).

Vonderhaar and Mueller (137) have prepared estradiol derivatives of polyvinyl (N-phenylenemaleimide) and of cellulose to attempt extraction of the estradiol receptor protein of uterus. Although the binding activity of the crude supernatant "disappeared" upon exposure to these materials, recovery of the binding protein was not possible. Similarly, Burstein (138) used a testosterone agarose support to "remove" the serum-binding activity, but he was unable to affect elution. Extensive studies in this laboratory (33) on the use of affinity chromatography for the purification of serum and uterine estradiol-binding proteins suggest that in the two studies quoted above, the disappearance of the binding activity resulted not from adsorption to the support but from release of free estradiol or testosterone, which then binds with extraordinary avidity to the protein free in solution and effectively "inactivates" its subsequent capacity to bind the small quantity of the radioactive hormone. This possibility was not examined in these studies, and it is very likely that the washing procedures were inadequate. It is necessary to wash the estradiol and testosterone adsorbents with extremely large volumes of solutions, including organic solvents, and over a prolonged period of time, to completely remove residual free steroid which is strongly adsorbed to the solid matrix. Even traces of such free estradiol will result in the erroneous impression of selective removal of the receptor protein. A number of

estradiol adsorbents prepared by coupling to derivatized agarose
(Fig. 17) and glass beads (Fig. 18) have been used successfully to
purify the serum-binding protein (33). The uterine receptor protein
can also be adsorbed, but the yields of elution have not yet been very
high.

Affinity chromatography can also be used to purify protein or
peptide inhibitors of certain enzymes. For example, a ribonuclease-
agarose column was useful in the purification of ribonuclease inhibitor
from liver (77).

Amyloid protein can be purified on columns of agarose that contain
Congo red dye (139), a compound that has extraordinary avidity for
this protein.

F. PURIFICATION OF COMPLEMENTARY AND SYNTHETIC PEPTIDES

Affinity chromatography applies also to cases in which a large
peptide or protein is immobilized in order to adsorb other peptides or
proteins capable of selectively associating with it. Some of the methodo-
logical considerations important in the attachment of protein
molecules to agarose have been presented in an earlier section. In
certain favorable cases, biologically active proteins may be cleaved
into fragments that are separately inactive but which reassociate with
the regeneration of much or all of the specific function. The well-known
association of RNase-S-protein with RNase-S-peptide, originally
described by Richards and Vithayathil (140), was used by Hofmann
and his colleagues (141) for the purification of RNase-S-peptide
synthesized by the classical methods of peptide chemistry. This
system was utilized by Kato and Anfinsen (142) to prepare useful
agarose adsorbents. Columns of RNase-S-protein, prepared by the
cyanogen bromide technique, showed a good capacity for RNase-S-
peptide and a relatively high degree of specificity in the selection of
active S-peptide from the crude preparation of this material resulting
from solid phase synthesis. Affinity chromatography made it possible
to demonstrate that synthetic material contained not only populations
of molecules that bound tightly to S-protein but yielded inactive
recombinants, but also molecules that yielded active RNase-S with
only very low affinity for S-protein.

Similar experiments have been carried out with another two-
component system, nuclease-T (143). When staphylococcal nuclease is

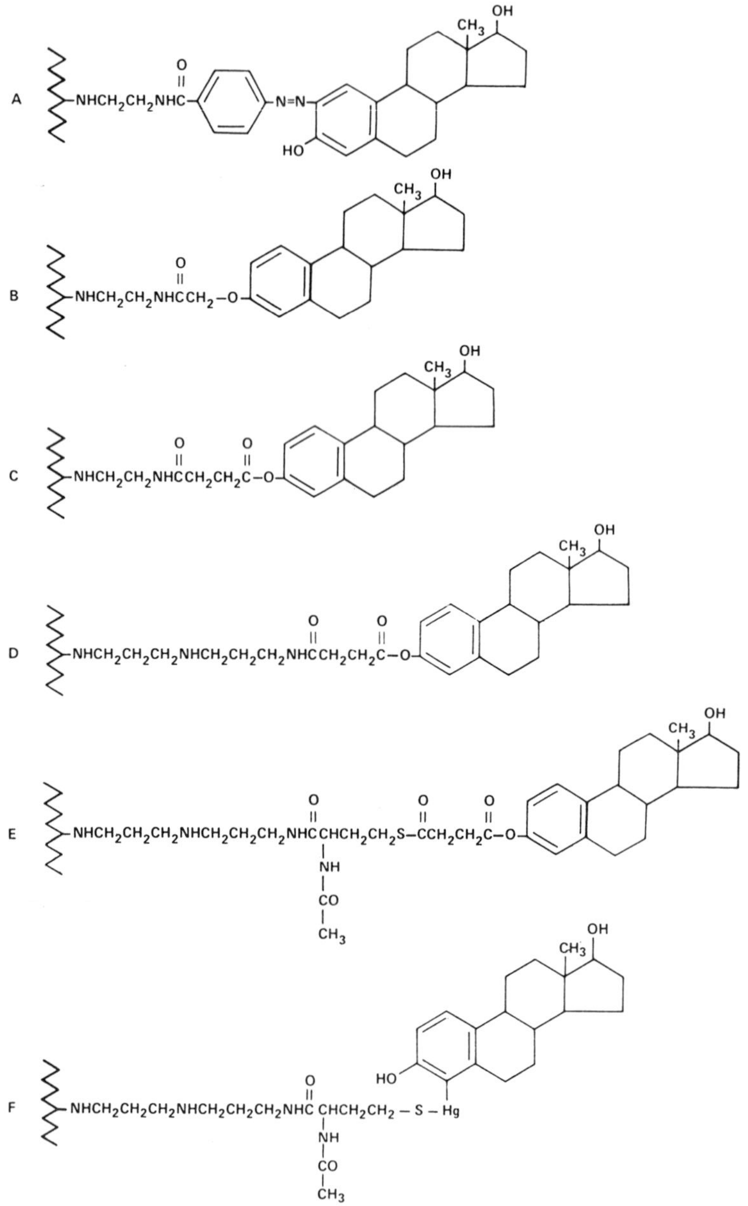

Fig. 17. Agarose derivatives of estradiol that have been used to study the purification of serum and uterine estradiol-binding proteins (33).

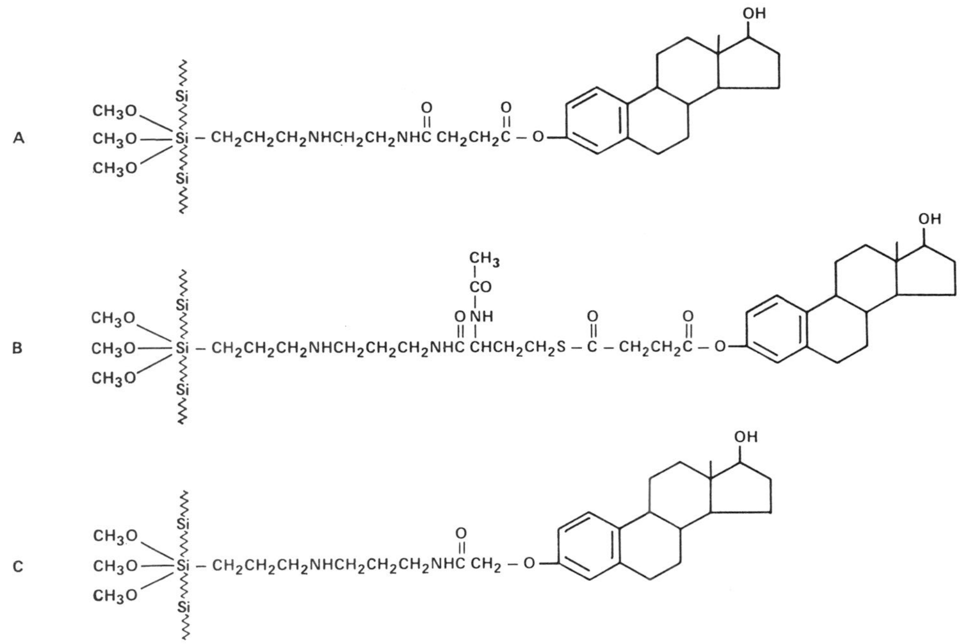

Fig. 18. Estradiol-glass derivatives used in the study of the purification of serum and uterine estradiol-binding proteins (33).

79

digested with trypsin in the presence of calcium ions and the tightly bound inhibitor pdTp, cleavage occurs at two positions in the 149 residue chain. A pentapeptide fragment, residues 1–5, is rapidly removed without effect on activity or physical structure, followed by cleavage between residues 48, 49, and 50. The two large fragments that result, 6–48 and 49 or 50–149, are structureless and inactive alone in solution but recombine noncovalently to yield nuclease-T, which has about 8–10% the activity of the native enzyme. An adsorbent for the 6–48 fragment was prepared by attachment of the larger fragment (residues 49, 50–149) to agarose (144). This material was then employed for the purification of the products of solid phase synthesis of the 6–47 sequence (residue 48 is dispensible for function) and of various analogues of this sequence in a study of the critical residues in the active site (144,145). The synthetic material contains a population of molecules with a spectrum of affinities for the 49, 50–149 fragment, yielding nuclease-T recombinants of varied specific activity. The technique has permitted the isolation of synthetic 6–47 having 30–50% of the biological activity of native material.

G. EXPLORATION OF ENZYME SUBUNIT INTERACTIONS

An instructive example of a special use of insoluble derivatives is the study of Hennig and Ginsburg on agarose-coupled glutamine synthetase subunits (146). The binding of ions, substrates, or inhibitors to the dissociated subunits is not possible because they cannot be maintained in the dissociated state in the absence of denaturants. The agarose-coupled subunits are catalytically inactive. If the sulfhydryl groups of these subunits are blocked with p-chloro-mercuriphenylsulfonate they cannot be adenylated by ATP: glutamine synthetase adenylytransferase. Adenylylation proceeds quite normally if the organ mercurial is removed with mercaptoethanol. The structure of the glutamine synthetase subunits attached to agarose is probably not very different from their structure in the native dodecameric molecule.

Insoluble derivatives of proteins can also be used advantageously to study effects of subunit aggregation and polymerization. For example, insulin, a polypeptide hormone with a great propensity for aggregation, is active biologically when insolubilized to agarose (64). Since these agarose derivatives are washed extensively with urea and guanidine before use, it is clear that it is the monomeric species of insulin which is the biologically active form of the hormone. This information must be carefully considered in interpreting structure–function features of this protein from the tertiary structure obtained on hexamer crystals by X-ray crystallography.

H. PURIFICATION OF ANTIGENS AND ANTIBODIES

The preparation and use of immunoadsorbents has been thoroughly discussed in other reviews (10,15–20). The advantages of agarose as a supporting solid carrier were presented in earlier sections. For these reasons, agarose derivatives have become increasingly popular over the past few years. Perhaps the most significant advantages of agarose over cellulose are the greater adsorptive capacity for proteins and the strength with which these proteins are adsorbed to the column. As described in the section on the coupling of proteins to agarose, superior protein–agarose adsorbents result if the protein is coupled by the fewest possible linkage points.

Agarose adsorbents have been used successfully to purify insulin antibodies (66); mouse myeloma IgA proteins, which bind nitrophenyl ligands (67); staphylococcal nuclease antibodies (69); a variety of antihapten antibodies (68); canine brucellosis antibodies (70); anti-polysaccharide antibodies from streptococcal Group A antiserum (71); and sperm-whale myoglobin (76). A variety of antigens have been purified by using specific immunoglobulin–agarose derivatives. Small quantities of insulin can be very efficiently extracted with agarose containing insulin antibodies (2,4,72). Antibody to human chorionic somato-mammotropin coupled to agarose has been used to purify labeled and unlabeled hormone for radioimmunoassay (73). Agarose containing antibodies to human placental lactogen can remove more than 99% of monkey growth hormone from homogenates of monkey pituitaries (82). Similarly, contaminants of an endopeptidase, protein-ase "c," can be removed from partially purified phaseolain preparations with an agarose adsorbent containing antibodies to the proteinase (147). Adsorbents having greater capacity are obtained if the protein is coupled to agarose at a pH between 6 and 7 (2,4,5).

I. APPLICATIONS TO NUCLEIC ACID BIOCHEMISTRY

Water-insoluble derivatives of nucleic acids have been used in a number of interesting studies. It has been shown that single-stranded DNA binds to nitrocellulose filters (148), and that this DNA can bind homologous RNA (149). This has been exploited to develop general procedures for isolation of gene-specific mRNA (150,151).

Nucleic acids have been covalently attached by the hydroxyl groups of oligonucleotides to acetylated phosphocellulose (phosphate ester bond) with dicyclohexylcarbodiimide in pyridine (152) or methanol (153). Coupling of oligonucleotides by a terminal phosphate to the hydroxyl group of cellulose has been done very successfully with special carbodiimide reagents, both in anhydrous (58,154–156) and in aqueous (60) media. These insoluble oligo- and polynucleotide celluloses have been used to separate, fractionate, and determine the structure of various nucleic acids (154,157,158). These studies are based on the relative stability of complexes formed between components in the aqueous mixture and complementary insoluble polynucleotide structures.

Water-soluble polynucleotide celluloses have also been useful in the study of enzyme mechanisms. Cozzarelli et al. attached a polynucleotide to cellulose by a phosphate terminus (155). Enzymes that catalyze the covalent linking of interrupted deoxyribonucleotide strands of a bihelix could then be assayed by determining the amount of H^3 poly dC (free in solution) that was linked covalently to the polynucleotide cellulose. Jovin and Kornberg (156) studied various polynucleotide cellulose derivatives as solid state primers and templates for DNA polymerases. Litman (159) has used deoxyRNA cellulose to purify deoxyRNA polymerase. Weith et al. prepared cellulose derivatives containing dihydroxyboryl groups that could form specific complexes with nucleic acid components, sugars, and polyols (160). Polynucleotide phosphorylase coupled to cellullose by the cyanogen bromide procedure has been used as a solid state synthetic enzyme to prepare nucleotide polymers (161).

Denbury and DeLuca described a specific method of purifying specific tRNA species by affinity chromatography (162). Purified isoleucyl-tRNA was coupled at a relatively low pH (64) to agarose with retention of 40% of the original enzymic activity in the PPi-ATP exchange reaction. Formation of ^{14}C-isoleucyladenylate on the insoluble enzyme could be demonstrated, and this occurred on about 40% of the bound enzyme. This isoleucyladenylate–enzyme–agarose complex could bind specific tRNA, demonstrating nicely the use of solid adsorbents in studying interactions between very large macromolecular complexes. The isoleucyl group on the immobilized enzyme could be transferred, under certain conditions, to the specific tRNA. The enzyme–agarose derivatives were relatively stable and could be used repeatedly.

Poonian, Schlaback, and Weissbach (163) have very recently described techniques for coupling nucleic acids to agarose by using the cyanogen bromide procedure described by Cuatrecasas (4). They demonstrated that single-stranded deoxy and ribonucleic acids are readily linked, whereas double-stranded ones are not. They described procedures by which single-stranded ends could be introduced into double-stranded molecules, thus permitting their covalent attachment to agarose. DNA polymerase from HeLa cells adsorbed to DNA agarose about 50-fold more efficiently than to DNA cellulose. The availability of these procedures should rapidly extend the usefulness of affinity chromatography to many areas of nucleic acid research.

J. PURIFICATION OF CELLS, RECEPTORS, AND OTHER COMPLEX STRUCTURES

The concepts and techniques of affinity chromatography can, in principle, be applied to the study and isolation of complex cellular structures, such as functional membrane structures, polyribosomal complexes, hormone and drug receptor structures, multienzyme complexes, repressor substances, and transport proteins (2,3). It is perhaps in this area that the application of the principles and techniques of affinity chromatography will find the greatest rewards, since it is here that the conventional purification methods are most difficult to apply. Until now, however, only a few but encouraging results have appeared in this area.

Miller, Cuatrecasas, and Thompson (34) have demonstrated that pyridoxamine phosphate–agarose, which adsorbs tyrosine aminotransferase of mouse heptatoma, can also be used to affect substantial purification of ribosomes capable of selectively synthesizing this enzyme. Presumably nascent protein, or unassembled protein subunits, present in the polyribosomal structures is capable of interacting strongly with the insoluble ligand. The ribosomes that do not stick to the column, in contrast to those that do and that are eluted with pyridoxal phosphate solutions, do not synthesize significant amounts of the enzyme.

Properly prepared insulin–agarose derivatives retain the biological properties of the hormone when tested on isolated fat cells, which indicates that interactions with the cell membrane are not destroyed and that the insulin receptor structures are probably located in the membrane (64). The insulin–agarose derivatives are also biologically active (stimulation of RNA synthesis) when tested on isolated mammary cells (81). Insulin-agarose columns retain isolated fat cell ghosts that can be eluted with solutions containing insulin. Such insulin-agarose derivatives can selectively adsorb receptor-containing membrane fragments from liver or fat cells. These adsorbents have been used to purify by a factor of about 250,000 the insulin receptor isolated from liver membranes (164).

Furthermore, the glucagon–agarose derivatives (Fig. 6) described in an earlier section can also selectively retain receptor-containing membranes (83) and they should be potentially very useful for receptor purification. Adrenocroticotropin coupled to agarose by the same procedures used to prepare glucagon–agarose (Fig. 6, B) can induce steroidogenesis in free adrenal cells in the same manner as the free hormone (80).

Wigzell and Andersson (51), Wigzell and Makela (32), and Evans et al. (165) have separated normal and immune lymphoid cells by antigen-(hapten-) coated glass and plastic beads. Truffa-Bachi and Wofsy (26) have used similar affinity columns by using polyacrylamide beads containing covalently linked hapten. In these columns the nonspecific attachment of cells is drastically reduced over that observed with the glass and plastic beads (31,32). Davie and Paul (166) have used hapten–agarose adsorbents for similar separations of immunocompetent cells. Edelman and colleagues (167) have recently described novel procedures for fractionating cells on fibers.

In general, cells that can potentially produce antibodies against the immobilized hapten bind to such affinity columns, whereas cells incapable of responding to that given immunogen are not retarded. The addition of hapten to the cells prevents their adsorption to the column. The results of all these studies are very encouraging, since they provide means for purifying all populations with characteristic membrane-receptor proteins. The studies indicate, furthermore, that preformed receptors are present in the membrane of the cell, and that these receptors have similar, if not identical, binding characteristics to the humoral antibody which can potentially be produced by that cell (32).

References

1. Cuatrecasas, P., Wilchek, M., and Anfinsen, C. B., *Proc. Nat. Acad. Sci. U.S.*, *61*, 636 (1968).
2. Cuatrecasas, P., In *Biochemical Aspects of Reactions on Solid Supports*, G. R. Stark, Ed., Academic Press, New York, 79 (1971).
3. Cuatrecasas, P., and Anfinsen, C. B., *Ann. Rev. Biochem.*, *40*, 259 (1971). Ed., Vol. 23, Academic Press, New York, p. 345 (1971).
4. Cuatrecasas, P., *J. Biol. Chem.*, *245*, 3059 (1970).
5. Cuatrecasas, P., and Anfinsen, C. B., In *Methods in Enzymology*, W. Jacoby, Ed., Vol. 23, Academic Press, New York, p. 345 (1971).
6. Cuatrecasas, P., *J. Agr. Food Chem.*, *19*, 600 (1971).
7. Jerina, D. M., and Cuatrecasas, P., *Proc. Int. Congr. Pharm.*, *4th*, *Basel, Switzerland*, Vol. 1, Schwabe, Basel/Stuttgart, 1970, p. 236.
8. Steers, E., Cuatrecasas, P., and Pollard, H., *J. Biol. Chem.*, *246*, 196 (1971).
9. Campbell, D. H., Leuscher, E., and Lerman, L. S., *Proc. Nat. Acad. Sci. U.S.*, *37*, 575 (1951).
10. Silman, I., and Katchalski, E., *Ann. Rev. Biochem.*, *35*, 873 (1966).
11. Lerman, L. S., *Proc. Nat. Acad. Sci. U.S.*, *39*, 232 (1953).
12. Arsenis, C., and McCormick, D. B., *J. Biol. Chem.*, *239*, 3093 (1964).
13. Arsenis, C., and McCormick, D. B., *J. Biol. Chem.*, *241*, 330 (1966).
14. Goldman, R., Goldstein, L., and Katchalski, E., In *Biochemical Aspects of Reactions on Solid Supports*, G. R. Stark, Ed., Academic Press, New York, p. 1 (1971).
15. Sehon, A. H., *Pure Appl. Chem.*, *4*, 483 (1962).
16. Sehon, A. H., *Brit. Med. Bull.*, *19*, 183 (1963).
17. Porter, R. R., and Press, E. M., *Ann. Rev. Biochem.*, *31*, 625 (1962).
18. Franklin, D. C., *Progr. Allergy*, *8*, 58 (1964).
19. Manecke, G., *Pure Appl. Chem.*, *4*, 507 (1962).
20. Katchalski, E., *Polyamino Acids, Polypeptides, Proteins, Proc. Int. Symp.*, University of Wisconsin Press, Madison, 1962, p. 283.
21. Hjerten, S., *Arch. Biochem. Biophys.*, *99*, 446 (1962).
22. Porath, J., Axén, R., and Ernbäck, S., *Nature*, *215*, 1491 (1967).
23. Axén, R., Porath, M., and Ernbäck, S., *Nature*, *214*, 1302 (1967).
24. Cuatrecasas, P., *Nature*, *228*, 1327 (1970).
25. Inman, J. K., Dintzis, H. M., *Biochemistry*, *8*, 4074 (1969).
26. Truffa-Bachi, P., and Wofsy, L., *Proc. Nat. Acad. Sci. U.S.*, *66*, 685 (1970).
27. Cuatrecasas, P., unpublished observations.
28. Krug, F., Desbuquois, B., and Cuatrecasas, P., *Nature New Biology*, *234*, 268 (1971).
29. Weetall, H. H., and Hersh, L. S., *Biochim. Biophys. Acta*, *185*, 464 (1969).
30. Weetall, H. H., *Science*, *166*, 615 (1969).
31. Wigzell, H., and Andersson, B., *J. Exp. Med.*, *129*, 23 (1969).
32. Wigzell, H., and Makela, O., *J. Exp. Med.*, *131*, 110 (1970).
33. Cuatrecasas, P., and Puca, G. A., unpublished observations.
34. Miller, J. V., Jr., Cuatrecasas, P., and Thompson, E. B., *Proc. Nat. Acad. Sci. U.S.*, *68*, 1014 (1971).

35. Berman, J. D., and Young, M., *Proc. Nat. Acad. Sci. U.S.*, *68*, 395 (1971).
36. Pressman, D., Campbell, D. H., and Pauling, L., *J. Immunol.*, *44*, 101 (1942).
37. Malley, A., and Campbell, D. H., *J. Amer. Chem. Soc.*, *85*, 487 (1963).
38. Talmage, D. W., Baker, H. R., and Akeson, W., *J. Infec. Dis.*, *84*, 199 (1954).
39. Gurvich, A. E., *Biochemistry*, *22*, 977 (1957).
40. Nezlin, R. S., *Biochemistry*, *24*, 282 (1959).
41. Gurvich, A. E., Kapner, R. B., and Nezlin, R. S., *Biochemistry*, *24*, 129 (1959).
42. Gurvich, A. E., Kuzovlena, O. B., and Tumanova, A. E., *Biochemistry*, *26*, 803 (1961).
43. Moudugal, N. R., and Porter, R. R., *Biochim. Biophys. Acta*, *71*, 185 (1963).
44. Gurvich, A. E., and Drizlikh, G. I., *Nature*, *203*, 648 (1964).
45. Webb, T., and Lapresle, C., *Biochem. J.*, *91*, 24 (1964).
46. Manecke, G., and Gillert, K. E., *Naturwissenschaften*, *42*, 212 (1955).
47. Gyenes, L., Rose, B., and Sehon, A. H., *Nature*, *181*, 1465 (1958).
48. Kent, L. H., and Slade, J. H. R., *Biochem. J.*, *77*, 12 (1960).
49. Gyenes, L., and Sehon, A. H., *Can. J. Biochem.*, *38*, 1235–1249 (1960).
50. Yagi, Y., Engel, K., and Pressman, D., *J. Immunol.*, *85*, 375 (1960).
51. Kent, L. H., and Slade, J. H. R., *Proc. Int. Congr. Biochem.*, *5th, 1961*, Vol. 9, p. 474, 1963.
52. Isliker, H. C., *Ann. N.Y. Acad. Sci.*, *57*, 225 (1953).
53. Jagendorf, A. T., Patchornik, A., and Sela, M., *Biochim. Biophys. Acta*, *78*, 516 (1963).
54. Robbins, J. B., Haimovich, J., and Sela, M., *Immunochemistry*, *4*, 11 (1967).
55. Levin, Y., Pecht, M., Goldstein, L., and Katchalski, E., *Biochemistry*, *3*, 1905 (1964).
56. Micheel, F., and Evers, J., *Makromol. Chem.*, *3*, 200 (1949).
57. Mitz, M. A., and Summaria, L. J., *Nature*, *189*, 576 (1961).
58. Gilham, P. T., *J. Amer. Chem. Soc.*, *84*, 1311 (1962).
59. Gilham, P. T., *J. Amer. Chem. Soc.*, *86*, 4982 (1964).
60. Gilham, P. T., *Biochemistry*, *7*, 2809 (1968).
61. Porath, J., *Nature*, *218*, 834 (1968).
62. Cuatrecasas, P., Wilchek, M., and Anfinsen, C. B., *J. Biol. Chem.*, *244*, 4316 (1969).
63. Cuatrecasas, P., and Illiano, G., *Biochem. Biophys. Res. Commun.*, *44*, 178 (1971).
64. Cuatrecasas, P., *Proc. Nat. Acad. Sci. U.S.*, *63*, 450 (1969).
65. Axen, R., and Porath, J., *Nature*, *210*, 367 (1966).
66. Cuatrecasas, P., *Biochem. Biophys. Res. Commun.*, *35*, 531 (1969).
67. Goetzel, E. J., and Metzger, H., *Biochemistry*, *9*, 1267 (1970).
68. Wofsy, L., and Burr, B., *J. Immunol.*, *103*, 380 (1969).
69. Omenn, G., Ontjes, D., and Anfinsen, C. B., *Nature*, *225*, 189 (1970).
70. Howe, C. W., Morisset, R., and Spink, W. W., *Fed. Proc.*, *29*, 830 Abstr. (1970).
71. Parker, D. C., and Briles, D., *Fed. Proc.*, *29*, 438 Abstr. (1970).
72. Akanuma, Y., Kuquya, T., and Hayashi, M., *Biochem. Biophys. Res. Commun.*, *38*, 947 (1970).

73. Weintraub, B., *Biochem. Biophys. Res. Commun.*, *39*, 83 (1970).
74. Axen, R., Heilbronn, E., and Winter, A., *Biochim. Biophys. Acta*, *191*, 478 (1969).
75. Feinstein, G., *FEBS Lett.*, *7*, 353 (1970).
76. Boegman, R. J., and Crumpton, M. H., *Biochem. J.*, *120*, 373 (1970).
77. Gribnau, A. A. M., Schoenmakers, J. G. G., van Kraaikamp, M., and Bloemendol, H., *Biochem. Biophys. Res. Commun.*, *38*, 1064 (1970).
78. Gabel, D., and v. Hofstein, B., *Eur. J. Biochem.*, *15*, 410 (1970).
79. Gryszkiewicz, J., *Folia Biologica*, *19*, 119 (1971).
80. Selinger, R. C., and Civen, M., *Biochem. Biophys. Res. Commun.*, *43*, 793 (1971).
81. Turkington, R. W., *Biochem. Biophys. Res. Commun.*, *41*, 1362 (1971).
82. Guyda, H. J., and Friesen, H. G., *Biochem. Biophys. Res. Commun.*, *41*, 1068 (1971).
83. Krug, F., Desbuquois, B., and Cuatrecasas, P., *Nature*, *234*, 268 (1971).
84. Cuatrecasas, P., Wilchek, M., and Anfinsen, C. B., *Science*, *162*, 1491 (1968).
85. Cuatrecasas, P., Wilchek, M., and Anfinsen, C. B., *Biochemistry*, *8*, 2277 (1969).
86. Cuatrecasas, P., Taniuchi, H., and Anfinsen, C. B., *Brookhaven Symp. Biol.*, *21*, 172 (1969).
87. Anfinsen, C. B., Cuatrecasas, P., and Taniuchi, H., in *The Enzymes*, 3rd ed. P. D. Boyer, Ed., Academic Press, New York, Vol. IV, p. 177, 1971.
88. Cuatrecasas, P., *J. Biol. Chem.*, *245*, 574 (1970).
89. Taniuchi, H., and Anfinsen, C. B., *J. Biol. Chem.*, *244*, 3864 (1969).
90. Erickson, R. P., and Steers, E., Jr., *Arch. Biochem. Biophys.*, *137*, 399 (1970).
91. Steers, E., Jr., and Shifrin, S., *Biochim. Biophys. Acta*, *133*, 454 (1967).
92. Perrin, D., *Cold Spring Harbor Symp. Quant. Biol.*, *28*, 529 (1963).
93. Perrin, D., Doctoral Dissertation, Institut Pasteur, Paris, France, 1965.
94. Shifrin, S., and Steers, E., Jr., *Biochim. Biophys. Acta*, *133*, 463 (1967).
95. Tomino, A., and Paigen, K., *The Lac. Operon*, D. Zipser and J. Beckwith, Eds., Cold Spring Harbor Laboratory, New York, 1970, p. 233.
96. Stevenson, K. J., and Landman, A., *Can. J. Biochem.*, *49*, 119 (1971).
97. Blumberg, S., Schechter, I., and Berger, A., *Eur. J. Biochem.*, *15*, 97 (1970).
98. Deutsch, D. G., and Mertz, E. T., *Science*, *170*, 1095 (1970).
99. Kalderon, M., Silman, I., Blumberg, S., and Dudai, Y., *Biochim. Biophys. Acta*, *207*, 560 (1970).
100. Cuatrecasas, P., and Wilchek, M., *Biochem. Biophys. Res. Commun.*, *33*, 235 (1968).
101. Green, N. M., *Biochem. J.*, *89*, 585 (1963).
102. McCormick, D. B., *Anal. Biochem.*, *13*, 194 (1965).
103. Reimann, E. M., Brostrom, C. O., Corbin, J. D., King, C. A., and Krebs, E. G., *Biochem. Biophys. Res. Commun.*, *42*, 187 (1971).
104. Cuatrecasas, C., and Parikh, I., unpublished observations.
105. Sprossler, B., and Lingens, F., *FEBS Lett.*, *6*, 232 (1970).
106. Schmit, J. C., Artz, S. W., and Zalkin, H., *J. Biol. Chem.*, *245*, 4019 (1970).
107. Erickson, J. S., and Mathews, C. K., *Biochem. Biophys. Res. Commun.*, *43*, 1164 (1971).

108. Pogell, B. M., in *Methods in Enzymology*, Vol. 9, W. A. Wood, Ed., Academic Press, New York, 1966, p. 9.
109. Starkenstein, E., *Biochem. Z.*, *24*, 210 (1910).
110. Fischer, E., and Stein, E. A., in *The Enzymes*, 2nd ed., P. D. Boyer, H. Lardy, and K. Myrback, IV, Eds., Academic Press, New York, p. 315. 1961.
111. Leloir, L. F., and Goldenberg, S. H., in *Methods in Enzymology*, Vol. 5, S. P. Colowick and N. O. Kaplan, Eds., Academic Press, New York, 1962, p. 145.
112. Leloir, L. F., Rongine De FeKete, M. A., Cardini, C. E., *J. Biol. Chem.*, *236*, 636 (1961).
113. Heppel, L. A., in *Methods in Enzymology*, Vol. 2, Eds., S. P. Colowick and N. O. Kaplan, Academic Press, New York, 1955, p. 576.
114. Pogell, B. M., *Biochem. Biophys. Res. Commun.*, *7*, 225 (1962).
115. Eley, J., *Biochemistry*, *8*, 1502 (1969).
116. Erlanger, B. F., *Biochim. Biophys. Acta*, *27*, 646 (1958).
117. Fritz, H., Schult, H., Neudecker, M., and Werle, E., *Angew. Chem.*, *78*, 775 (1966).
118. Chan, W. C., and Takahashi, M., *Biochem. Biophys. Res. Commun.*, *33*, 235 (1969).
119. Lingens, F., Goebel, W., and Vesseler, H., *Eur. J. Biochem.*, *1*, 363 (1967).
120. Chua, G. K., and Bushuk, W., *Biochem. Biophys. Res. Commun.*, *37*, 535 (1969).
121. Wilchek, M., and Gorecki, M., *Eur. J. Biochem.*, *11*, 491 (1969).
122. Holohan, P. D., Mahazan, K., and Fondy, T. P., *Fed. Proc.*, *29*, 888 Abstr. (1970) and personal communication.
123. Kristiansen, T., Einarsson, M., Sundberg, M., and Porath, J., *FEBS Lett.*, *7*, 294 (1970).
124. Shaper, J. H., Barker, R., and Hill, R. L., *Fed. Proc.*, *30*, 1265 abstr. (1971).
125. Claeyssens, M., Kersters-Hilderson, H., Van Wauwe, J. P. and DeBruyne, K., *FEBS Lett.*, *11*, 336 (1970).
126. Poillon, W. N., *Fed. Proc.*, *30*, 1086 Abstr. (1971).
127. Whitney, P. L., *Fed. Proc.*, *30*, 1291 Abstr. (1971).
128. Salvatore, N., and Cuatrecasas, P., unpublished observations.
129. Sluyterman, L. A. E., and Wijdenes, J., *Biochim. Biophys. Acta*, *200*, 593 (1970).
130. Shainoff, J. R., *J. Immunol.*, *100*, 187 (1968).
131. Eldjarn, L., and Jellum, E., *Acta Chem. Scand.*, *17*, 2610 (1963).
132. Givol, D., Weinstein, Y., Gorecki, M., and Wilchek, M., *Biochem. Biophys. Res. Commun.*, *38*, 825 (1970).
133. Wilchek, M., *FEBS Lett.*, *7*, 161 (1970).
134. Pensky, J., and Marshall, J. S., *Arch. Biochem. Biophys.*, *135*, 304 (1969).
135. Pages, R. A., Cahnmann, H. J., and Robbins, J., personal communication.
136. Marchesi, V., personal communication (1971).
137. Vonderharr, B., Mueller, G. C., *Biochim. Biophys. Acta*, *176*, 626 (1969).
138. Burstein, S. H., *Steroids*, *14*, 2 (1969).
139. Cuatrecasas, P., and Glenner, G., unpublished observations.
140. Richards, F. M., and Vithayathil, P. J., *J. Biol. Chem.*, *234*, 1459 (1959).

141. Hofmann, K., Smithers, M. J., and Finn, F. M., *J. Amer. Chem. Soc.*, *88*, 4107 (1966).
142. Kato, I., and Anfinsen, C. B., *J. Biol. Chem.*, *244*, 1004 (1969).
143. Taniuchi, H., Anfinsen, C. B., and Sodja, A., *Proc. Nat. Acad. Sci. U.S.*, *58*, 1235 (1967).
144. Ontjes, D., and Anfinsen, C. B., *J. Biol. Chem.*, *244*, 6316 (1969).
145. Anfinsen, C. B., Ontjes, D., and Chaiken, I., *Eur. Peptide Symp.*, *10th*, Sept. 1969, in press.
146. Hennig, S. E., and Ginsburg, A., personal communication.
147. Carey, W. F., and Wells, J. R. E., *Biochem. Biophys. Res. Commun.*, *41*, 574 (1970).
148. Aliapoulios, M. A., Savery, A., and Munson, P. L., *Fed. Proc.*, *24*, 322 (1965).
149. Goldhaber, P., *Science*, *147*, 407 (1965).
150. Bautz, E. K. F., and Reilly, E., *Science*, *151*, 328 (1966).
151. Nyggard, A. P., and Hall, B. D., *Biochem. Biophys. Res. Commun.*, *12*, 98 (1963).
152. Alder, A. J., and Rich, A., *J. Amer. Chem. Soc.*, *84*, 3977 (1962).
153. Bautz, E. L. F., and Holt, B. D., *Proc. Nat. Acad. Sci. U.S.*, *48*, 400 (1962).
154. Gilham, P. T., and Robinson, W. E., *J. Amer. Chem. Soc.*, *86*, 4985 (1964).
155. Cozzarelli, N. R., Melachen, M. E., Jovin, T. M., and Kornberg, A., *Biochem. Biophys. Res. Commun.*, *28*, 578 (1967).
156. Jovin, T. M., and Kornberg, A., *J. Biol. Chem.*, *243*, 250 (1968).
157. Sander, E. G., McCormick, D. G., and Wright, L. D., *J. Chromatogr.*, *21*, 419 (1966).
158. Erhan, S. L., Northrup, L. G., and Leach, F. R., *Proc. Nat. Acad. Sci. U.S.*, *53*, 646 (1965).
159. Litman, R., *J. Biol. Chem.*, *243*, 6222 (1968).
160. Weith, H. L., Wiebers, J. L., and Gilham, P. T., *Biochemistry*, *9*, 4396 (1970).
161. Hoffman, C. H., Harris, E., Chodroff, S., Michelson, S., Rothrock, J. W., Peterson, E., and Reuter, W., *Biochem. Biophys. Res. Commun.*, *41*, 710 (1970).
162. Denburg, J., and DeLuca, M., *Proc. Nat. Acad. Sci. U.S.*, *67*, 1057 (1970).
163. Poonian, M. S., Schlaback, A. J., and Weissbach, A., *Biochemistry*, *10*, 424 (1971).
164. Cuatrecasas P., *Proc. Nat. Acad. Sci. U.S.*, In Press (May 1972).
165. Evans, W. H., Wage, M. G., and Peterson, E. A., *J. Immunol.*, *102*, 899 (1969).
166. Davie, J. M., Paul, W. E., *Cellular Immunol.*, *1*, 404 (1970).
167. Edelman, G. M., Rutishauser, U. and Millette, C. F., *Proc. Nat. Acad. Sci. U.S. 68*, 2153 (1971).

BIOCHEMISTRY OF α-GALACTOSIDASES

By P. M. DEY and J. B. PRIDHAM, *London, England*

CONTENTS

1. Introduction

In 1895 Bau (1) and Fischer and Lindner (2) isolated enzyme preparations (melibiases) from bottom yeast which hydrolyzed the disaccharide melibiose. The name melibiase was later changed to α-galactosidase by Weidenhagen (3,4), who studied the specificity of action of the enzyme using a variety of sugars possessing nonreducing terminal α-D-galactosyl residues.

α-Galactosidases [α-D-galactoside galactohydrolases (E.C. 3.2. 1.22)] catalyze the following reaction:

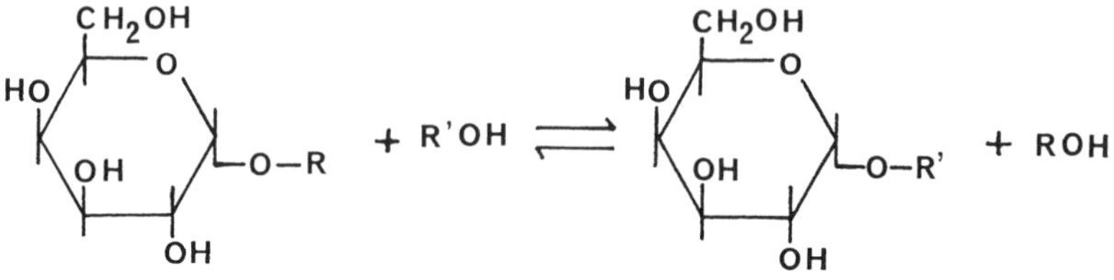

The hydroxylic acceptor molecule, R'OH, is commonly water, although R and R' can be aliphatic or aromatic groups. This means that the enzyme may hydrolyze a variety of simple α-D-galactosides as well as more complex molecules, such as oligo- and polysaccharides. In addition, O-transfer of α-D-galactosyl residues to various alcohol derivatives may be effected. Under special conditions, de novo synthesis can occur using D-galactose (R=H) as the donor.

Studies with α-galactosidases have mostly been carried out using relatively crude enzymes; however, the kinetics and specificity of highly purified preparations from Vicia faba (5), V. sativa (6), and Mortierella vinacea (7) have recently been investigated. Interest has centered around the mode of action and physiological significance of these enzymes, and their use as tools for structural studies with biological molecules.

A. OCCURRENCE

α-Galactosidases have been reported to occur widely in micro-organisms, plants, and animals (Table I).

Regarding tissue and cellular locations, a survey of the organs of the rat, using a histochemical method, has revealed highest activities in the cytoplasm of epithelial cells of Brunner's glands in the intestine. In addition, the distal segments of the proximal tubes of the kidney, and the thyroid and parathyroid glands also possess high activities. Blood cells and bone marrow of some animals also appear to contain some α-galactosidase (8–10). α-Galactosidase in cells from most organisms is present in the soluble fraction. However, Debris, Courtois, and Petek (145) have detected a particulate renal enzyme in pig; the

TABLE I
Sources of α-Galactosidases

Organism	Reference	Organism	Reference
Microorganisms		*Coffea* sp. (seeds)	17,44,71–82
		Coniferous seeds (various)	83,84
A bacterial strain from guinea pig faeces (unidentified)	11	*Cyamopsis tetragonolobus* (seeds)	85
Actinomyces spp.	12,13	*Gleditschia ferox* (seeds)	44
Aerobacter aerogenes	14	*Gossypium* sp. (seeds)	86
Agaricus bisporus	15	*Helianthus annuus* (seeds)	68
Aspergillus spp.	16–21	*Hordeum* sp. (seeds)	87
Calvatia cyanthiformis	22,23	*Medicago* sp. (seeds)	17,44,88
Clostridium spp.	24,25	Papain (commercial preparation)	89
Colletotrichum rindemuthianum	26	*Phaseolus* spp. (seeds)	68,90–94
Diplococcus pneumoniae	27,28	*Pisum sativum* (seeds)	67
Epidinium ecaudatum	29	*Plantago* spp. (seeds)	44,97,98
Escherichia coli	30–38	*Populus tremuloides* (phloem)	87
Eudiplodinium maggii	39	*Porphyra umbilicalis* (whole organism)	99
Lactobacillus spp.	40,41	*Prunus amygdalus* (seeds)	17,48,49, 100–107
Mortierella vinacea	7,42	*Prunus armeniaca* (seeds)	16
Ophryoscolex caudatus	39	*Raphanus sativus* (seeds)	68
Penicillium spp.	43,45	*Ricinus communis* (seeds)	68
Polyplaston multivesculatum	39	*Spinacia oleracea* (leaves)	108
Pseudomonas saccharophila	46	*Trigonellum foenum graecum* (seeds)	44,109,110
Saccharomyces spp.	1–4,16, 47–59	*Triticum* spp. (seeds)	111
Streptococcus bovis	60,61	*Vicia* spp. (seeds)	5,6,67, 112–116
Streptomyces spp.	62,63		
Trichomonas foetus	24,64–66		
Plants			
Acer pseudoplatanus (seeds)	67		
Brassica oleracea (seeds)	68		
Cajanus indicus (seeds)	67		
Canavalia ensiformis (seeds)	69		
Citrulus vulgaris (seeds)	70		

TABLE I (*Contd.*)

Organism	Reference	Organism	Reference
Animals		Tissues (various mammals)	
Astacus fluviatilis	117	dog	10
Helix pomatia and	16,118–120	human	10
other snails (digestive		ox	139,140
tract, liver, pancreas)		pig	16,141,142
Insects (various)	121–135	rabbit	10,145,146
Lumbricus terrestris	136	rat	8,10,146–
Sperm and seminal	137,138		149
plasma (various			
mammals)			

enzyme was solubilized with trypsin (cf. refs. 150,151). The enzyme has also been found in a rat-brain mitochondrial fraction (149), in the plasma membrane of bovine liver (139), and in chloroplast, mitochondrial, and microsomal fractions of spinach leaves (108).

B. INDUCTION

In *E. coli* (152) and *Aerobacter aerogenes* (14) α-galactosidase is not constitutive but can be induced. Several α-D-galactosides (32–34, 152,153) and β-D-galactosides (152,153) will induce the formation of α-galactosidase in *E. coli*. Melibiose can induce β-galactosidase (154–159), as well as α-galactosidase, simultaneously (152,153,160). Phenyl- and *o*-nitrophenyl α-D-galactosides only induce the formation of α-galactosidase (14,34). Other studies on melibiose-induced formation of α-galactosidase in *E. coli* have revealed that a galactoside transport system (permease) and α-galactosidase control the utilization of melibiose (158). The two permeases which are known to transport melibiose (95,158) have been termed thiomethylgalactoside (TMG) permease I (96) and TMG permease II (158). Mutation experiments and demonstration of coordinate induction of α-galactosidase and TMG permease II by several α-galactosides and D-galactose suggest that these two enzymes may be components of a common operon (37).

II. Detection and Measurement of Activity

A. METHODS OF ASSAY

Melibiose and raffinose, presumed natural substrates for plant α-galactosidases, are commonly used to determine enzyme activity (5,104,115,161). Following incubation, the extent of hydrolysis is measured in terms of the hexose liberated. This can be achieved by the measurement of increased reducing power (261,162) or by enzymic methods (163–165). Substituted phenyl α-D-galactosides (166) can be used as convenient assay substrates. The release of galactose from these compounds can be measured as stated above, or alternatively, the appearance of aglycon can be estimated (5,167–174).

In some cases, inclusion of cofactors in the assay medium may be necessary for maximum enzyme activity. For example, it has been demonstrated (36,38) that α-galactosidase activity in cell-free extracts of $E.$ $coli$ is dependent upon Mn^{2+} ions and NAD^+. The presence of both manganese ions and reducing agents, such as glutathione or Cleland's reagent (175) stabilizes activity during the enzyme assay. In addition, K^+ ions have been shown to activate α-galactosidase from $V.$ $faba$ seeds (176).

Galactosyl transferase activity, as opposed to hydrolysis, is normally assayed by carrying out the enzymic reaction in the presence of a suitable galactose donor and an acceptor, fractionating the resulting mixture by a chromatographic procedure and then determining the transfer products. This is conveniently achieved by the use of a ^{14}C-labeled galactose donor (177). Transferase activity can also be determined by calculation after measuring the amounts of galactose and the glycosidic substituent of the original donor in the reaction mixture (177).

B. HISTOCHEMICAL DETECTION

The location of the enzyme in cells and tissues may be effected by the use of 4-bromo- or 6-bromo-2-naphthyl α-D-galactosides as substrates. Hydrolysis of these compounds yields a water-insoluble aglycon that can subsequently be coupled with Fast blue B, resulting in the formation of a pigment (8,9,178–181).

III. Isolation

α-Galactosidases from various sources can be isolated by conventional methods of extraction. Commonly, α-galactosidases occur in the cell in association with various other glycosidases and it has often proved difficult to fractionate these activities. The normal techniques of isolation used include ammonium sulphate (7,13,20,27,68–70,85,94,103, 104,108,115,116,140,148,176) and organic solvent (22,68,78,103,104, 114,115) fractionations, heat treatment (7,27), acidification (103,104, 114,115), ion exchange (13,20,21,27,69,70,84,94,103,104,140,148,176) and gel chromatography (6,7,20,21,36,67,69,85,104,108,114–116,148), and isoelectric focusing (21). The principle of affinity chromatography has been used to separate sweet almond α-galactosidase from other accompanying glycosidases: the enzyme was specifically adsorbed on an insoluble matrix of poly (p-hydroxystyrene α-D-galactoside) (102). One of the steps in purification of pneumococcal α-galactosidase involved adsorption of the enzyme on human red blood cells. This again is really an application of affinity chromatography, where the enzyme shows an affinity for the α-D-galactosyl residues of blood group B substance. Shibata and Nisizawa (182) have shown that it is possible to separate α-galactosidase from β-glucosidase in tannin-precipitated emulsin by paper electrophoresis.

Removal of nucleic acids in the initial stage of purification may be important in order to achieve higher degrees of purity at later steps

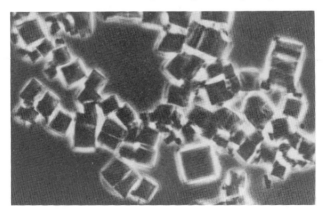

Fig. 1. Crystals of α-galactosidase isolated from *M. vinacea*, magnified 800 times (7).

(103,104,115). Highly purified and apparently homogeneous α-galactosidases have only been isolated in a few cases (6,7,85,104,115, 116). The first crystalline form of the enzyme has been obtained from the fungus *Mortierella vinacea* (7). (See Fig. 1.)

IV. Physical Properties

A. MULTIMOLECULAR FORMS

Petek and his collaborators (78,80,98) first reported the existence of multimolecular forms of α-galactosidase; they separated two forms from the seeds of both *Coffea* sp. and *Plantago ovata* by chromatography on alumina columns. Dey and Pridham (114–116) later showed that dormant *V. faba* seeds also possessed two α-galactosidases (I and II), which differed in their molecular weights. The separation was accomplished by Sephadex gel filtration. The two forms detected by Petek and Dong (78) in coffee are not apparent when extracts of the seed are examined by this latter technique (67). Using gel filtration, two forms of the enzyme have been observed in seven other species of dormant seeds (67). (See Fig. 2.)

α-Galactosidase II from *V. faba* has recently been further resolved into two active fractions, $II_{(1)}$ and $II_{(2)}$, with approximately similar molecular weights using CM–cellulose chromatography (176) [Fig. 3(a)]. Enzyme I, however, could not be further resolved by this treatment. Suzuki et al. (7), using DEAE-Sephadex columns, have detected three forms of α-galactosidases in *M. vinacea* [Fig. 3(b)]; the authors have crystallized one of the components. It is not known, however, whether the three forms of the enzyme have similar molecular weights. Multimolecular forms of α-galactosidase may have very similar properties and hence may be difficult to resolve. For example, the enzyme obtained from *A. niger* was homogeneous, as judged by chromatography on Sephadex G-200, Bio-Gel P-200, DEAE-Sephadex, and DEAE–cellulose (20,21); but when passed through a CM–cellulose column it was resolved into three active forms (21). The existence of several forms was also confirmed by isoelectric focusing (21).

Some evidence for the interconversion of multimolecular forms of α-galactosidase *in vitro* has been obtained in the case of the *V. faba* enzymes (176). A preparation of the lower molecular weight form, II, from green seeds (see p. 122) was prepared by pH precipitation and acetone and ammonium sulphate fractionations, and then stored at pH 5.5 and 4°. Over a period of days the specific activity of the solution increased rapidly; examination by Sephadex-gel filtration revealed that the activity of enzyme II was decreasing at the expense of a higher molecular weight enzyme with an elution pattern identical to that of enzyme I. The interconversion did not occur if the preparation of enzyme II had previously been passed through Sephadex G-100. It has not been proved conclusively that the higher molecular weight enzyme produced in this reaction is in fact enzyme I. The picture is complicated by the recent discovery that enzyme II consists of two components [see Fig. 3(a)].

Some dormant seeds appear to contain enzyme of only one molecular size (see Table II); however, it is important to note that the number of

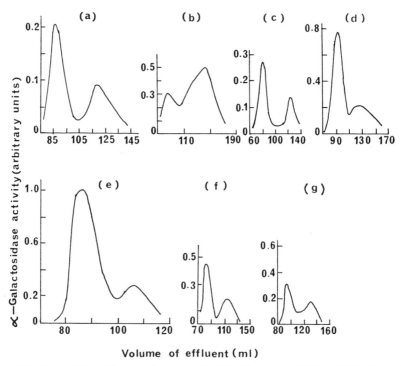

Fig. 2. α-Galactosidase patterns obtained by Sephadex G-100 gel filtration of extracts from (a) *Acer pseudoplatanus*, (b) *Helianthus annuus*, (c) *Phaseolus aureus*, (d) *Phaseolus vulgaris*, (e) *Vicia sativa*, (f) *Vicia dumitorum*, and (g) *Pisum sativum*.

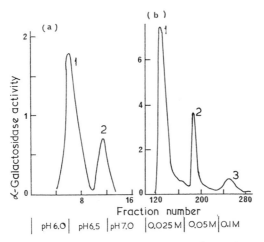

Fig. 3. Separation of α-galactosidase isoenzymes on ion-exchange columns. (a) CM-cellulose chromatography of α-galactosidase II from *V. faba*. The column was equilibrated with McIlvaine buffer (pH 4.0) and eluted stepwise with the same buffer at varying pH values (176). (b) DEAE-Sephadex A-50 chromatography of α-galactosidase from *M. vinacea*. The column was equilibrated with 0.01 M-phosphate buffer (pH 7.0) and eluted with buffer of increasing ionic strength (7).

forms of α-galactosidase may be related to the physiological state of the seed (67).

B. MOLECULAR WEIGHTS

Table II indicates the apparent molecular weights of α-galactosidases from various organisms. In most cases these have only been estimated by gel filtration techniques (183). It is of interest to observe that many dormant seeds contain two forms of the enzyme, one form apparently having a substantially higher molecular weight (2–6 times) than the other. There is every reason to suppose, however, that many of the forms listed in Table II may be mixtures of active proteins with similar molecular weights but different ionic properties.

In addition, there is evidence that α-galactosidase I, the high molecular weight form, from *V. faba* has a subunit structure. Dey and Pridham (115) have shown that this enzyme is dissociated into six inactive protein fractions when it is passed through a Sephadex G-100 column in the presence of 6 *M*-urea. The high molecular weight enzymes from other plant species also probably possess quaternary structures.

TABLE II

Molecular Weights of α-Galactosidases from Various Seeds

Source of α-galactosidase	Molecular weight	Reference
Acer pseudoplatanus[a]		67
I	167,000	
II	50,000	
Cajunus indicus[a]	87,000	67
Coffea sp.[a]	26,000	67
Cyamopsis tetragonolobus[b]	25,000	85
Helianthus annuus[a]		67
I	159,000	
II	23,000	
Medicago sativa[a]	38,000	67
Phaseolus aureus[a]		67
I	209,000	
II	38,000	
Phaseolus vulgaris[a]		67
I	125,900	
II	39,800	
Pisum sativum[a]		67
I	121,600	
II	32,300	
Prunus amygdalus[a]	32,400	67
	33,000	104
Vicia dumitorum[a]		67
I	195,000	
II	57,000	
Vicia faba[a]		114,115
I	209,000	
II	38,000	
Vicia sativa[a]		67
I	166,000	
II	77,000	
Vicia sativa[b]	30,000[d]	6

[a] Dormant seeds.
[b] Germinated seeds.
[c] Composed of $II_{(1)}$ and $II_{(2)}$; see Fig. 3(a).
[d] By the method of sucrose density gradient.

V. Chemical Analysis

Few details of the compositions of α-galactosidases are known. The crystalline enzyme from *M. vinacea* appears to be a glycoprotein containing 2.7% D-glucosamine and 10.8% hexoses (7). α-Galactosidase I from *V. faba* is also reported to contain bound carbohydrate (114,115). In the case of α-galactosidase from germinated *V. sativa* seeds, the full amino acid analysis has been published (6). The protein has a high acidic amino acid content and contains little cysteine. Treatment with 1-dimethylaminonaphthalene-5-sulphonylchloride (dansyl chloride) or 1-fluoro-2,4-dinitrobenzene shows that the enzyme has a single polypeptide chain with L-alanine as the *N*-terminal group.

VI. Specificity

A. HYDROLASE ACTIVITY

In general, change of configuration of hydrogen and hydroxyl groups on any single carbon atom of a glycoside substrate is sufficient to reduce the rate or completely inhibit the hydrolytic action of the corresponding hydrolase. In the case of α-galactosidases, two main factors govern the rate of hydrolysis of the substrate. First, the ring structure must be pyranoid, and second, the configuration of —H and —OH on carbon atoms 1, 2, 3, and 4 must be similar to that on α-D-galactose. As in the case of other glycosidases, namely, β-galactosidase (50,184, 185), β-glucosidase (50,185), and α-mannosidase (50), changes at C-6 of the glycosyl moiety of the substrate are normally tolerated by α-galactosidases. Hence β-L-arabinosides (see Fig. 4) are hydrolyzed by enzymes from several sources (7,50,105,115,185,186). However, α-galactosidases from *Streptococcus bovis* (61), *Epidinium ecaudatum* (29), *Diplococcus penumoniae* (27), and *Calvatia cyanthiformis* (22) cannot use arabinosides as substrates. Dey and Pridham (5) have shown that α-galactosidases can hydrolyze *p*-nitrophenyl α-D-fucoside; this compound also has the D-galactose configuration (see Fig. 4). Pigman (187) postulated that D-glycero-D-galactoheptosides (see Fig. 4) would serve as substrates for galactosidases. He partially confirmed this using phenyl D-glycero-β-D-galactoheptoside and sweet almond β-galactosidase (188). However, the α-galactosidases from almond (188) and yeast (50) are not able to catalyze the hydrolysis of the corresponding α-isomer.

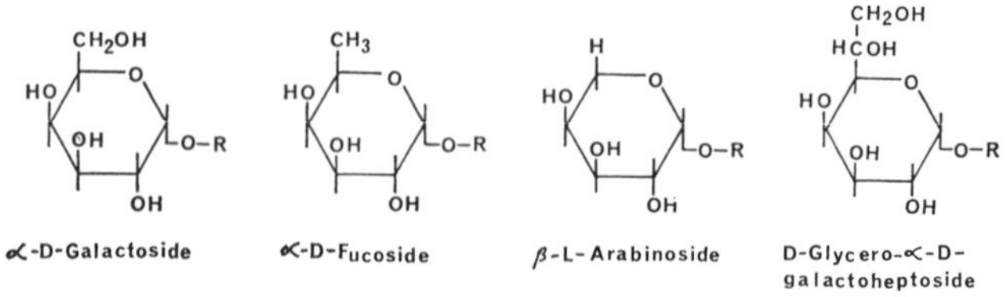

Fig. 4. α-D-Galactopyranoside and related glycosides.

The quantitative evaluation of glycon specificity has been carried out with α-galactosidases from sweet almond (105) and V. faba (5); the V_{max} and K_m values are compared in Table III. The hydrolyzability (V_{max}) of the glycosides shows an apparent random variation. However, the affinity ($1/K_m$) of the enzymes for the substrates seems to depend largely on the structural changes in the glycon moiety and follows the order: α-D-galactoside > α-D-fucoside > β-L-arabinoside. This suggests that one of the specific points of attachment of the substrate to the enzyme is through the primary alcohol group of the galactose structure.

It is not known whether the replacement of the anomeric oxygen atom of an α-D-galactoside by a sulphur atom affects the hydrolyzability. In the case of E. coli β-galactosidase, o- and p-nitrophenyl-1-thio β-D-galactosides have affinities for the enzyme that are similar to those of the corresponding O-galactosides, although o-nitrophenyl β-D-galactoside is hydrolyzed 7×10^5 times faster than the corresponding sulphur analogue (154,168,171,189,190).

The aglycon group of a substrate may or may not have a marked effect on hydrolysis by glycosidases. Normally, the group does not completely inhibit hydrolysis. The following naturally occurring and synthetic α-D-galactosides are known to be hydrolyzed by various α-galactosidases:

Galactosides: Methyl-, ethyl- (14), n-propyl- (105), phenyl-, o-nitrophenyl- (35), m-nitrophenyl- (105), p-nitrophenyl- (103), o-cresyl- (105), m-cresyl-, p-cresyl- (16), m-chlorophenyl- (105), 1-naphthyl-, 2-naphthyl- and 6-bromo-2-naphthyl α-D-galactosides (173), 1-O- and 2-O-α-D-galactosyl glycerols (99), galactinol (5,7), digalactosyl glycerol (149), and α-D-galactosyl fluoride (174).

Oligosaccharides: Melibiose, epimelibiose, (167), O-α-D-gal-(1 → 4)-D-gal (7,29,61), melibiitol (29,60,61), melibionic acid (49,167),

TABLE III
Substrate Specificity of α-Galactosidases[a]

	α-Galactosidases from					
	V. faba (5) pH 4.0				Sweet almonds (105) pH 5.5	
	I		II			
Substrate	V_{max}	K_m	V_{max}	K_m	V_{max}	K_m
Glycon and stereospecificity						
p-Nitrophenyl α-D-galactoside	25.53	0.38	2.39	0.45	27.00	0.53
p-Nitrophenyl α-D-fucoside	24.10	4.76	6.96	5.88	—	—
p-Nitrophenyl β-L-arabinoside	16.40	14.30	2.39	12.50	5.00	33.30
p-Nitrophenyl β-D-galactoside	b	b	b	b	b	b
p-Nitrophenyl α-D-glucoside	b	b	b	b	b	b
p-Nitrophenyl β-D-glucoside	b	b	b	b	—	—
Sucrose	b	b	b	b	—	—
Aglycon specificity						
Methyl α-D-galactoside	1.66	7.13	0.29	14.30	0.59	10.90
Ethyl α-D-galactoside	1.66	8.93	0.28	8.00	0.62	6.25
n-Propyl α-D-galactoside	2.20	6.13	0.27	5.88	1.08	6.25
Phenyl α-D-galactoside	20.30	1.11	4.36	1.25	22.70	5.00
o-Cresyl α-D-galactoside	26.00	1.33	2.90	0.78	32.20	4.54
m-Cresyl α-D-galactoside	24.30	1.38	2.80	2.00	32.70	8.33
p-Cresyl α-D-galactoside	23.00	1.54	2.43	1.00	40.00	4.76
p-Aminophenyl α-D-galactoside	26.60	0.95	2.72	0.87	—	—
m-Chlorophenyl α-D-galactoside	20.60	0.83	3.30	1.17	32.70	8.33
o-Nitrophenyl α-D-galactoside	42.10	1.14	2.80	0.69	43.00	0.33
m-Nitrophenyl α-D-galactoside	5.86	10.00	0.31	2.50	23.60	1.57
p-Nitrophenyl α-D-galactoside	25.53	0.38	2.39	0.45	27.00	0.53
6-Bromo-2-naphthyl α-D-galactoside	22.70	1.80	1.66	0.62	—	—
α-D-Galactose-1-phosphate	c	—	c	—	b	b
Galactinol	2.15	0.13	0.72	0.69	—	—
Melibiose	2.54	0.96	0.41	0.77	1.61	2.24
Raffinose	28.40	4.00	4.18	5.00	11.80	12.50
Stachyose	9.00	7.50	1.36	5.26	—	—

[a] V_{max} is expressed as micromoles of substrate hydrolyzed per minute per milligram of enzyme and K_m as mM.

[b] Not hydrolyzed.

[c] Hydrolyzed.

raffinose, umbelliferose (191), planteose (17,145,167), O-α-D-gal-$(1 \rightarrow 3)$-β-D-fru-$(2 \rightarrow 1)$-α-D-glc (192), O-α-D-gal-$(1 \rightarrow 1)$-β-D-fru-$(2 \rightarrow 1)$-α-D-glc (193), manninotriose, manninotriitol (29,60,61), mannino-trionic acid (167). Stachyose, verbascose, and higher homologues (17,29,60,61,91), lychnose, isolychnose, and higher homologues (193–195).

Polysaccharides: Galactomannans are attacked by α-galactosidases from various sources (20,22,61,80,81,85,94). These polysaccharides normally have a basic structure consisting of a backbone of β-1,4-linked D-mannosyl residues to which D-galactosyl residues are attached by α-1,6-linkages. The galactose content varies, depending on the plant source. Not all α-galactosidases hydrolyze galactomannans; those that do, appear to remove terminal galactose residues only.

Some α-galactosidases can also remove terminal α-D-galactosyl residues from blood group B substance (77,196).

The results in Table III show that, in general, aryl α-D-galactosides are better substrates than alkyl derivatives or disaccharides (cf. refs. 16,22,43,109,148,168). Relationships between K_m and V_{max} values are highly irregular with changing aglycons, and high affinity of the enzyme for the substrate does not necessarily parallel high V_{max} values, and *vice versa*. The nature of the substituent on the phenyl ring of aromatic galactosides does not have a large effect on V_{max} in the case of the enzymes from *V. faba* but the affinity $(1/K_m)$ shows some tendency to increase when electron-attracting substituents are present (5). It has been reported that factors affecting the affinity are probably complex and include the position and size of the aromatic substituent, its electronic effect, and the degree of hydration (5). The V_{max} values with α-galactosidase from sweet almond have been shown to increase with increasing electron-releasing and electron-withdrawing properties of the benzene ring substituent of the substrate (105).

Regarding galactose-containing oligosaccharides, such as melibiose and manninotriose, reduction of the terminal reducing group (producing melibiitol and manninotriitol, respectively) decreases the rate of enzymic hydrolysis (29,60,61). Oxidation of the reducing group, as in the case of the conversion of melibiose to melibionic acid, does not appear to affect the rate of hydrolysis (49).

In a homologous series of α-D-galactosides, the rate of hydrolysis seems to be reduced by an increase in the chain length (5,6,17,22,27,60,

61,109,124,145,148), but in two microbial enzymes the reverse is reported to occur (43,61).

In addition to hydrolyzing terminal galactosyl residues, almond α-galactosidase is also able to split the internal galactosidic linkage of stachyose, forming galactobiose and sucrose (198,199). On the other hand, coffee α-galactosidase can only cleave stachyose (73,74) and tetra-O-D-galactosyl sucrose (17) in a stepwise fashion starting from the nonreducing end. α-Galactosidases from *Streptococcus bovis* (60,61) and *Epidinium ecaudatum* (29) acting on verbascose and verbascotetraose behave in a similar manner.

B. TRANSGALACTOSYLATION

Transgalactosylase activity was first observed with yeast α-galactosidase by Blanchard and Albon (197), who showed that galactose could be transferred from melibiose to a second melibiose acceptor molecule with the formation of manninotriose (1,208). The transferase properties of α-galactosidases have since been studied extensively with respect to the effect of various parameters, such as galactosyl donor and acceptor specificity, acceptor concentration, pH, temperature, and the source of enzyme (80,109,167,168). Transgalactosylation reactions to aliphatic hydroxyl groups are normally accompanied by a pronounced hydrolytic activity. Tanner and Kandler (177), however, have reported an enzyme in *Phaseolus vulgaris* seeds that transfers galactose from galactinol to raffinose with the formation of stachyose, which produces only slight hydrolysis. An α-galactosidase from wheat bran with a high transfer/hydrolysis ratio has also been described by Wohnlich (111).

Table IV shows that although hexoses serve as galactosyl acceptors, this property so far has not been demonstrated with the pentoses. Y. T. Li and Shetlar (28) have shown with pneumococcal α-galactosidase that phosphorylation, reduction or oxidation of either C-1 or C-6 of D-glucose or D-galactose destroys the acceptor properties of these hexoses. They have further reported that 2-deoxy D-galactose, but not 2-deoxy D-glucose, is an acceptor and methylation of C-3 of D-glucose also destroys its acceptor properties.

In most cases a long incubation of the reaction mixture results in the disappearance of the transfer products (168), which clearly shows that the α-D-galactosyl configuration is retained in the transfer products: this has normally been confirmed by chemical analysis. Incubation of

TABLE IV
Specificity of Transgalactosylation by α-Galactosidases from Different Sources

Acceptor	Donor*	Transfer product	Source of α-galactosidase	Reference
Methanol	a, c	Methyl α-D-galactoside	*Coffea* sp.	75
Methanol	d, e	Methyl α-D-galactoside, verbascose and ajugose	*Coffea* sp.	72
D-Glucose, D-galactose, gentianose, gentiobiose, lactose, and maltose	a	Unidentified products	*Coffea* sp.	75, 78
Mannose	a	Epimelibiose	*Coffea* sp.	75, 76, 78
Cellobiose	a	α-D-gal(1 → 3)-D-glc (4 ← 1)-β-D-glc; α-D-gal(1 → 6)-β-D-glc(1 → 4)-D-glc; α-D-gal(1 → 3)-β-D-glc(1 → 4)-D-glc	*Coffea* sp.	75, 78, 79
Glycerol	c	Floridoside	*Coffea* sp.	72
Sucrose	a, c	Raffinose, stachyose and planteose	*Coffea* sp.	75
L-Arabinose, D-ribose, D- and L-xylose, D-fructose, L-sorbose, L-rhamnose, glycerol, mannitol, meso-inositol, α-L-glycerophosphate, D-glucosamine, trehalose, methyl α- and β-D-glucosides, methyl α-D-mannoside, and amygdaloside	a	No products formed	*Coffea* sp.	75
Methanol, ethanol, n-propanol, n-butanol, isopropanol, isopentanol, glycerol, mannitol, sorbitol, meso-inositol, D-glucose, sucrose, and trehalose	a	Unidentified products	*Trigonellum foenum*	109, 110
Methanol, n-propanol, and glycerol	a	Unidentified products	*Hordeum* sp., *Brassica oleracea*, *Helianthus annuus*	68
D-Galactose		Di- and trigalactosides (all with α-1 → 6 linkages)	*Raphanus sativus*, *Phaseolus vulgaris*, *Ricinus communis*	68

TABLE IV (*cont'd*)

Acceptor	Donor*	Transfer product	Source of α-galactosidase	Reference
Sucrose Raffinose	f	Raffinose Stachyose	*Phaseolus vulgaris*	177
D-Glucose, D-mannose, and D-galactose	a	Unidentified products	*Plantago ovata, Vicia sativa*	6, 97, 98
Sucrose	a	Raffinose	*Plantago ovata, Vicia sativa*	6, 97, 98
Melibiose	c	Two nonreducing sugars (one containing only D-galactose and the other with D-galactose and D-glucose), galactobiose, galactotriose, manninotriose, verbascotetraose and one isomer each of the latter three oligosaccharides	*Diplococcus pneumoniae*	27, 28 (also see 7, 148)
Raffinose	d	Galactobiose, galactotriose, stachyose, verbascose and two sucrose-containing oligosaccharides		
D-Glucose	c	Melibiose		
D-Glucose	d	Incorporation of D-glucose into raffinose		
Stachyose	e	Unidentified product		
D-Glucose and D-galactose	d	No products formed	*Streptococcus bovis*	61
Melibiose	c	Manninotriose		
Sucrose	c	Raffinose		
Raffinose	c	Stachyose		
Sucrose	d	Planteose	*Triticum* sp.	111
Sucrose	c	Raffinose and planteose	*Vicia faba*	113
o- and p-Nitrophenyl α-D-galactosides	b	A galactobiose and two unidentified sugars	*Calvatia cyanthiformis*	22
N-Acetyl-D-glucosamine	a	6-O-α-D-Galactosyl-N-acetyl-D-glucosamine	*Saccharomyces* sp.	58
N-Acetyl-D-glucosamine	a	No products formed	*Coffea* sp.	58
N-Acetyl-D-glucosamine	c	6-O-α-D-Galactosyl-N-acetyl-D-glucosamine	*Trichomonas foetus*	65
N-Acetyl-D-galactosamine	c	6-O-α-D-galactosyl-N-acetyl-D-galactosamine		
D-Ribose and D-fructose	a	No products formed	*Vicia sativa*	6
Cellobiose	a	Three α-D-galactosides of cellobiose		
Raffinose	d	Stachyose		
Galactinol	f	Digalactosidoinositol		
D-Galactose	b	Galactobiose, galactotriose and galactotetraose (all with α-1,6-linkages)	*Prunus amygdalus*	104,107

TABLE IV (cont'd)

Acceptor	Donor*	Transfer product	Source of α-galactosidase	Reference
D-Glucose	b	Melibiose		
Sucrose	b	Raffinose		
Maltose	b	α-D-Gal(1 → 6)-D-glc- (4 ← 1)-α-D-glc and one unidentified product		
Melibiose	c	Three unidentified products		

* Donors: a, phenyl α-D-galactoside; b, p-nitrophenyl α-D-galactoside; c, melibiose; d, raffinose; e, stachyose; f, galactinol.

α-galactosidase from *Calvatia cyanthiformis* with o- or p-nitrophenyl α-D-galactosides gives rise to two oligosaccharides containing only D-galactose, which continue to accumulate over a 5-hr incubation period. Y. T. Li and Shetlar (22), who made this observation, showed that the oligosaccharides were not hydrolyzed by pneumococcal α-galactosidase, thus suggesting that α-D-galactosidic linkages were absent. They inferred that the transgalactosylation reaction was accompanied by inversion of configuration at C-1 of the glycon group (cf. refs. 200,201).

Courtois and Percheron (109) have calculated the efficiency of various sugars as galactose acceptors, using phenyl α-D-galactoside as donor, in terms of percent transfer, that is:

$$\frac{\text{phenol liberated (mole)–galactose liberated (mole)}}{\text{phenol liberated (mole)}} \times 100$$

(Also see ref. 111). In general, it appears that α-galactosidases preferentially transfer galactosyl groups to the primary alcoholic groups of acceptor molecules.

C. DE NOVO SYNTHESIS

Galactosidases are known to synthesize oligosaccharides when incubated with high concentrations of monosaccharides (202,203), and this procedure has been used for the preparation of several glucose (204–206) and galactose (207) derivatives. For example yeast α-galactosidase (59,208) in a solution of ~17% D-galactose polymerizes 7.5% of the sugar and about 60% of this amount appears as 6-O-α-D-galactosyl-D-galactose (55,59). Other products formed include α-1,3-, α-1,4-, and α-1,5-galactobioses. Clancy and Whelan (58) have also produced 3-*O*- and 6-*O*-α-D-galactosyl-*N*-acetyl-D-glucosamine by allowing a mixture of D-galactose and *N*-acetyl-D-glucosamine to react in the presence of yeast α-galactosidase.

VII. Kinetic Properties

Detailed kinetic investigations with α-galactosidases have been limited because few enzymes are available in purified forms. This section will be concerned largely with studies on these purified enzymes.

A. EFFECT OF SUBSTRATE CONCENTRATION

p-Nitrophenyl α-D-galactoside has been shown to be inhibitory at high concentrations with both molecular forms (I and II) of α-galactosidase from *V. faba* [Fig. 5(*a*)]. In contrast, galactose-containing oligosaccharides, such as melibiose and raffinose, do not show any inhibitory effect (105,115). In the case of enzyme I, where a high concentration of the nitrophenyl galactoside completely inhibits activity [Fig. 5(*a*)], it is possible that two molecules of substrate associate with the enzyme, giving rise to an inactive complex (5):

$$\text{E} + \text{S} \underset{k-1}{\overset{k+1}{\rightleftharpoons}} \underset{\text{(active)}}{\text{ES}} \overset{k+2}{\longrightarrow} \text{E} + \text{P}$$

$$\text{ESS (inactive)}$$

Hence, at a high substrate concentration, the rate of product formation would approach zero (cf. refs. 209,210). It has also been shown that when reaction rate is plotted against $\log_{10}$ [*p*-nitrophenyl α-D-galactoside] (5) a symmetrical, bell-shaped curve results, with the major inhibition occurring with substrate concentrations greater than 0.75 mM. Further, when D-galactose was added in increasing amounts to reaction mixtures containing initial substrate concentrations of 0.75 mM, the resulting curve closely paralleled the curve produced by substrate inhibition (5). D-Galactose behaves as a competitive inhibitor (5), and hence it was postulated that *p*-nitrophenyl α-D-galactoside at high concentrations was also competitive.

In the case of sweet almond α-galactosidase, the rate of hydrolysis of the above substrate did not approach zero at a high substrate

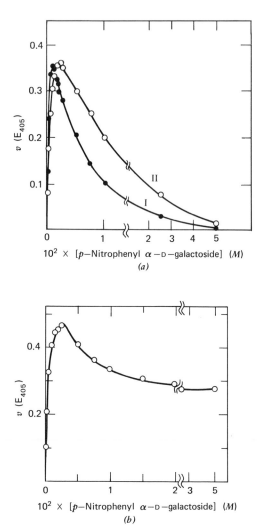

Fig. 5. Effect of substrate concentration on the initial rate of hydrolysis of *p*-nitrophenyl α-D-galactoside by (*a*) α-galactosidase I (●) and α-galactosidase II (○) from *V. faba* (5). (*b*) α-Galactosidase from sweet almond (105).

concentration [Fig. 5(b)] (See refs. 104,105). It, therefore, appears that in this case the complex ESS is not completely inactive and that it decomposes to products at a lower rate than ES (105).

B. EFFECT OF TEMPERATURE

α-Galactosidases display varying degrees of stability depending on their origin. The cell-free crude enzyme from *E. coli* is extremely unstable, whereas the purified enzyme can be lyophilized and stored at 4° for up to 2 months without loss of activity (36). Similarly, the enzymes from *A. niger* (20,21), *S. oleracea* (108), *P. amygdalus* (104), *C. ensiformis* (69), and beef liver (140) can be stored at low temperatures but the enzyme from *V. faba* is completely inactivated at temperatures below zero (211). The ionic strength of α-galactosidase solutions may also be an important factor when considering stability (27). The thermal stabilities of α-galactosidases from various sources are summarized in Table V.

TABLE V

Thermal Stability of Some α-Galactosidases

Source of α-galactosidase	Temperature (°C)	Time (min)	Loss of activity (%)	Reference
Aspergillus niger	55	60	35	21
Beef liver	55	5	20	140
	60	5	Complete	
Canavalia ensiformis	62	5	20	69
Diplococcus pneumoniae	50	60	30	27
Helix pomatia	70	30	Complete	118
Prunus amygdalus	60	20	15	104
Spinacia oleracea	43	10	50	108
Streptomyces olivaceus	55	15	90	13
Vicia faba				115
I	60	30	42	
II	60	30	81	
Vicia sativa	75	40	16	6

Most α-galactosidases behave normally in that the rate of the enzyme-catalyzed reaction increases to an optimum value with increasing temperature until inactivation of the enzyme occurs. Dey and co-workers (5,106) have followed the effect of temperature on both K_m and V_{max} values for α-galactosidases from sweet almonds and *V. faba;*

changes in K_m do not parallel the changes in V_{max}. This supports the results obtained by substrate specificity studies (see p. 104), that is, the nature of the substrate affects K_m and V_{max} differently. Therefore, it is likely that K_m is independent of V_{max} and represents the dissociation of the enzyme–substrate complex only, that is, $K_m \simeq K_s$ (5,106). In Arrhenius plots, strict linear relationships between the kinetic constant and $1/T$ suggest that K_m and V_{max} are simple constants rather than complex functions of several velocity constants, and that the breakdown of a single intermediate is rate-determining (5,106). Dey and Pridham (5) have calculated the entropy values from the effect of temperature on K_m and suggest that the high negative figures are caused by considerable conformational changes in the enzyme protein during the formation of the ES complex.

The energies of activation for substrate hydrolysis by various α-galactosidases are given in Table VI.

TABLE VI

Activation Energies for Substrate Hydrolysis by α-Galactosidases from Various Sources

Source of α-galactosidase	pH	Substrate*	Activation energy (Kcal/mole)	Reference
Aspergillus niger	4.0	a	16.4	21
Cyamopsis tetragonolobus	?	?	7.2	85
Escherichia coli (intact cells)	?	b	14.1	35
Mortierella vinacea	5.8?	b	12.4	7
Phaseolus vulgaris	~6.6	a	13.6	94
Prunus amygdalus	6.0	a	19.0	106
	6.0	c	12.4	
Vicia faba				5,114
I	4.0	a	15.3	
II	4.0	a	27.2	

* Substrates: a, *p*-Nitrophenyl α-D-galactoside; b, *o*-nitrophenyl α-D-galactoside; c, melibiose.

C. EFFECT OF pH

The pH optima of α-galactosidases vary to a considerable extent (see Table VII). In several cases the enzymes show two peaks, but in general, most α-galactosidases exhibit single broad optima.

TABLE VII
Optimum pH of α-Galactosidases from Various Sources

Source of α-galactosidase	Substrate*	Optimum pH	Reference
Aspergillus niger	b	3.8–4.5	20, 21
	j	4.2–4.8	21
Aspergillus paxillus	a	4.6	43
Calvatia cyanthiformis	f	3.0–5.0	22
Canavalia ensiformis (seeds)	b	4.0–5.0	69
Citrullus vulgaris (seeds)	c, d, e	4.2	70
Coffea sp. (seeds)	c, d, k	3.6–4.2 (broad)	17
Coffea sp. (seeds)			78
I	a	5.3	
II	a	6.0	
Diplococcus pneumoniae	f	5.6–6.0	27
Epidinium ecaudatum	c	5.0–5.5	29
Escherecia coli	f	7.5	37
Eudiplodinium maggii	c	5.1	39
Helix pomatia	a	3.2–3.8	118
Malolontha melolontha	a	6.0	126
Mortierella vinacea	c, f	4.0–6.0	7
	d	3.0–5.0	
Ophryoscolex caudatus	c	5.6	39
Phaseolus vulgaris (seeds)	b	6.5–6.7	94
Pig liver	a	5.2	145
Pisum sativum (seeds)			67
I	b	3.0–6.5 (broad)	
II	b	3.0–3.5 and 5.0–5.5	
Plantago ovata (seeds)			98
I	a	5.9	
II	a	5.3	
Polyplastron multivesculatus	c	5.3	39
Prunus amygdalus (seeds)	b, c	5.5–5.7	105
Rat brain	b	4.9	149
Rat kidney	h	4.0–4.5	172
Rat uterus	f	5.2	148
Saccharomyces sp.	c	3.5–5.3	50
	g, h, i	3.5–4.5	173
Spinacia oleracea (leaves)	b	5.3	108
Streptococcus bovis	c	5.6–6.3	61
Streptomyces olivaceus	c	5.4	13
Trigonellum foenum graecum (seeds)	a	3.2 and 4.6	109

TABLE VII (cont'd)

Source of α-galactosidase	Substrate*	Optimum pH	Reference
Vicia faba (seeds)			115
I	b	3.0–3.5 and 6.0–6.5	
	d	3.5–5.5	
II	b	2.5–3.5 and 5.0–5.5	
	d	3.5–4.5	
Vicia sativa (seeds)	a	6.3	6

* Substrates: a, phenyl α-D-galactoside; b, p-nitrophenyl α-D-galactoside; c, melibiose; d, raffinose; e, stachyose; f, o-nitrophenyl α-D-galactoside; g, 1-naphthyl α-D-galactoside; h, 2-naphthyl α-D-galactoside; i, 6-bromo-2-naphthyl α-D-galactoside; j, melibiitol; k, planteose.

The effect of pH on K_m and V_{max} has been studied only with α-galactosidases from sweet almond (106) and V. faba (5). The results obtained with α-galactosidase I from V. faba (5) are presented in Figure 6 in the form recommended by Dixon (212). In the pK_m-pH plots, the straight line portions (with slopes $+1$, 0, -1) of the curves intersect when they are extended. An analysis of the results according to Dixon (212,213) indicates the possible participation of a carboxyl and a histidine imidazolium group in the enzyme-catalyzed hydrolysis of both substrates (5).

In a similar study on sweet almond α-galactosidase (106), using melibiose and p-nitrophenyl α-D-galactoside as substrates, it was shown that ionized carboxyl and imidazolium groups were required at the active site of the enzyme. The role of the carboxyl group was less pronounced in the enzyme–p-nitrophenyl α-D-galactoside complex (V_{max} was pH-independent below pH 5.0) than in the enzyme–melibiose complex (V_{max} fell continuously below pH 6.0).

D. EFFECT OF INHIBITORS

1. Group Specific Reagents

Hogness and Battley (14) have shown that α-galactosidase from Aerobacter aerogenes can be inhibited by "sulphydryl reagents," such as p-chloromercuribenzoate, N-ethyl maleimide, and iodoacetamide. Sulphydryl reagents also inhibit α-galactosidases from Diplococcus pneumoniae (27) and Streptomyces olivaceus (13). On the other hand, α-galactosidases from Calvatia cyanthiformis (22), spinach leaves (108),

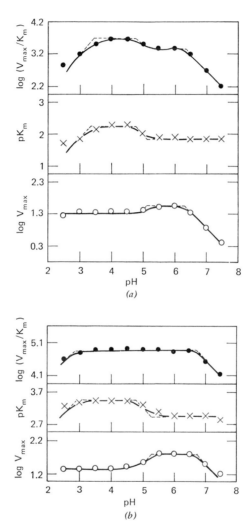

Fig. 6. K_m and V_{max} values as a function of pH for the hydrolysis of (a) raffinose and (b) p-nitrophenyl α-D-galactoside by α-galactosidase I from *V. faba* (5).

115

sweet almond (107), *V. faba* (5), and *Mortierella vinacea* (7) are not specifically inhibited by such reagents. Thus not all α-galactosidases require —SH groups for activity.

Photooxidation in the presence of methylene blue inactivated enzymes from sweet almond (107) and *V. faba* (5). Although photooxidation under these conditions is said to be specific for the destruction of histidine residues, possible reaction with other residues, such as tryptophan, tyrosine, and thiol groups cannot be ruled out (214–220). With sweet almond α-galactosidase (107), tryptophan, tyrosine (221), and thiol groups (222) were determined, both before and after photooxidation. The results presented in Table VIII show that these aminoacid residues are not affected by photooxidation in this instance.

TABLE VIII

Effect of Photooxidation on Some Amino Acid Residues of
Sweet Almond α-Galactosidase (107).

Amino acid residue	Native enzyme	Photooxidized enzyme
Tyrosine	16.9 ± 1.2	17.0 ± 1.3
Tryptophan	9.2 ± 1.0	9.4 ± 1.2
Free thiol	Nil	Nil
Masked thiol		
(Urea denatured enzyme)	7.5 ± 0.2	7.5

The amino acid residues are expressed as mole/mole enzyme (mol. wt. = 33,000).

2. Metal Ions

The inhibitory effect of some metal ions on various α-galactosidases is summarized in Table IX. A wide variety of effects by other metal ions on α-galactosidases has also been observed (6,13,22,49,107,108).

With *V. faba* enzyme I, the Ag^+ inhibition (5) may be attributed to combination with carboxyl and/or histidine residues (cf. refs. 213,223). The K_i (competitive) values at pH 4.0 and 6.0 were 4.0 μM and 0.59 μM respectively. The decrease in K_i with the rise in pH suggests that the ions were binding groups that lost protons over this pH range, that is, groups that were not markedly unprotonated at pH 4.0. However, the fact that the decrease was much less than 100-fold suggests that these groups did not have pK values greater than 6.

TABLE IX
Inhibition of various α-galactosidases by Ag^+, Cu^{2+}, and Hg^{2+}

Metal ion	Source of α-galactosidase	Concentration of metal ion (M)	Inhibition (%)	Nature of inhibition	K_i (M)	Reference
Ag^+	Calvatia cyanthiformis	10^{-3}	Nil	—	—	22
	Prunus amygdalus (seeds)	5.7×10^{-6}	50	Competitive	2.46×10^{-6} (pH 5.5)	107
	Spinacia oleracea (leaves)	2×10^{-7}	50	—	—	108
	Streptomyces olivaceus	2×10^{-6}	81	—	—	13
	Vicia faba (seeds)	10^{-6}	23	Competitive	5.9×10^{-7} (pH 6.0)	5
Cu^{2+}	Vicia sativa (seeds)	10^{-6}	100	—	—	6
	Calvatia cyanthiformis	10^{-3}	Nil	—	—	22
	Diplococcus pneumoniae	2×10^{-4}	100	—	—	27
	Prunus amygdalus (seeds)	1.2×10^{-3}	50	Competitive	1.1×10^{-3} (pH 5.5)	107
Hg^{2+}	Streptomyces olivaceus	10^{-3}	50	—	—	13
	Vicia faba (seeds)	2.5×10^{-2}	46	—	—	5
	Aspergillus niger	3×10^{-3}	95	—	—	21
	Calvatia cyanthiformis	10^{-4}	100	—	—	22
	Diplococcus pneumoniae	2×10^{-4}	100	—	—	27
	Mortierella vinacea	10^{-7}	~23	Noncompetitive	—	7
	Prunus amygdalus (seeds)	4.7×10^{-3}	50	Competitive	6.5×10^{-4} (pH 5.5)	107
	Streptomyces olivaceus	2×10^{-7}	63	—	—	13
	Vicia faba (seeds)	10^{-7}	43	Competitive	3×10^{-7} (pH 6.0)	5

The Ag^+ inactivated $V.$ $faba$ enzyme I regained most of its activity on dialysis against McIlvaine buffer (pH 4.0 or 6.0), particularly in the presence of cysteine. It was further shown that the inhibitory effect of Ag^+ decreased when low concentrations of galactose were present. Such protection was observed only when galactose was added before the addition of Ag^+. This, therefore, confirms that Ag^+ reacts with the active site (5).

Dey (106) has shown that the Ag^+ inhibition of sweet almond α-galactosidase is also competitive. An examination of the effect of pH on K_i values according to Dixon (212,213) has further indicated that a histidine imidazolium group present in the enzyme active site is strongly affected by the binding of Ag^+.

The metal ion Hg^{2+} is a potent inhibitor of several α-galactosidases (see Table IX), and this usually suggests reaction with thiol groups. However, Webb (224) has discussed the coordination of Hg^{2+} between carboxyl and amino groups. Mercuric ions are also known to react with amino and imidazolium groups of histidine (225) and with peptide linkages (226). Using α-galactosidase I from $V.$ $faba$, Dey and Pridham (5) have shown that Hg^{2+} inhibition is weaker than that of Ag^+ at pH 4.0 and that the former was noncompetitive (K_i 75 μM). This discounts the possibility that Hg^{2+} reacts with the carboxyl or histidyl residues at the active site of the enzyme. On the other hand, inhibition by Hg^{2+} at pH 6.0 (K_i 0.3 μM) was much greater and competitive. In the latter case, therefore, reaction may have occurred with the histidine at the active site.

3. Sugars and Their Derivatives

D-Galactose is a powerful and competitive inhibitor of α-galactosidases (5,7,13,16,21,22,34,207,208). The structural analogues of D-galactose, that is, L-arabinose (5,7,21,107) and D-fucose (5), also inhibit the enzymes, whereas their enantiomers are ineffective (5). 2-Deoxy-D-galactose, D-glucose, D-mannose, D-fructose, D-xylose, and D-ribose do not produce inhibition of the α-galactosidases that have so far been examined (5,7,22,34,107). Li and Shetlar (22) have pointed out that for attachment of the sugar to the enzyme, the D-galactose configuration is required and C-1, C-2, C-4, and C-6 are involved in binding. The open chain derivatives of galactose, and sodium and calcium galactonates do not inhibit the enzyme (16,107). Sharma (227) has shown that a highly specific inhibition of α-galactosidases occurs with myo-inositol (cf. ref. 5). He has attributed this to the similarity in the orientation of —OH groups at C-2, C-3, and C-4 in myo-inositol to those at C-4, C-3, and C-2 or C-1, C-2, and C-3 of the α-D-galacto-pyranosyl residue of the substrate. Similar conclusions have been made by Kelemen and Whelan (19) and Heyworth and Walker (228), who noted configurational similarities between substrates and polyol-inhibitors of various glycosidases.

D-Galactal was shown to inhibit α-galactosidase from *Aspergillus niger* (21). The planar half-chair conformation of this cyclitol could be responsible for its inhibitory action (cf. ref. 229). Half-chair conformations of monosaccharides are postulated to occur as intermediates in acid hydrolysis of glycosides (230,231) and in lysozyme-catalyzed hydrolysis of aminosugar-containing oligosaccharides (232–234). The inhibitory action of aldono-(1 → 5)-lactones on glycosidases can also be expected on the basis of the preferred half-chair conformations of these compounds (235).

Dey and co-workers (5,106) have shown that various α-D-galactosides are competitive inhibitors of α-galactosidases. The K_i values of the galactosides are very close to their Michaelis constants when these compounds are used as substrates.

VIII. Mechanism of Action

Few concrete facts are available regarding the mechanism of action of α-galactosidases (5,106,107) because of insufficient knowledge of the chemistry and kinetics of the enzymes from most sources.

So far there have been no studies on bond fission by α-galactosidases, although by analogy with other glycosidases (168,236–239) it is likely that the galactose-oxygen bonds of substrates are cleaved. Nuclear magnetic resonance and polarimetry studies have clearly shown with *Cajanus indicus* and sweet almond α-galactosidases that the liberated galactosyl residues possess the same anomeric configuration as the substrate (107,240; cf. ref. 22).

Acid-base catalyses operate in enzyme systems such as fumarase (241), maltase (242), yeast invertase (243), β-galactosidase (168), and so forth. The specificity studies on sweet almond α-galactosidase (105) with aryl α-D-galactosides show that the electronic nature of the aglycon has a noticeable influence on the rate of enzymic hydrolysis. Two sets of straight lines ($\rho = -0.054$ and -1.5) are obtained by plotting $\log_{10} V_{max}$ against Hammett constants (σ) for substituent groups on the aromatic ring. The straight lines intercept at a point corresponding to phenyl α-D-galactoside ($\sigma = 0$). This resembles alkaline and acid hydrolyses of aryl glucosides (244) and, therefore, may be attributed to the presence of basic and acidic groups at the active site. These groups were identified by kinetic studies as carboxyl (deprotonated) and imidazolium (protonated), respectively (106). Photooxidation in the presence of methylene blue and inhibition by Ag$^+$ ions, as discussed in Section VII, D also support these conclusions. A similar state of affairs seems to exist with *V. faba* enzyme I (5). It has further been shown with the almond enzyme that binding of *p*-nitrophenyl α-D-galactoside to the active site lowers the pK of the

Two-step mechanism

group dissociating on the acidic side and raises the value of the group dissociating on the alkaline side of the pH optimum (106). A similar shift of pK values was observed with chitotriose bound to lysozyme (245). It seems that the changes (presumably conformational) induced by the substrate in the enzyme molecule enlarge the effective pH range of the enzyme [i.e. induced change favors the catalytic reaction (246)].

On the basis of those results a "two-step" mechanism has been postulated for the action of sweet almond α-galactosidase (107). Here the aglycon is cleaved by the concerted action of carboxyl and imidazolium groups. This is followed by reaction with an acceptor molecule (R′OH), which may be water or an aliphatic alcohol, resulting in hydrolysis or transfer products. It is possible that the electrophilic attack of the imidazolium group alone is sufficient to cleave the glycosyl–oxygen bond with the formation of a carbonium ion at C-1 of the galactose moiety. In the complete two-step mechanism, two Walden inversions probably occur, resulting in the retention of the anomeric configuration in the final product. However, formation of the carbonium ion intermediate need not necessarily lead to racemization; the configuration could conceiviably be stabilized by a specific binding of the intermediate to the enzyme (cf. refs. 233,247).

An alternative "one-step" mechanism of action has also been suggested (107):

One - step mechanism

This involves the formation of a ternary complex of the enzyme, substrate, and acceptor with the carboxyl and the imidazolium groups playing a similar role, as in the two-step mechanism. A front-side attack (cf. ref. 236) on the anomeric carbon atom of the galactose moiety leads to a product with retention of configuration. An examination of molecular models of phenyl α-D-galactoside shows that the C1 and 1C chair conformations of this compound (cf. ref. 248) would not allow a front-side attack (168,249). This steric hindrance is not apparent in the B2 and 3B boat conformations; the latter is the theoretically preferred boat structure with fewer interacting groups than the B2 form. Hence, in the one-step mechanism, the conformation of the galactose moiety may change from C1 to 3B during enzyme–substrate complex formation.

IX. Physiological Significance

In this section no attempt has been made to present a comprehensive review of the large number of publications concerned with the physiological roles of α-galactosidases. Only some of the more recent studies in this field will be mentioned.

In the plant kingdom galactose-containing oligo- and polysaccharides and lipids are ubiquitous (250,251); in the tissues they are commonly accompanied by α-galactosidase. In many maturing seeds, there is a concomitant synthesis of galactosylsucrose derivatives and an apparent increase in α-galactosidase activity (68,112,252–254). On germination the enzyme is undoubtedly involved in the hydrolysis of these oligosaccharides, which serve as a soluble and a readily metabolizable energy reserve (e.g., ref. 255 and references therein).

In order that galactosyl sucrose derivatives may accumulate during seed maturation, there must be some mechanism that prevents the interaction of enzyme and substrate; perhaps this is effected by compartmentalization or by an endogenous inhibitor. A related problem has been revealed by Shiroya (86), who examined the breakdown of raffinose and stachyose in germinating cotton seeds. It would appear that oxygen is somehow involved in the α-galactosidase-catalyzed hydrolysis of these sugars *in vivo*, as this is inhibited if moistened seeds are kept under a reduced oxygen tension. Shiroya (86) noted that moistened seeds that had lost viability were also unable to degrade endogenous oligosaccharides. In the case of *V. faba*, the physiological state of the seed is reflected in changing patterns of the different molecular forms of α-galactosidases (67). Green, immature seeds only possess activity corresponding to enzyme II; as development occurs, enzyme I is formed (254). The *in vitro* experiments (see Section IV) suggest that enzyme I is derived from enzyme II. In dormant bean seeds, the activity of I is normally much greater than that of II (cf. Fig. 2), but the degree of hydration of the tissues appears to be related to the activity ratio (67). Similar enzyme patterns occur in dormant seeds from several other plant species (67). It is perhaps physiologically significant that in dormant *V. faba* seeds, it is the enzyme with high specific activity which is predominant. Thus at the onset of germination a maximum rate of breakdown of the reserve oligosaccharides can occur.

The germination of *V. faba* seeds appears to be a reversal of maturation, in that the level of enzyme I rapidly decreases (67). Despite this, the total α-galactosidase activity of the tissues increases during the first few days of germination (254).

α-Galactosidase may be involved in the metabolism of plant galactolipids. Sastry and Kates (256), for example, have shown that runner bean leaves possess all the necessary enzymes, including α-galactosidase, for the complete breakdown of these compounds to fatty acids, glycerol, and galactose. Chloroplast membrane galactolipids may also be affected by α-galactosidases. Bamberger and Park (93), studying the Hill reaction of isolated spinach chloroplasts, noted that a crude preparation containing galactolipases and galactosidases changed the physiological activity of the organelles. In this connection it is interesting to note that α-galactosidase has been detected in spinach (108) and sugar cane chloroplasts (257).

In the animal kingdom, few observations appear to have been made on the role of α-galactosidase. The enzyme does occur in brain tissues; *in vivo* it may be involved in the hydrolysis of digalactosyl diglycerides (149).

References

1. Bua, A. *Chem. Ztg., Chem. App.*, *19*, 1873 (1895).
2. Fischer, E., and Lindner, P., *Ber. Bunsenges. Phys. Chem.*, *28*, 3034 (1895).
3. Weidenhagen, R., *Z. Zuckerind.*, *77*, 696 (1927); *78*, 99 (1928).
4. Weidenhagen, R., and Renner, A., *Z. Zuckerind.*, *86*, 22 (1936).
5. Dey, P. M., and Pridham, J. B., *Biochem. J.*, *115*, 47 (1969).
6. Petek, F., Villarroya, E., and Courtois, J. E., *Eur. J. Biochem.*, *8*, 395 (1969).
7. Suzuki, H., Li, Su-Chen, and Li, Yu-Teh., *J. Biol. Chem.*, *245*, 781 (1970).
8. Morris, B., Tsou, Kwan-Chung, and Seligman, A. M., *J. Histochem. Cytochem.*, *11*, 653 (1963).
9. Szmigielski, S., *J. Lab. Clin. Med.*, *67*, 709 (1966).
10. Morris, B., and Wasserkrug, H., *Histochemie*, *10*, 363 (1967).
11. Carrère, C., Lambin, S., and Courtois, J. E., *Compt. Rend. Soc. Biol.*, *154*, 1747 (1960).
12. Suzuki, H., and Tanabe, O., *Nippon Nogei Kagaku Kaishi*, *37*, 623 (1963).
13. Suzuki, H., Ozawa, Y., and Tanabe, O., *Agr. Biol. Chem.* (*Tokyo*), *30*, 1039 (1966).
14. Hogness, D. S., and Battley, E. H., *Fed. Proc.*, *16*, 197 (1957).
15. Yakulis, V., Costea, N., and Heller, P., *Proc. Soc. Exp. Biol. Med.*, *121*, 812 (1966).
16. Wakabayashi, K., and Nishizawa, K., *Seikagaku*, *27*, 662 (1955).
17. Courtois, J. E., Wickstrom, A., and Dizet, P. L., *Bull. Soc. Chim. Biol.*, ·*38*, 851 (1956).
18. Kelemen, M. V., and Whelan, W. J., *Biochem. J.*, *100*, 5P (1966).
19. Kelemen, M. V., and Whelan, W. J., *Arch. Biochem. Biophys.*, *117*, 423 (1966).
20. Bahl, O. P., and Agrawal, K. M. L., *J. Biol. Chem.*, *244*, 2970 (1969).
21. Lee, Y. C., and Wacek, V., *Arch. Biochem. Biophys.*, *138*, 264 (1970).
22. Li, Yu-Teh, and Shetlar, M. R., *Arch. Biochem. Biophys.*, *108*, 523 (1964).
23. Bulmer, G. S., and Li, Yu-Teh, *Mycologia*, *58*, 555 (1966).

24. Watkins, W. M., and Morgan, W. T. J., *Nature*, *175*, 676 (1955).
25. Furukawa, K., Yamamoto, S., and Iseki, S., *Igaku to Seibutsugaku*, *53*, 142 (1959).
26. English, P. D., and Albersheim, P., *Plant Physiol.*, *44*, 217 (1969).
27. Li, Yu-Teh, Li, Su-Chen, S., and Shetlar, M. R., *Arch. Biochem. Biophys.*, *103*, 436 (1963).
28. Li, Yu-Teh, and Shetlar, M. R., *Arch. Biochem. Biophys.*, *108*, 301 (1964).
29. Bailey, R. W., and Howard, B. H., *Biochem. J.*, *87*, 146 (1963).
30. Hoeckner, E., *Z. Hyg. Infektionskrankh.*, *129*, 519 (1949).
31. Sheimin, R., and Crocker, B. F., *Congr. Intern. Biochim.*, *Resumes Communs.*, *3 Congr.*, *Brussels*, 91 (1955).
32. Lester, G., and Bonner, D. M., *J. Bacteriol.*, *73*, 544 (1957).
33. Sheimin, R., and McQuillen, K., *Biochim. Biophys. Acta*, *31*, 72 (1959).
34. Sheimin, R., and Crocker, B. F., *Can. J. Biochem.*, *39*, 63 (1961).
35. Sheimin, R., and Crocker, B. F., *Can. J. Biochem.*, *39*, 55 (1961).
36. Schmitt, R., and Rotman, B., *Biochem. Biophys. Res. Commun.*, *22*, 473 (1966).
37. Schmitt, R., *J. Bacteriol.*, *96*, 462 (1968).
38. Burstein, C., and Kepes, A., *Biochim. Biophys. Acta*, *230*, 52 (1971).
39. Howard, B. H., *Biochem. J.*, *89*, 90P (1963).
40. Ruttloff, H., Taeufel, A., Haenel, H., and Taeufel, K., *Nahrung*, *11*, 47 (1967).
41. Hofmann, E., *Biochem. Z.*, *272*, 133 (1934).
42. Suzuki H., Ozawa, Y., Oota, H., and Yoshida, H., *Agr. Biol. Chem.*, *(Tokyo)*, *33*, 506 (1969).
43. Courtois, J. E., Carrerem, C., and Petek, F., *Bull. Soc. Chim. Biol.*, *41*, 1251 (1959).
44. Courtois, J. E., and Dizet, P. L., *Bull. Soc. Chim. Biol.*, *45*, 743 (1963).
45. Suzuki, H., Ozawa, Y., and Tanabe, O., *Chem. Abstr.*, *65*, P9700g (1965).
46. Doudoroff, M., *J. Biol. Chem.*, *157*, 699 (1945).
47. Isaiev, V. I., *Chem. Listy*, *21*, 101, 141, 191 (1927).
48. Cattaneo, C., *Bull. Soc. Ital. Biol. Sper.*, *11*, 902 (1936).
49. Cattaneo, C., *Arch. Sci. Biol. (Italy)*, *23*, 472 (1937).
50. Adams, M., Richtmyer, N. K., and Hudson, C. S., *J. Amer. Chem. Soc.*, *65*, 1369 (1943).
51. Winge, O., and Roberts, C., *Nature*, *177*, 383 (1956).
52. Losada, M., *Compt. Rend. Trav. Lab. Carlsberg Ser. Physiol.*, *25*, 460 (1957).
53. Winge, O., and Roberts, C., *Compt. Rend. Trav. Lab. Carlsberg Ser. Physiol.*, *25*, 420 (1957).
54. Rosa, M., and Barta, J., *Ind. Aliment. et Agr. (Paris)*, *12*, 889 (1957).
55. Clancy, M. J., and Whelan, W. J., *Biochem. J.*, *76*, 22P (1960).
56. Maria, J. S., *J. Gen. Microbiol.*, *28*, 375 (1962).
57. Haupt. W., and Alps, H., *Arch. Mikrobiol.*, *45*, 179 (1963).
58. Clancy, M. J., and Whelan, W. J., *Arch. Biochem. Biophys.*, *118*, 730 (1967).
59. Clancy, M. J., and Whelan, W. J., *Arch. Biochem. Biophys.*, *118*, 724 (1967).
60. Bailey, R. W., *Nature*, *195*, 79 (1962).
61. Bailey, R. W., *Biochem. J.*, *86*, 509 (1963).

62. Suzuki, H., Ozawa, Y., and Tanabe, O., *Nippon Nogei Kagaku Kaishi, 38,* 542 (1964).
63. Suzuki, H., Ozawa, Y., and Tanabe, O., *Hakko Kyokaishi, 22,* 455 (1964).
64. Watkins, W. M., *Biochem. J., 54,* 33 (1953).
65. Watkins, W. M., *Nature, 181,* 117 (1958).
66. Watkins, W. M., Zermitz, M. L. and Kabat, E. A., *Nature, 195,* 1204 (1962).
67. Barham, D., Dey, P. M., Griffiths, D, and Pridham, J. B., *Phytochemistry, 10,* 1759 (1971).
68. Lechevallier, D., *Compt. Rend. Soc. Biol., 255,* 3211 (1962).
69. Cabezas, J. A., and Vazquez-Pernas, R., *Rev. Espan. Fisiol., 25,* 147 (1969).
70. Ahmed, M. U., and Cook, F. S., *Can. J. Biochem., 42,* 605 (1964).
71. Helferich, B., and Vorsatz, F., *Z. Physiol. Chem., 237,* 254 (1935).
72. Courtois, J. E., and Petek, F., *Bull. Soc. Chim. Biol., 36,* 1115 (1954).
73. Courtois, J. E., and Anagnostopoulos, C., and Petek, F., *Compt. Rend. Soc. Biol., 238,* 2020 (1954).
74. Courtois, J. E., and Anagnostopoulos, C., *Enzymologia, 17,* 69 (1954).
75. Anagnostopoulos, C., Courtois, J. E., and Petek, F., *Arch. Sci. Biol. (Italy), 39,* 631 (1955).
76. Courtois, J. E., and Petek, F., *Bull. Soc. Chim. Biol., 39,* 715 (1957).
77. Zarmitz, M. L., and Kabat, E. A., *J. Amer. Chem. Soc., 82,* 3953 (1960).
78. Petek, F., and Dong, T., *Enzymologia, 23,* 133 (1961).
79. Petek, F., and Courtois, J. E., *Bull. Soc. Chim. Biol., 46,* 1093 (1964).
80. Courtois, J. E., and Petek, F., *Methods Enzymol., 8,* 565 (1966).
81. Courtois, J. E., and Dizet, P. L., *Carbohyd. Res., 3,* 141 (1966).
82. Shadaksharaswami, M., and Ramachandra, G., *Phytochemistry, 7,* 715 (1968).
83. Hattori, S., and Shiroya, T., *Botan. Mag. (Tokyo), 64,* 137 (1951).
84. Hasegawa, M., Takayama, M., and Yoshida, S., *J. Japan Forest Soc., 35,* 156 (1953).
85. Lee, S. R., *Soul Taehakkyo Nonmunjip, Chayon Kwahak Saengnougge, 17,* 7 (1966).
86. Shiroya, T., *Phytochemistry, 2,* 33 (1963).
87. Pridham, J. B., *Biochem. J., 76,* 13 (1960).
88. Villarroya, E., Petek, F., Courtois, J. E., and Lanchec, C., *Abstr. 6th Fed. Spanish. Biochem. Soc., Madrid,* p. 930 (1969).
89. Zoch, E., *J. Chromatogr., 4,* 21 (1960).
90. Lechevallier, D., *Compt. Rend. Soc. Biol., 250,* 2825 (1960).
91. Cooper, R. A., and Greenshields, R. N., *Biochem. J., 81,* 6P (1961).
92. Lechevallier, D., *Compt. Rend. Soc. Biol., 258,* 5519 (1964).
93. Bamberger, E. S., and Park, R. B., *Plant Physiol., 41,* 1591 (1966).
94. Agrawal, K. M. L., and Bahl, O. P., *J. Biol. Chem., 243,* 103 (1968).
95. Prestidge, L. S., and Pardee, A. B., *Biochem. Biophys. Acta, 100,* 591 (1965).
96. Jacob, F., and Monod, J., *Symp. Soc. Exptl. Biol., 12,* 75 (1958).
97. Courtois, J. E., Petek, F., and Dong, T., *Bull. Soc. Chim. Biol., 43,* 1189 (1961).
98. Courtois, J. E., Petek, F., and Dong, T., *Bull. Soc. Chim. Biol., 45,* 95 (1963).
99. Peat, S., and Rees, D. A., *Biochem. J., 79,* 7 (1961).

126   P. M. DEY AND J. B. PRIDHAM

100. Neuberg, C., *Biochem. Z.*, *3*, 519 (1907).
101. Zechmeister, L., Toth, G., and Balant, M., *Enzymologia*, *5*, 302 (1938).
102. Helferich, B., and Jung, K. H., *Z. Physiol. Chem.*, *311*, 54 (1958).
103. Malhotra, O. P., and Dey, P. M., *Biochem. Z.*, *340*, 565 (1964).
104. Malhotra, O. P., and Dey, P. M., *Biochem. J.*, *103*, 508 (1967).
105. Malhotra, O. P., and Dey, P. M., *Biochem. J.*, *103*, 739 (1967).
106. Dey, P. M., and Malhotra, O. P., *Biochem. Biophys. Acta*, *185*, 402 (1969).
107. Dey, P. M., *Biochem. Biophys. Acta*, *191*, 644 (1969).
108. Gatt, S., and Baker, E. A., *Biochem. Biophys. Acta*, *206*, 125 (1970).
109. Courtois, J. E., and Percheron, F., *Bull. Soc. Chim. Biol.*, *43*, 167 (1961).
110. Percheron, F., and Guilloux, E., *Bull. Soc. Chim. Biol.*, *46*, 543 (1964).
111. Wohnlich, J., *Bull. Soc. Chim. Biol.*, *45*, 1171 (1963).
112. Bourne, E. J., Pridham, J. B., and Walter, M. W., *Biochem. J.*, *82*, 44P (1962).
113. Pridham, J. B., and Walter, M. W., *Biochem. J.*, *92*, 20P (1964).
114. Dey, P. M., and Pridham, J. B., *Phytochemistry*, *7*, 1737 (1968).
115. Dey, P. M., and Pridham, J. B., *Biochem. J.*, *113*, 49 (1969).
116. Dey, P. M., and Pridham, J. B., *Abstr. 6th Fed. Spanish Biochem. Soc. Madrid*, p. 931 (1969).
117. Kooiman, P., *J. Cell. Comp. Physiol.*, *63*, 197 (1964).
118. Nagaoka, T., *Tohoku J. Exp. Med.*, *51*, 137 (1949).
119. Bierry, H., *Biochem. Z.*, *44*, 446 (1912); *Compt. Rend. Soc. Biol.*, *156*, 265 (1913).
120. Utsushi, M., Huji, K., Matsumoto, S., and Nagaoka, T., *Tohoku J. Exp. Med.* *50*, 175 (1949).
121. Koike, H., *Zool. Mag. Tokyo*, *63*, 228 (1954).
122. Evans, W. A. L., *Exp. Parasitol.*, *5*, 191 (1956).
123. Newcomer, W. S., *Physiol. Zool.*, *29*, 157 (1956).
124. Webber, L. G., *Aust. J. Zool.*, *5*, 164 (1957).
125. Courtois, J. E., Chararas, C., and Debris, M. M., *Compt. Rend. Soc. Biol.*, *252*, 2608 (1961).
126. Courtois, J. E., Petek, F., and Zanouzi, M. A. K., *Compt. Rend. Soc. Biol.*, *156*, 565 (1962).
127. Bhatnagar, P., *Indian J. Entomol.*, *24*, 19 (1962).
128. Ehrhardt, P., and Voss, G., *J. Insect Physiol.*, *8*, 165 (1962).
129. Banks, W. M., *Science*, *141*, 1191 (1963).
130. Evans, W. A. L., and Payne, D. W., *J. Insect Physiol.*, *10*, 657 (1964).
131. Bhatnagar, P., *Naturwissenschaften*, *51*, 17 (1964).
132. Devi, R. V., *Physiol. Zool.*, *38*, 158 (1965).
133. Srivastava, P. N., *Enzymologia*, *30*, 127 (1966).
134. Saxena, K. N., and Ghandi, J. R., *Comp. Biochem. Physiol.*, *17*, 765 (1966).
135. Khan, M. R., and Ford, J. B., *J. Insect Physiol.*, *13*, 1619 (1967).
136. Li, Yu-Teh, and Shetlar, M. R., *Comp. Biochem. Physiol.*, *14*, 275 (1965).
137. Conchi, J., and Mann, T., *Nature*, *179*, 1190 (1957).
138. Roston, C. P. J., Caygill, J. C., and Jevons, F. R., *Life Sci. (Oxford)*, *5*, 535 (1966).
139. Fleischer, B., and Fleischer, S., *Biochem. Biophys. Acta*, *183*, 265 (1969).

140. Langley, T. J., and Jevons, F. R., *Arch. Biochem. Biophys.*, *128*, 312 (1968).
141. Iseki, S., and Yamamoto, H., *Proc. Jap. Acad.*, *44*, 269 (1968).
142. Yamamoto, H., *Igaku to Seibutsugaku*, *76*, 286 (1968).
143. Van Hoof, F., and Hers, H. G., *Eur. J. Biochem.*, *7*, 34 (1968).
144. Romeo, G., and Migeon, B. R., *Science*, *170*, 180 (1970).
145. Debris, M. W., Courtois, J. E., and Petek, F., *Bull. Soc. Chim. Biol.*, *44*, 291 (1962).
146. Suzuki, I., Kushida, H., and Shida, H., *Seikagaku*, *41*, 334 (1969).
147. Monis, B., Goldberg, J. D., and Diamond, E. L., *Proc. Soc. Exp. Biol. Med.*, *116*, 580 (1964).
148. Coleman, R. L., *Biochem. Biophys. Acta*, *159*, 192 (1968).
149. Subba Rao, K., and Pieringer, R. A., *J. Neurochem.*, *17*, 483 (1970).
150. Borgstrom, B., and Dahlqvist, A., *Acta Chem. Scand.*, *12*, 1997 (1958).
151. Dahlqvist, A., *Biochem. Biophys. Acta*, *50*, 55 (1961).
152. Koppel, J. L., Porter, C. J., and Crocker, B. F., *J. Gen. Physiol.*, *36*, 703 (1953).
153. Porter, C. J., Holmes, R., and Crocker, B. F., *J. Gen. Physiol.*, *37*, 271 (1953).
154. Monod, J., Cohen-Bazire, G., and Cohn, M., *Biochem. Biophys. Acta*, *7*, 585 (1951).
155. Landman, O. E., and Bonner, D. M., *Arch. Biochem. Biophys.*, *52*, 93 (1954).
156. Pardee, A. B., *J. Bacteriol.*, *69*, 233 (1955).
157. Rickenberg, H. V., and Lester, G., *J. Gen. Microbiol.*, *13*, 279 (1955).
158. Pardee, A. B., *J. Bacteriol.*, *73*, 376 (1957).
159. Landman, O. E., *Biochem. Biophys. Acta*, *23*, 558 (1959).
160. Lester, G., *Arch. Biochem. Biophys.*, *40*, 390 (1952).
161. Hestrin, S., Feingold, D. S., and Schramm, M., in *Methods in Enzymology*, Vol. 1, S. P. Colowick and N. O. Kaplan, Eds., Academic Press, New York, 1955, p. 231.
162. Nelson, N., *J. Biol. Chem.*, *153*, 375 (1944).
163. Martinsson, A., *J. Chromatogr.*, *24*, 487 (1966).
164. Johnson, J., and Fusaro, R., *Anal. Biochem.*, *18*, 107 (1967).
165. Wallenfels, K., and Kurz, G., in *Methods in Enzymology*, Vol. 9, S. P. Colowick and N. O. Kaplan, Eds., Academic Press, 1966, p. 112.
166. Dey, P. M., *Chem. Ind. (London)*, 1637 (1967).
167. Courtois, J. E., *Proc. Intern. Congr. Biochem.*, *4th Congr. Vienna*, *1*, 140 (1959).
168. Wallenfels, K., and Malhotra, O. P., *Advan. Carbohyd. Chem.*, *16*, 239 (1961).
169. Levvy, G. A., Hay A. J., and Conchie, J., *Biochem. J.*, *91*, 378 (1964).
170. Furth, A. J., and Robinson, D., *Biochem. J.*, *97*, 59 (1965).
171. Wallenfels, K., Lehmann, J., and Malhotra, O. P., *Biochem. Z.*, *333*, 209 (1960).
172. Gillman, S. M., Tsou, K. C., and Seligman, A. M., *Arch. Biochem. Biophys.*, *138*, 264 (1970).
173. Tsou, K. C., and Su, H. C. F., *Anal. Biochem.*, *8*, 415 (1964).
174. Barnett, J. E. G., Jarvis, W. T. S., and Munday, K. A., *Biochem. J.*, *105*, 669 (1967).
175. Cleland, W. W., *Biochemistry*, *3*, 480 (1964).

176. Dey, P. M., Khaleque, A., and Pridham, J. B., *Biochem. J.*, *124*, 27p (1971).
177. Tanner, W., and Kandler, O., *Eur. J. Biochem.*, *4*, 233 (1968).
178. Cohen, R. B., Tsou, K. C., Rutenburg, S. H., and Seligman, A. M., *J. Biol. Chem.*, *195*, 239 (1952).
179. Seligman, A. M., Tsou, K. C., Rutenburg, S. H., and Cohen, R. B., *J. Histochem. Cytochem.*, *2*, 209 (1954).
180. Rutenburg, A. M., and Seligman, A. M., *J. Histochem. Cytochem.*, *3*, 455 (1955).
181. Rutenburg, A. M., Rutenburg, S. H., Monis, B., Teague, R., and Seligman, A. M., *J. Histochem. Cytochem.*, *6*, 122 (1958).
182. Shibata, Y., and Nisizawa, K., *Arch. Biochem. Biophys.*, *109*, 516 (1965).
183. Andrews, P., *Biochem. J.*, *91*, 222 (1964).
184. Bridel, M., and Béguin, C., *Compt. Rend. Soc. Biol.*, *182*, 812 (1926).
185. Helferich, B., and Appel, H., *Z. Physiol. Chem.*, *205*, 231 (1932).
186. Helferich, B., Winkler, S., Gootz, R., Peters, O., and Gunther, E., *Z. Physiol. Chem.*, *208*, 91 (1932).
187. Pigman, W. W., *J. Amer. Chem. Soc.*, *62*, 1371 (1940).
188. Pigman, W. W., *J. Res. Nat. Bur. Stand.*, *A26*, 197 (1941).
189. Cohn, M., *Bacteriol. Revs.*, *21*, 140 (1957).
190. Wallenfels, K., and Fischer, J., *Z. Physiol. Chem.*, *321*, 223 (1960).
191. Wickstrom, A., and Svendsen, A. B., *Acta Chem. Scand.*, *10*, 1190 (1956).
192. Courtois, J. E., Dizet, P. L., and Petek, F., *Bull. Soc. Chim. Biol.*, *41*, 1261 (1959).
193. Courtois, J. E., and Ariyoshi, U., *Bull. Soc. Chim. Biol.*, *42*, 737 (1960).
194. Courtois, J. E., *Bull. Soc. Chim. Biol.*, *42*, 1451 (1960).
195. Archambault, A., Courtois, J. E., Wickstrom, A., and Dizet, P. L., *Bull. Soc. Chim. Biol.*, *38*, 1133 (1956).
196. Watkins, W. M., in *Methods in Enzymology*, Vol. 8, S. P. Colowick and N. O. Kaplan, Eds., Academic Press, New York, 1966, p. 700.
197. Blanchard, P. H., and Albon, N., *Arch. Biochem.*, *29*, 220 (1950).
198. Neuberg, C., and Lachman, J., *Biochem. Z.*, *24*, 171 (1910).
199. French, D., Wild, G. M., and James, W. J., *J. Amer. Chem. Soc.*, *75*, 3664 (1953).
200. Fitting, C., and Doudoroff, M., *J. Biol. Chem.*, *199*, 153 (1952).
201. Ayers, W. A., *J. Biol. Chem.*, *234*, 2819 (1959).
202. Bourquelot, E., and Aubry, A., *Compt. Read. Acad. Sci. (Paris)*, Ser. *163*, 60 (1916).
203. Bourquelot, E., and Aubry, A., *Compt. Read. Acad. Sci. (Paris)*, Ser. *164*, 443, 521 (1917).
204. Helferich, B., and Leete, J. F., *Organic Syntheses*, Collective Vol. 3, Wiley, New York, 1955.
205. Peat, S., Whelan, W. J., and Hinson, K. A., *Nature*, *170*, 1056 (1952).
206. Peat, S., Whelan, W. J., and Hinson, K. A., *Chem. Ind. (London)*, *385* (1955).
207. Stephen, A. M., Kirkwood, S., and Smith, F., *Can. J. Chem.*, *40*, 151 (1962).
208. French, D., *Advan. Carbohyd. Chem.*, *9*, 158 (1954).
209. Haldane, J. B. S., *Enzymes*, Longmans London, 1930.

210. Laidler, K. J., *The Chemical Kinetics of Enzyme Action*, Clarendon Press, Oxford, England, 1958.
211. Dey, P. M., Khaleque, A., and Pridham, J. B., unpublished results.
212. Dixon, M., *Biochem. J.*, *55*, 161 (1953).
213. Dixon, M., and Webb, E. C., *Enzymes*, 2nd ed. Longmans, London, 1964.
214. Weil, L., and Burchert, A. R., *Arch. Biochem. Biophys.*, *34*, 1 (1951).
215. Weil, L., James, S., and Burchert, A. R., *Arch. Biochem. Biophys.*, *46*, 266 (1953).
216. Weil, L., and Seibles, T. S., *Arch. Biochem. Biophys.*, *54*, 368 (1955).
217. Mountler, L. A., Alexander, H. C., Tuck, K. D., and Dien, L. T. H., *J. Biol. Chem.*, *226*, 86 (1957).
218. Barnard, E. A., and Stein, W. D., *Advances in Enzymology*, Vol. 20, F. F. Nord, Ed., Interscience, New York, 1958, p. 51.
219. Koshland, D. E., *Advances in Enzymology*, Vol. 22, F. F. Nord, Ed. Interscience, New York, 1960, p. 45.
220. Nakatani, M., *J. Biochem. (Tokyo)*, *48*, 633 (1960).
221. Goodwin, T. W., and Morton, R. A., *Biochem. J.*, *40*, 628 (1946).
222. Boyer, P. D., *J. Amer. Chem. Soc.*, *76*, 4331 (1954).
223. Myrback, K., *Ark. Kemi*, *11*, 471 (1957).
224. Webb, J. L., *Enzyme and Metabolic Inhibitors*, Vol. 2, Academic, New York, 1966.
225. Simpson, R. B., *J. Amer. Chem. Soc.*, *83*, 4711 (1961).
226. Haarmann, W., *Biochem. J.*, *314*, 1 (1943).
227. Sharma, C. B., *Biochem. Biophys. Res. Commun.*, *43*, 572 (1971).
228. Heyworth, R., and Walker, P. G., *Biochem. J.*, *83*, 331 (1962).
229. Lee, Y. C., *Biochem. Biophys. Res. Commun.*, *35*, 161 (1969).
230. Edwards, J. T., *Chem. Ind. (London)*, 1102 (1955).
231. Bamford, C., Capon, B., and Overend, W. G., *J. Chem. Soc.*, 5138 (1962).
232. Phillips, D. C., *Proc. Nat. Acad. Sci. U.S.*, *57*, 484 (1967).
233. Vernon, C. A., *Proc. Roy. Soc. Ser. B*, *167*, 389 (1967).
234. Rupley, J. A., Gates, V., and Bilbrey, R., *J. Amer. Chem. Soc.*, *90*, 5633 (1968).
235. Leaback, D. H., *Biochem. Biophys. Res. Commun.*, *32*, 1025 (1968).
236. Koshland, D. E., *Mechanism of Enzyme Action*, McElroy, W. D., and Glass, B., Eds., Johns Hopkins Press: Baltimore, 1954.
237. Mayer, F. C., and Larmer, J., *J. Amer. Chem. Soc.*, *81*, 188 (1959).
238. Halpern, M., and Leibowitz, J., *Biochem. Biophys. Acta*, *36*, 29 (1959).
239. Koshland, D. E., and Stein, S. S., *J. Biol. Chem.*, *208*, 139 (1954).
240. Dey, P. M., Gillies, D. G., Lake-Bakaar, D. M., Pridham, J. B., and Weston, A. F., *Abstr. Chem. Soc.*, *Autumn meeting*, *F16*, (1970).
241. Massey, V., and Alberty, R. A., *Biochem. Biophys. Acta*, *13*, 354 (1954).
242. Larner, J., and Gillespie, R. E., *Arch. Biochem. Biophys.*, *58*, 252 (1955).
243. Myrback, K., and Willstadt, E., *Ark. Kemi*, *12*, 203 (1958).
244. Nath, R. L., and Rydon, H. N., *Biochem. J.*, *57*, 1 (1954).
245. Dahlqvist, F. W., Jao, L., and Raftery, M., *Proc. Nat. Acad. Sci. U.S.*, *56*, 26 (1966).
246. Koshland, D. E., *Proc. Nat. Acad. Sci. U.S.*, *44*, 98 (1958).

247. Blake, C. C. F., Johnson, L. N., Mair, G. A., North, A. C. T., Phillips, D. C., and Sarma, V. R., *Proc. Roy. Soc., Ser. B, 167,* 378 (1967).
248. Reeves, R. E., *J. Amer. Chem. Soc., 72,* 1499 (1950).
249. Bartlett, P. D., and Rosen, L. J., *J. Amer. Chem. Soc., 64,* 543 (1942).
250. Carter, H. E., Johnson, P., and Weber, E. J., *Ann. Rev. Biochem., 34,* 109 (1965).
251. Courtois, J. E., *Bull. Soc. Bot. Fr., 115,* 309 (1968).
252. Korytnyk, W., and Metzler, E., *Nature, 195,* 616 (1962).
253. Gould, M. F., and Greenshields, R. N., *Nature, 202,* 108 (1964).
254. Dey, P. M., Khaleque, A., Palan, P. R., and Pridham, J. B., unpublished results.
255. Pridham, J. B., Walter, M. W., and Worth, H. G. J., *J. Exp. Bot., 20,* 317 (1969).
256. Sastry, P. S., and Kates, M., *Biochemistry, 3,* 1271 (1964).
257. Bourne, E. J., Davies, D. R., and Pridham, J. B., unpublished results.

ENZYMATIC BASIS FOR BLOOD GROUPS IN MAN

By VICTOR GINSBURG, *Bethesda, Maryland*

CONTENTS

I. Introduction

The antigens present on human erythrocytes differ among individuals, and their presence or absence is the basis for the various blood groups into which people can be classified. Since the discovery of the ABO system by Landsteiner in 1900, over 250 red cell antigens have been described, and the number is rapidly increasing. This chapter will be limited to the ABO system involving the antigens A, B, and H[1] and the closely related Lewis system involving the antigens Le^a and Le^b. These antigens are the best understood of all from a chemical standpoint, and the biochemical basis for their inheritance has recently been established.

[1] The H antigen is the structure found on group O cells, and to a lesser extent on most other cells, that is responsible for their agglutination by specific reagents of human, animal, or plant origin. People who belong to blood group O often secrete a glycoprotein that inhibits agglutination by these reagents. However, the glycoprotein is clearly not a product of the *O* gene, as it is also secreted by most people including those of genotype AB who do not possess an *O* gene. To avoid confusion the glycoprotein is called H substance instead of O substance (1). The reagents that react with it, as well as with the corresponding H determinants on red cells, are anti-H reagents.

II. The ABH and Lewis Antigens

The A, B, H, and Lewis antigens are carbohydrates whose structure has been worked out in the last 20 years by Morgan and Watkins and their coworkers at the Lister Institute in London, and by Kabat and his coworkers at Columbia University College of Physicians and Surgeons in New York (for recent reviews, see Refs. 2,3). Their studies mainly involved "soluble blood group substances," which are glycoproteins found in secretions such as saliva, gastric juice, or ovarian cyst fluid and have A, B, H, and Lewis specificities. The seriologic specificity of these molecules is determined by the nature and linkage of the monosaccharides at the nonreducing ends of their carbohydrate chains. There are two kinds of nonreducing ends, Type 1 and Type 2. Type 1 contains a $Gal\beta1$–$3GlcNAc^1$ sequence, while Type 2 contains a $Gal\beta1$–$4GlcNAc$ sequence. To these structures are attached fucose, galactose, and N-acetylgalactosamine to form determinants of A, B, H, Le^a, and Le^b specificity, as shown in Table I.

The A, B, and H determinants can arise from either Type 1 chains or Type 2 chains; Le^a and Le^b determinants can only be formed from Type 1 chains, since both structures contain a $Fuc\alpha1$–$4GlcNAc$ sequence, a sequence not possible in Type 2 chains because the 4 position of N-acetylglucosamine is already occupied by galactose. The most important sugars for each specificity are known as immunodominant sugars. The immunodominant sugar for H specificity is fucose as $Fuc\alpha1$–$2Gal$; for A specificity, N-acetylgalactosamine as $GalNAc\alpha1$–$3Gal$; for B specificity, galactose as $Gal\alpha1$–$3Gal$; and for Le^a specificity, fucose as

$$\cdots GlcNAc \cdots$$
$$\overset{4}{\underset{|}{}}$$
$$Fuc\alpha\text{-}1$$

Both fucoses, as

$$Gal\beta1\text{--}3GlcNAc \cdots$$
$$\overset{2}{\underset{|}{}} \quad \overset{4}{\underset{|}{}}$$
$$Fuc\alpha\text{-}1 \quad Fuc\alpha\text{-}1$$

are required for Le^b specificity.

---

[1] Abbreviations used in this paper include: Glc for glucose; Gal for galactose; Fuc for fucose; Man for mannose; Gm for glucosamine; GlcNAc for N-acetylglucosamine; and GalNAc for N-acetylgalactosamine. Unless specifically mentioned all monosaccharides except fucose have a D configuration; fucose has an L configuration.

TABLE I

Structures Responsible for A, B, H, Le[a], and Le[b] Specificities (2)

Structure

Specificity	Type 1	Type 2
H	$\text{Gal}\beta1\text{-}3\text{GlcNAc}\cdots$ $\quad\overset{\displaystyle\mid}{2}$ $\text{Fuc}\alpha1$	$\text{Gal}\beta1\text{-}4\text{GlcNAc}\cdots$ $\quad\overset{\displaystyle\mid}{2}$ $\text{Fuc}\alpha1$
A	$\text{GalNAc}\alpha1\text{-}3\text{Gal}\beta1\text{-}3\text{GlcNAc}\cdots$ $\quad\quad\quad\quad\overset{\displaystyle\mid}{2}$ $\quad\quad\quad\text{Fuc}\alpha1$	$\text{GalNAc}\alpha1\text{-}3\text{Gal}\beta1\text{-}4\text{GlcNAc}\cdots$ $\quad\quad\quad\quad\overset{\displaystyle\mid}{2}$ $\quad\quad\quad\text{Fuc}\alpha1$
B	$\text{Gal}\alpha1\text{-}3\text{Gal}\beta1\text{-}3\text{GlcNAc}\cdots$ $\quad\quad\quad\overset{\displaystyle\mid}{2}$ $\quad\quad\text{Fuc}\alpha1$	$\text{Gal}\alpha1\text{-}3\text{Gal}\beta1\text{-}4\text{GlcNAc}\cdots$ $\quad\quad\quad\overset{\displaystyle\mid}{2}$ $\quad\quad\text{Fuc}\alpha1$
Le[a]	$\text{Gal}\beta1\text{-}3\text{GlcNAc}\cdots$ $\quad\quad\quad\quad\overset{\displaystyle\mid}{4}$ $\quad\quad\quad\text{Fuc}\alpha1$	
Le[b]	$\text{Gal}\beta1\text{-}3\text{GlcNAc}\cdots$ $\quad\overset{\displaystyle\mid}{2}\quad\quad\overset{\displaystyle\mid}{4}$ $\text{Fuc}\alpha1\quad\text{Fuc}\alpha1$	

Lloyd and Kabat (4) have recently proposed that both Type 1 and Type 2 groupings shown in Table I occur as part of the same carbohydrate chain. The skeleton of this chain is common to all soluble blood group substances, regardless of serologic specificity, and has the following structure:

Galβ1–4GlcNAcβ-1
|
6
Galβ1–3GlcNAcβ1–3Galβ1–3GlcNAcβ1–3Galβ1–3GalNAc · · ·

Because this chain is common, all soluble blood group substances, regardless of their A, B, H, or Lewis specificity, cross-react to varying degrees with horse antitype XIV peumococcal serum, which is directed against both the Type 1 and the Type 2 groupings (6). Many chains are attached to the same polypeptide by glycosidic bonds between the proximal N-acetylgalactosamine residue and the hydroxyl groups of serine or threonine. There is considerable heterogeneity among the chains, however, and chains at various stages of completion and with variable fucose content have been isolated (7). This heterogeneity can explain why single glycoprotein molecules possess different specificities: glycoproteins from A or B individuals, for example, may have variable amounts of H, Le[a], or Le[b] determinants in addition to A or B and Type XIV pneumococcal determinants. Precipitation experiments with specific antisera indicate that the various determinants are present on the same molecule (2). Another cause of heterogeniety among the carbohydrate chains of glycoproteins may be the acceptor specificity of transglycosylases as will be discussed later.

Family studies over the past 60 years have shown that the inheritance of ABO(H) and Lewis blood types, and hence the ability to form the structures in Table I, is controlled by the action of genes at four independent loci: the *ABO*, the *Lele*, the *Hh*, and the *Sese* loci (8). When the chemical relationship between the structures responsible for the different blood types was emerging, Watkins and Morgan (9) proposed that these genes were not responsible for the production of the entire determinant but only for the formation of enzymes that attach the outer or immunodominant sugars to a central "precursor" chain common to all phenotypes. These enzymes acted sequentially, the product of one being the substrate for the next, and as each sugar is added a new specificity emerges as the underlying specificity is suppressed. A similar scheme based on serological and genetic evidence obtained from family studies was proposed by Ceppellini (10). Recent studies have provided abundant proof for the correctness of these proposals; the presence or absence of four specific glycosyltransferases in individuals belonging to different blood groups agrees with their predicted occurrence based on genotype.

The *ABO* locus is responsible for two enzymes; an *N*-acetylgalacto-saminyltransferase specified by the *A* gene and a galactosyltransferase specified by the *B* gene. Both enzymes transfer the appropriate hexose to galactose and form αl-3 linkages. The third gene at the *ABO* locus, the *O* gene, is inactive and does not produce a functional transgly-cosylase.

The *Lele* locus produces one enzyme, a fucosyltransferase specified by the *Le* gene that transfers fucose to *N*-acetylglucosamine and forms αl-4 linkages. The *le* gene is inactive.

The *Hh* locus is responsible for a second fucosyltransferase that transfers fucose to galactose to form αl-2 linkages. The *H* gene specifies this fucosyltransferase, while its allele, the *h* gene, is inactive.

The *Sese* locus does not specify an enzyme but in an unknown way is necessary for the expression of the *H* gene (in this sense, the formation of the fucosyltransferase specified by the *H* gene) in certain organs but not in others. The presence or absence of the fucosyltransferase specified by the *H* gene in secretory organs is the basis for the ability to secrete glycoproteins that exhibit A, B, or H specificities. As shown in Table I, the determinants of all three specificities contain Fucαl–2Gal groupings. The 80% of the population whose secretions contain glycoproteins with A, B, or H activity possess an *Se* gene and are known as "secretors." The remaining 20%, the "nonsecretors," have the genotype *sese;* their secreted glycoproteins lack A, B, or H specificity. The *Se* gene does not control the formation of the fucosyltransferase specified by the *H* gene in hematopoietic tissue, and its absence does not affect the formation of A, B, or H determinants of erythrocytes. The *Se* gene does affect Lewis blood groups. Unlike the A, B, or H determinants, the determinants of the Lewis system on red cells are not integral parts of the erythrocyte membrane but are sphingoglycolipids acquired from the serum, where they circulate associated with serum lipoproteins (11). The origin of serum sphingoglycolipids is unknown, but clearly the organ in which they are produced is affected by the *Se* gene because of the association between the secretor status and Lewis blood type (12): all individuals in the Le(a—b+) blood group are secretors, while nonsecretors never belong to the Le(a— b+) group (8). As shown in Table I, the fucosyltransferase specified by the *H* gene is required for the synthesis of the Le^{b} determinant.

The *Se* gene may be analogous in action to a gene in mice described by Ganschow and Paigen (13) that affects the intracellular location of hepatic β-glucuronidase. The enzyme is specified by a single structural gene and occurs within lysosomes, as well as in the microsomal membranes of the liver cells of DBA/2 and C3H/HeHa mice. A gene distinct from the structural gene has been found in another strain of mice (strain YBR) that affects this distribution: hepatic β-glucuronidase of mice that inherit the new gene occurs only in lysosomes and not in microsomal membranes. The gene is tissue-specific and seems to affect only β-glucuronidase, as microsomal glucose 6-phosphatase and other microsomal proteins appear unaltered. One explanation offered by the authors is that the gene in YBR mice is an altered form of a gene in DBA/2 and C3H/HeHa mice responsible for the formation of specific binding sites for β-glucuronidase in microsomal membranes. The product of the altered gene would not bind the enzyme, and as a result β-glucuronidase would not be incorporated into the membrane. Possibly, the *se* gene in a similar way affects the distribution of the fucosyltransferase specified by the *H* gene so that it does not become a part of the synthetic machinery of certain secretory organs.

Whether the A, B, and H determinants on erthrocytes are sphingoglycolipids, glycoproteins or both is still unsettled. Sphingoglycolipids with A, B, and H activity have been isolated from erythrocytes (14–17), but their levels appear too low to account for all the antigenic sites; it is likely that membrane-bound glycoproteins are mainly responsible for the A, B, and H reactivity of erythrocytes (18,19). However, uncertainty as to the nature of the erythrocyte determinants is not important for the purposes of this chapter: the glycosyltransferases specified by the genes that determines blood type are involved in the synthesis of the carbohydrate chains of both glycolipids and glycoproteins. For example, the glycosyltransferase specified by the *A* gene forms the GalNAcα1–3Gal sequence shown in Table I. In people who have blood type A, this sequence occurs in glycolipids (16,20,21), glycoproteins (2,22), and also in the free oligosaccharides of milk (23) and urine (24). The same sequence is not found in similar material from people with blood type B or O. Since blood type A is determined by one gene (8), it follows that the same glycosyltransferase, the one specified by the A gene, synthesizes GalNAcα1–3Gal structures in both glycolipids and glycoproteins as well as in free oligosaccharides.

III. Nucleotide-linked Sugars

As in the synthesis of other complex carbohydrates, the formation of blood group determinants involves the stepwise transfer of mono-saccharides from their activated derivatives, the nucleotide-linked sugars. Pathways leading to these derivatives are summarized in Figure 1. Only glucose, mannose, and N-acetylglucosamine arise directly from fructose 6-phosphate; the remaining sugars are formed

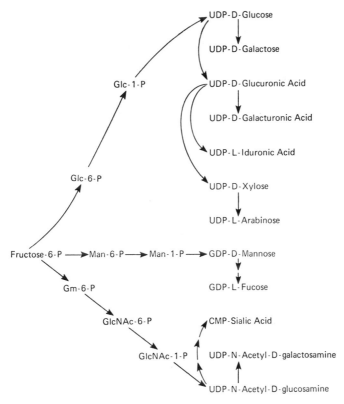

Fig. 1. Formation of nucleotide-linked sugars involved in the synthesis of mammalian complex carbohydrates. For reference to individual reactions, see (25).

from these three monosaccharides after their incorporation into the appropriate nucleotide. In animals, each sugar is usually associated with only one base; for example, fucose is only carried by GDP and galactose only by UDP. However, exceptions have been reported (26–28). In contrast, bacteria and plants often have the same sugar carried by different bases. Glucose, for instance, is found linked to UDP, GDP, ADP, CDP, and TDP, and there are enzymes specific for each glucose derivative: enzymes from various salmonella will convert glucose attached to CDP to dideoxyhexoses, glucose attached to TDP to L-rhamnose, and glucose attached to UDP to galactose; and enzymes from plants will form different polymers of glucose depending on whether the glucose is attached to UDP, GDP, or ADP (29).

What factors govern the choice of a particular base for a given reaction is not apparent, but it is clear that the use of different bases sharply separates pathways of synthesis. This separation may be advantageous for several reasons. One advantage is that it offers the organism a means for the independent control of the various pathways. For example, ADP–glucose gives rise to α1-4 glucans in bacteria and plants (glycogen and starch, respectively); the synthesis of these polymers may be regulated by the activation and inhibition of ADP–glucose pyrophosphorylase by various metabolities (30). Regulation at this step in the biosynthetic sequence implies not only that the pyrophosphorylase reaction is rate limiting, but also that the ADP–glucose that is produced in the reaction serves only for the formation of these products and is not an intermediate in the synthesis of other sugars or a glycosyl donor for the synthesis of other complex saccharides. The use of ADP as a carrier for the glucose allows the pyrophosphorylase step to be unique to the synthesis of α1-4 glucans and eminently suited for a regulatory role. In animals UDP-glucose gives rise to glycogen, and, possibly because UDP–glucose has so many other functions, the control of glycogen synthesis is at the level of the glycosyltransferase. In addition to the above example, there are several instances of control by negative feedback in which nucleotide-linked sugars regulate their own concentration by inhibiting an enzyme catalyzing the formation of a precursor (29). Again, separation of pathways by the use of different bases creates reactions unique to the synthesis of certain intermediates, and therefore effective sites for control.

Another possible advantage for using different bases is suggested by the mechanism of synthesis of complex saccharides. Complicated structures like the A, B, H, and Lewis determinants arise by the ordered and stepwise addition of glycosyl residues from nucleotide-linked sugars catalyzed by glycosyltransferases. Clearly, the structures that are formed depend on the inherent specificities of the glycosyl-transferase; the more specific these enzymes are, the more accurately will each structure be synthesized. However, enzyme specificity is a relative concept and glycosyltransferases make mistakes: the glycogen of chickens fed galactose contains some galactose (31) and the glycogen of rats fed glucosamine contains some glucosamine (32). Glycogen synthetase, whose normal substrate is UDP–glucose, will use as glycosyl donors the UDP–galactose and UDP–glucosamine that accumulates in the liver of these animals. As glycosyltransferases exhibit specificity for both base and sugars, carrying monosaccharides on different bases may increase the accuracy of synthesis by decreasing the probability of error. For example, the fucosyltransferase specified by the *H* gene transfers fucose from GDP-fucose and is less likely to make a mistake and transfer galactose from UDP–galactose than galactose from GDP–galactose.

IV. Oligosaccharides of Milk

Oligosaccharides isolated from human milk have long been useful in studies on the biochemical basis of blood type. Originally, these sugars were used in hapten inhibition studies, playing a key role in elucidating the structures of the H, Le^a, and Le^b determinants (2). Some of the sugars are shown in Table II. Several of them are found in sphingoglycolipids: lactose occurs in lactosyl ceramide (cytolipin H), a glycolipid that accounts for most of the serological differences between normal and tumor tissues of humans (44); lacto-*N*-tetraose and lacto-N-*neo*tetraose occur in glycolipids of human spleen (45); and lacto-*N*-fucopentaose III occurs in a glycolipid that accumulates in human adenocarcinomas (46).

The oligosaccharides shown in Table II do not occur intact in glyco-proteins but are found minus their glucose residues in the distal portions of the carbohydrate chains of glycoproteins. For example, the branched protion of lacto-*N*-hexaose (43) is identical to the branched portion of the carbohydrate chain of soluble blood group substances

TABLE II
Some Oligosaccharides of Human Milk

Trivial name	Structure	Reference
Lactose	Galβ1–4Glc	33

2'Fucosyllactose	Galβ1–4Glc
	  2
	  |
	Fucα1
		34

3-Fucosyllactose	Galβ1–4Glc
	      3
	      |
	    Fucα1
		35

Lacto-difucotetraose	Galβ1–4Glc
	  2 3
	  | |
	Fucα1 Fucα1
		36

Lacto-N-tetraose	Galβ1–3GlcNAcβ1–3Galβ1–4Glc	37
Lacto-N-neotetraose	Galβ1–4GlcNAcβ1–3Galβ1–4Glc	38

Lacto-N-fucopentaose I	Galβ1–3GlcNAcβ1–3Galβ1–4Glc
	  2
	  |
	Fucα1
		39

Lacto-N-fucopentaose II	Galβ1–3GlcNAcβ1–3Galβ1–4Glc
	         4
	         |
	       Fucα1
		40

Lacto-N-fucopentaose III	Galβ1–4GlcNAcβ1–3Galβ1–4Glc
	         3
	         |
	       Fucα1
		41

Lacto-N-difucohexaose I	Galβ1–3GlcNAcβ1–3Galβ1–4Glc
	  2 4
	  | |
	Fucα1 Fucα1
		42

Lacto-N-difucohexaose II	Galβ1–3GlcNAcβ1–3Galβ1–4Glc
	        4 3
	        | |
	     Fucα1 Fucα1
		42

Lacto-N-hexaose	Galβ1–4GlcNAcβ1
	                   6
	                   |
	Galβ1–3GlcNAcβ1–3Galβ1–4Glc
		43

shown in Section II, and pieces of the other milk sugars are also found in these glycoproteins (see Table I and ref. 2).

Recently, milk oligosaccharides have been useful in assaying glycosyltransferases by serving as acceptors. The reaction products are relatively simple and can be characterized chemically. In addition, glycosyltransferases specified by the genes that determine blood type occur in soluble form in milk, a convenient source for their isolation and study.

V. Glycosyltransferases Specified by Genes That Determine Blood Groups

1. The H Gene Enzyme

The H gene specifies a fucosyltransferase that catalyzes the following reaction:

$$\text{GDP–fucose} + \text{Gal} \cdots \rightarrow \text{Fuc}\alpha\text{1–2Gal} \cdots + \text{[GDP]}$$

The enzyme has been detected in milk (47) and submaxillary gland preparations (48) from secretors but not in similar preparations from nonsecretors, as both organs are under the control of the Se gene. The presence or absence of this enzyme in secretory organs is the biochemical basis for secretor status. Most individuals have an H gene. The few who do not (the hh genotype) belong to the rare "Bombay" blood group. Presumably, these individuals are unable to form the enzyme in any tissue and, for this reason, are unable to synthesize any of the structures in Table I except the Le^a-active structure, and then only if they possess the Le gene (2).

2. The Le Gene Enzyme

The Le gene specifies a fucosyltransferase that catalyzes the following reaction:

$$\text{GDP–fucose} + \cdots \text{GlcNAc} \cdots \rightarrow \cdots \underset{\substack{4 \\ | \\ \text{Fuc}\alpha 1}}{\text{GlcNAc}} \cdots + \text{[GDP]}$$

The enzyme has been found in milk (47,49) and is conveniently assayed by using lacto-N-fucopentaose I as an acceptor. Its presence is absolutely dependent on the Lewis blood group of the donor as shown in Table III: it is present in milk from donors belonging to the blood group Le(a + b—) or Le(a — b+) and is absent from milk of donors who belong to neither group, the "Lewis negative" phenotype (Le(a — b—)). Since both Le^a and Le^b determinants contain Fucα1–4GlcNAc groupings (see Table I), it follows that individuals who are "Lewis negative" [about 7% of the population (8)] would be unable to make either structure.

The milk enzyme can transfer fucose to a glycoprotein without blood group activity to yield a product precipitable with anti-Le^a sera (50). The enzyme has also been detected in submaxillary gland preparations (48).

3. The A Gene Enzyme

The A gene specifies an N-acetylgalactosaminyltransferase that catalyzes the following reaction:

$$\text{UDP–GalNAc} + \underset{\underset{\text{Fuc}\alpha1}{\overset{|}{2}}}{\text{Gal}} \cdots \rightarrow \underset{\underset{\text{Fuc}\alpha1}{\overset{|}{2}}}{\text{GalNAc}\alpha1\text{–3Gal}} \cdots + [\text{UDP}]$$

As shown in Table IV, the enzyme is present in milk from donors with an A or AB blood type and is absent from milk of donors with B or O blood type. The enzyme has a strict acceptor requirement: it will transfer N-acetylgalactosamine to galactose only if the galactose is substituted on the 2 position with fucose. Of the oligosaccharides in Table II that were tested, only 2-fucosyllactose and lacto-N-fucopentaose I were able to accept N-acetylgalactosamine (23,51). The

TABLE III

Occurrence of the Fucosyltransferase Specified by the
Le Gene in Human Milk (49)

Donor	Lewis group	Secretor status	Fucose transferred to lacto-N-fucopentaose I $\mu\mu M/(5 \text{ hr})(30 \mu l \text{ milk})$
J. S.	Le(a + b—)	Nonsecretor	90
D. R.	Le(a + b—)	Nonsecretor	7
S. L.	Le(a — b+)	Secretor	9
L. N.	Le(a — b+)	Secretor	120
J. C.	Le(a — b+)	Secretor	120
G. H.	Le(a — b+)	Secretor	30
E. K.	Le(a — b—)	Secretor	0
C. R.	Le(a — b—)	Secretor	0

TABLE IV

Occurrence of the N-Acetylgalactosaminyltransferase Specified by the A Gene and the Galactosyltransferase Specified by the B Gene in Human Milk (23,51,59)

Blood Type				Sugar transferred to lacto-N-fucopentaose I	
Donor	ABO	Lewis	Secretor status	N-acetylgalactosamine	galactose
				$\mu\mu\mu$M/(5 hr)(30 μl milk)	
J. M.	A_1	Le(a − b+)	Secretor	27	0
M. K.	A_1	Le(a − b+)	Secretor	12	0
D. C.	A_1	Le(a − b+)	Secretor	15	0
J. B.	A_2	Le(a − b+)	Secretor	5	0
E. K.	A_2	Le(a − b−)	Secretor	67	0
J. C.	A_1B	Le(a − b+)	Secretor	40	4.5
L. M.	A_2	Le(a + b−)	Nonsecretor	16	0
B. B.	A_1	Le(a + b−)	Nonsecretor	7	0
G. H.	B	Le(a − b+)	Secretor	0	1.7
R. S.	B	Le(a − b+)	Secretor	0	4.5
N. C.	B	Le(a − b+)	Secretor	0	4.2
M. M.	B	Le(a − b+)	Secretor	0	7.1
F. W.	B	Le(a + b−)	Nonsecretor	0	3.8
L. N.	O	Le(a − b+)	Secretor	0	0
C. D.	O	Le(a − b+)	Secretor	0	0
J. S.	O	Le(a + b−)	Nonsecretor	0	0
D. R.	O	Le(a + b−)	Nonsecretor	0	0

same sugars without fucose (lactose and lacto-N-tetraose, respectively) were not acceptors. This stringent specificity explains why the Se gene not only controls the formation of H-active glycoproteins but also of A-active (and B-active) glycoproteins as well. Individuals with blood type A who do not have the Se gene (donors LM and BB in Table IV) have normal levels of the A enzyme. Nevertheless, they are not able to synthesize A-active soluble glycoproteins because the necessary Fucα1–2Gal grouping required by the A enzyme for the acceptor is missing from their secreted glycoproteins. Similarly, the absence of an H gene can "modify" the expression of the A or B gene: individuals belonging to the Bombay blood group cannot make A or B determinants on their red cells, even though they may possess an A or B gene.

Other inactive acceptors for the milk enzyme include 3-fucosyllactose lacto-difucotetraose, lacto-N-neotetraose, lacto-N-fucopentaose II, lacto-N-fucopentaose III, lacto-N-difucohexaose I, and lacto-N-difucohexaose II. Especially interesting is the inability of lacto-difucotetraose and lacto-N-difucohexaose I to accept N-acetyl-galactosamine, even though their penultimate galactosyl residues are substituted on the 2 position with fucose. Presumably, the second fucose residue on the adjacent sugar sterically hinders the enzyme. Lacto-difucotetraose and lacto-N-difucohexaose I resemble the Le^b determinant (see Table I) and are potent haptenic inhibitors of the agglutination of Le^b erythrocytes by anti-Le^b serum (2). It is probable that once Le^b-active structures are formed *in vivo*, they will persist, as they cannot act as acceptors for further addition of sugars. If this is true, the formation of the structure proposed for the A determinant (2),

would involve a strict sequence of addition of the three terminal sugars as dictated by enzyme specificities. First, fucose would be added to the galactose; then N-acetylgalactosamine would be added to the same galactose; and finally, the second fucose would be added to N-acetyl-glucosamine.

The enzyme specified by the A gene has been found in the submaxillary glands (52) and gastric mucosa of humans (53), as well as in the organs of pigs (54–56), and can convert O erthrocytes into A erythrocytes *in vitro* (57).

The difference between blood group A_1 and A_2 is not yet known. Possibly the A_1 and A_2 genes specify N-acetyl-galactosaminyltransferases with slightly different acceptor requirements, resulting in more complete conversion of H-active structures to A-active structures by the A_1 enzyme in contrast to the A_2 enzyme. However, no difference has yet been found between the A_1 or A_2 enzymes *in vitro* (23,51,52), including their ability to add N-acetylgalactosamine to oligosaccharides with the following structures (58): Type I,

$$\text{Gal}\beta1\text{–}3\text{GlcNAc} \cdots$$
$$2$$
$$|$$
$$\text{Fuc}\alpha1$$

and Type II,

$$\text{Gal}\beta1\text{–}4\text{GlcNAc} \cdots$$
$$2$$
$$|$$
$$\text{Fuc}\alpha1$$

4. The B Gene Enzyme

The B gene specifies a galactosyltransferase that catalyzes a reaction identical to that catalyzed by the enzyme specified by the A gene, except that galactose is transferred in place of N-acetylgalactosamine as follows:

$$\text{UDP–galactose} + \underset{\underset{\text{Fuc}\alpha1}{|}}{\overset{2}{\text{Gal}}} \cdots \longrightarrow \underset{\underset{\text{Fuc}\alpha1}{|}}{\overset{2}{\text{Gal}\alpha1\text{–3Gal}}} \cdots + \text{[UDP]}$$

The distribution of the B gene enzyme in milk is shown in Table IV. It is present in milk from donors of B or AB blood type and is absent from the milk of donors with A or O blood type. Its occurrence, like that of the A gene enzyme, is independent of secretor status (see donor F. W. in Table IV). The B gene enzyme has also been detected in gastric mucosa and submaxillary glands of humans and baboons (60,61) and can transform O erythrocytes into B erythrocytes *in vitro* (57).

Like the A gene enzyme, the B gene enzyme requires the galactose acceptor to be substituted with fucose on the 2 position; its acceptor specificity for other milk oligosaccharides is the same (59–61); and "modification" of the expression of the B gene by the *Se* gene or the *H* gene can be explained in the same way. As the A gene and the B gene are alleles, the primary structure of the two enzymes may be quite similar. The third allele at the *ABO* locus, the *O* gene, is inactive, presumably producing a nonfunctional A or B "enzyme."

VI. Summary

The action of the four enzymes described above are summarized in Figure 2. The carbohydrate structures that they form are depicted, along with their associated serologic activity. Clearly, these determinants of blood type are secondary gene products, in that the primary gene products are the enzymes and it is these enzymes, working in concert, that determine which structures are formed. This mechanism of synthesis has an important biologic consequence: it provides a biochemical explanation for antigens produced by "gene interaction"—that is to say, antigens present in a hybrid that are not found in either parent. The Le[b] antigen is one of these. From family studies, Ceppellini proposed that the Le[b] antigen was an interaction product of two genes (62). In the scheme of Watkins and Morgan (2), these

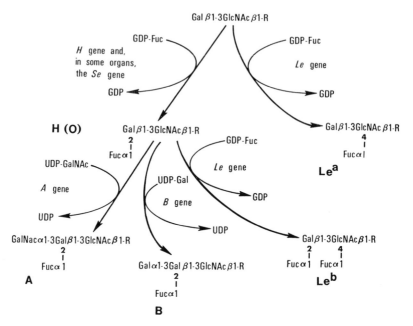

Fig. 2. Glycosyltransferases specified by the genes that determine blood type (59).

genes would be the *Le* gene and the *Se* gene. As shown in Figure 2, both genes are necessary to produce the two fucosyltransferases required for the synthesis of the Le^b antigen. A child who inherited an *Le* gene from one parent who happened to have the genotype *sese* and an *Se* gene from the other parent who happened to have the genotype *lele* would synthesize a structure that neither parent could make alone.

There are several reports of interaction of other genes resulting in new cell antigens; interactions, for example, within the Rh system or interactions of the *I* gene with the *A*, *B*, or *H* genes (8). The formation of unique carbohydrate structures by the concerted action of glycosyltransferases provided by both parents may be a factor in the expression of individuality. These structures would reflect not only the various glycosyltransferases that are present but, because the reactions are competitive, they would also reflect the relative activities of these enzymes as well (25). For example, as mentioned above, the

A gene enzyme and the *B* gene enzyme probably cannot use as acceptors those carbohydrate chains that contain Fucα1–4GlcNAc groupings. If the level of the *Le* gene enzyme that makes this grouping is high relative to the *H* gene enzyme, it is clear from Figure 2 that more Le^a and Le^b determinants would be formed than A, B, or H determinants. If the *Le* gene enzyme is less active than the *H* gene enzyme, the reverse would be true. Coupled with the enzyme specificity of subsequent steps, competition of this type would tend to generate heterogeneity among the carbohydrate chains of glycoproteins and, along with the incomplete synthesis of these chains, would contribute to the multiple blood group specificities that are exhibited by single glycoprotein molecules.

References

1. Morgan, W. T. J., and Watkins, W. M., *Brit. J. Exp. Pathol.*, *29*, 159 (1948).
2. Watkins, W. M., *Glycoproteins*, A. Gottschalk, Ed., Elsevier, London-New York, 1966, p. 462.
3. Marcus, D. M., *New Eng. J. Med.*, *280*, 944 (1969).
4. Lloyd, K. O., and Kabat, E. A., *Proc. Nat. Acad. Sci. U.S.*, *61*, 1470 (1968).
5. Kabat, E. A., *Blood Group Substances*, Academic, New York, 1956.
6. Watkins, W. M., and Morgan, W. T. J., *Vox Sanquinus*, *7*, 129 (1962); Kabat, E. A., *Arch. Biochem. Biophys.*, Supplement 1, 181 (1962).
7. Lloyd, K. O., Kabat, E. A., and Licerio, E., *Biochemistry*, *7*, 2976 (1968).
8. Race, R. R., and Sanger, R., *Blood Groups in Man*, 5th ed., Blackwell, Oxford, 1968.
9. Watkins, W. M., and Morgan, W. T. J., *Vox Sanguinus*, *4*, 97 (1959).
10. Ceppellini, R., in *Symposium on the Biochemistry of Human Genetics*, G. E. W. Wolstenholme and C. M. O'Connor, Eds., Churchill, London, 1959, p. 242.
11. Marcus, D. M., and Cass, L. E., *Science*, *164*, 553 (1969).
12. Grubb, R., *Acta. Path. Microbiol. Scand.*, *28*, 61 (1951).
13. Ganschow, R., and Paigen, K., *Proc. Nat. Acad. Sci.*, *U.S.*, *58*, 939 (1967).
14. Yamakawa, T., Irie, R., and Iwanaga, M., *J. Biochem. (Tokyo)*, *48*, 490 (1960).
15. Handa, S., *Jap. J. Exp. Med.*, *33*, 347 (1963).
16. Hakomori, S., and Strycharz, G. D., *Biochemistry*, *7*, 1279 (1968).
17. Koscielak, J., Piasek, A., and Gorniak, H., *Blood and Tissue Antigens*, D. Aminoff, Ed., Academic Press, New York, 1970, p. 163.
18. Poulik, M. D., and Bron, C., in *Red Cell Membranes*, G. A. Jamieson and T. J. Greenwalt, Eds., Lippincott, Philadelphia, 1969, p. 131.
19. Whittemore, N. B., Trabold, N. C., Reed, C. F., and Weed, R. K., *Vox Sanguinus*, *17*, 289 (1969).
20. Yamakawa, T., and Suzuki, S., *J. Biochem. (Tokyo)*, *39* (1952).
21. Koscielak, J., and Zakrzewski, K., *Nature*, *187*, 516 (1960).

148   VICTOR GINSBURG

22. Aminoff, D., Morgan, W. T. J., and Watkins, W. M., *Biochem. J.*, *46*, 426 (1950).
23. Kobata, A., and Ginsburg, V., *J. Biol. Chem.*, *245*, 1484 (1970).
24. Lundblad, A., *Biochem. Biophys. Acta*, *148*, 151 (1967).
25. Ginsburg, V., in *Advances in Enzymology*, Vol. 26, F. F. Nord, Ed., John Wiley, New York, 1964, p. 35.
26. Denamur, R., Fauconneau, G., and Guntz, G. J., *Ann. Biol. Animale Biochim. Biophys.*, *1*, 74 (1961).
27. Carlson, D. M., and Hansen, R. G., *J. Biol. Chem.*, *237*, 1260 (1962).
28. Cantore, M. L., Leoni, P., Leveroni, A. F., and Recondo, E. F., *Biochim. Biophys. Acta*, *230*, 423 (1971).
29. Neufeld, E. F., and Ginsburg, V., *Ann. Rev. Biochem.*, *34*, 297 (1965).
30. Preiss, J., in *Current Topics in Cellular Regulation*, B. L. Horecker and E. R. Stadtman, Eds., Academic Press, New York, 1970.
31. Nordin, J. H., and Hansen, R. G., *J. Biol. Chem.*, *238*, 489 (1963).
32. Maley, F., McGarrahan, J. F., and DelGiacco, R., *Biochem. Biophys. Res. Commun.*, *23*, 85 (1966).
33. Whittier, E. O., *Chem. Rev.*, *2*, 84 (1925–1926).
34. Kuhn, R., Baer, H. H., and Gauhe, A., *Chem. Ber.*, *88*, 1135 (1955).
35. Monteruil, J., *Compt. Rend.*, *242*, 192 (1956).
36. Kuhn, R., and Gauhe, A., *Ann. Chem.*, *611*, 249 (1958).
37. Kuhn, R., and Baer, H. H., *Chem. Ber.*, *89*, 504 (1956).
38. Kuhn, R., and Gauhe, A., *Chem. Ber.*, *95*, 518 (1962).
39. Kuhn, R., Baer, H. H., and Gauhe, A., *Chem. Ber.*, *89*, 2514 (1956).
40. Kuhn, R., Baer, H. H., and Gauhe, A., *Chem. Ber.*, *91*, 364 (1958).
41. Kobata, A., and Ginsburg, V., *J. Biol. Chem.*, *244*, 5496 (1969).
42. Kuhn, R., and Gauhe, A., *Chem. Ber.*, *93*, 647 (1960).
43. Kobata, A., and Ginsburg, V., *J. Biol. Chem.*, in press.
44. Rapport, M. M., and Graf, L., *Progr. Allergy*, *13*, 273 (1969).
45. Wiegandt, H., and Bücking, H. W., *Eur. M. Biochem.*, *15*, 287 (1970).
46. Yang, H., and Hakomori, S., *J. Biol. Chem.*, *246*, 1192 (1971).
47. Shen, L., Grollman, E. F., and Ginsburg, V., *Proc. Nat. Acad. Sci. U.S.*, *59*, 224 (1968).
48. Chester, M. A., and Watkins, W. M., *Biochem. Biophys. Res. Commun.*, *34*, 835 (1969).
49. Grollman, E. F., Kobata, A., and Ginsburg, V., *J. Clin. Invest.*, *48*, 1489 (1969).
50. Jarkovsky, Z., Marcus, D. M., and Grollman, A. P., *Biochemistry*, *9*, 1123 (1970).
51. Kobata, A., Grollman, E. F., and Ginsburg, V., *Arch. Biochem. Biophys.*, *124*, 609 (1968).
52. Hearn, V. M., Smith, Z. G., and Watkins, W. M., *Biochem. J.*, *109*, 315 (1968).
53. Tuppy, H., and Schenkel-Brunner, H., *Vox Sanguinus*, *17*, 139 (1969).
54. Carlson, D. M., *IV Int. Conference on Cystic Fibrosis of the Pancreas*, E. Rossi and E. Stoll, Eds., S. Karger, Basel-New York, 1968, p. 304.
55. Tuppy, H., and Schenkel-Brunner, H., *Eur. J. Biochem.*, *10*, 152 (1969).

56. McGuire, E. J., *Blood and Tissue Antigens*, D. Aminoff, Ed., Academic Press, New York, 1970, p. 461.
57. Schenkel-Brunner, H., and Tuppy, H., *Nature*, *223*, 1272 (1969).
58. Kobata, A., Ginsburg, V., and Kabat, E. A., unpublished experiments.
59. Kobata, A., Grollman, E. F., and Ginsburg, V., *Biochem. Biophys. Res. Commun.*, *32*, 272 (1968).
60. Race, C., Ziderman, D., and Watkins, W. M., *Biochem. J.*, *107*, 733 (1968).
61. Race, C., and Watkins, W. M., *FEBS Lett.*, *10*, 279 (1970).
62. Ceppellini, R., *Proc. 5th Int. Cong. of Blood Transfusion*, Paris, 1954, p. 207.

THE INHIBITION OF GLYCOSIDASES
BY ALDONOLACTONES

By G. A. LEVVY and SYBIL M. SNAITH, *Aberdeen, Scotland*

CONTENTS

I. Introduction

The purpose of this review is to trace the consecutive steps in the development of a single theme in research. No attempt has been made to compile corroboratory evidence or to deal with side issues, such as applications in related fields.

The term glycosidase is a general one, covering those hydrolytic enzymes that split off sugar residues, attached by way of their reducing group to the rest of the substrate molecule. Glycuronidase is the generic term for enzymes that liberate the uronic acids in which the primary alcohol group of the sugar has been converted to a carboxyl group. In aldonic acids the reducing group in the sugar has been oxidized to a carboxylic acid and the derived lactones are internal esters. Since a glycuronic acid already has a carboxyl group, it gives rise to a dicarboxylic acid, or glycaric acid. Such an acid can form lactones involving either carboxyl group.

Under normal conditions most glycosidases are completely specific for water as the acceptor molecule, and fission occurs at the glycosyl–oxygen bond (as in eq. 1):

| β-D-Glucoside | Glucose | Aglycone |

$$+ H_2O \longrightarrow \quad H, OH + ROH \qquad (1)$$

The discovery that a glycosidase is inhibited by the aldonolactone derived from the same sugar as the substrate arose from a study of the inhibition of β-glucuronidase by solutions of glucaric acid (24). This prompted an investigation of the effect of the appropriate aldonolactones upon other glycosidases (9). Although its significance was not realized at the time, the first mention of inhibition of a glycosidase by an aldonolactone was made in 1940. Gluconic and galactonic acids were employed as inhibitors in an attempt to distinguish between different types of β-glucosidase and β-galactosidase activities (12,18). Glucono-1:4-lactone and galactono-1:4-lactone were included among a number of other sugar derivatives that were examined. The effect of lactonization upon the apparent inhibitory action of the aldonic acids at different pH values was not realized and this makes the results difficult to interpret (cf. ref. 34). It is now known that aldonic acids as such do not inhibit glycosidases.

Inhibition of glycosidases by aldonolactones is competitive in character and highly specific. One might anticipate this from the close resemblance between the inhibitors and the substrates in their chemical structure. The relative affinities of inhibitors and substrates* are best compared as reciprocals of K_i and K_m, the latter being assumed to provide a good approximation to K_s. Because it could be more easily measured than K_i, the concentration of a compound required for 50% inhibition of enzyme activity was often determined. The 50% figure is directly proportional to K_i, but it is dependent upon the K_m and the concentration of the substrate employed. Where K_m is known, K_i can of course be calculated from the other data. Aldonolactones have a very high affinity for the glycosidases that they inhibit, in comparison with the substrates.

---

* The affinity for substrate or inhibitor may vary with the source of the enzyme.

Sugars and aldonolactones are shown in the following pages in Fischer projection, Haworth ring, or conformational formulas, according to whichever is most convenient for the purposes of illustration. Thus β-D-glucopyranose may be shown variously as:

Fischer   Haworth   Conformation

It is important to note that the conformation shown is only one of several possibilities. It is the preferred "chair" conformation for β-D-glucose.

Similarly, glucono-1:5-lactone may appear in any one of the following structural formulas:

In this case the preferred conformation is the "half-chair" shown.

II. Structural Relations

Aldonolactones are powerful, competitive, and highly specific inhibitors of enzymes that catalyze the hydrolysis of glycosides derived from the same aldoses. All β-glycosidases, but only certain α-glycosidases (glucosidase and mannosidase), are subject to inhibition. Although they can be effective under almost any conditions, one must understand the transformations that aldonolactones can undergo in aqueous solution, in order to make most use of their inhibitory action and interpret it correctly. The composition of an aldonolactone solution is dependent upon its age, temperature, and acidity. For purposes of illustration, let us consider a solution of D-gluconic acid (1) in water. Unless it is neutralized, the acid undergoes rapid inter-conversion with D-glucono-1:5-lactone (2) and D-glucono-1:4-lactone (3). The solutions inhibit the enzymic hydrolysis of α-D-glucosides (4) and β-D-glucosides (5), to an extent that is governed by the relative proportions of the two lactones and the aldonic acid. In this particular instance, both the 1:4- and the 1:5-lactone have been isolated as crystalline solids. Gluconolactone, however, is the exception rather than the rule, because the existence of most other 1:5-lactones, unlike the 1:4-lactones, can only be inferred from physical measurements made in aqueous solutions.

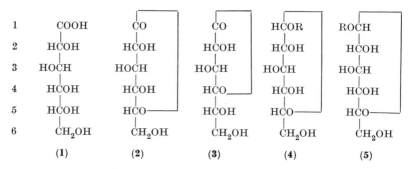

(1)   (2)   (3)   (4)   (5)

A. β-GLUCURONIDASE AND GLUCARIC ACID SOLUTIONS

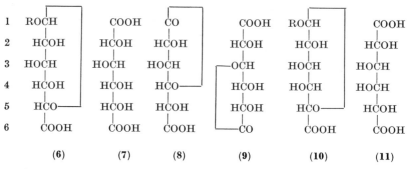

(6)   (7)   (8)   (9)   (10)   (11)

β-Glucuronidase hydrolyzes all β-D-glucuronides of the general formula (6); R can be an alkyl, aryl, alicyclic, or acyl group. This enzyme also hydrolyzes β-D-galacturonides (10), that is, it is not specific for C4 in the substrate (33). The substrate most commonly employed for β-glucuronidase is phenolphthalein β-glucuronide. It is also one of the substrates with the highest affinity for the enzyme (K_m about 100 μM).

In 1949, Karunairatnam and Levvy (21) reported that glucaric acid (formerly known as saccharic acid, 7) is an efficient competitive inhibitor of β-glucuronidase, with a K_i of about 100 μM. Soon after, it was shown that β-glucuronidase is very much more powerfully inhibited by glucaro-1:4-lactone (saccharo-1:4-lactone; 8), with a K_i of less than 1 μM (24). With a purified enzyme preparation, a value of 0.1 μM was later obtained (29). It was concluded (24) that the inhibitory action of glucaric acid can be entirely accounted for by traces of the 1:4-lactone present in aqueous solutions of the acid, or formed from it under the conditions of enzyme assay. Only when the solution is completely neutralized is lactonization of glucaric acid avoided. Most β-glucuronidase preparations are active at a slightly acid pH. For instance, mammalian β-glucuronidase has optima at pH 4.5 and 5.2.

Potassium hydrogen glucarate is the salt of glucaric acid most commonly employed; in dilute aqueous solution it gives a pH of 4.2. When such a solution is boiled, its inhibitory power reaches a maximum after 30 min, at a figure corresponding to one-third conversion into the 1:4-lactone. This solution is a convenient source of inhibitor for β-glucuronidase (24). The solutions also contain glucaro-3:6-lactone (L-gularo-1:4-lactone; 9), which is not an inhibitor for β-glucuronidase (5).

Boiled solutions of galactaric acid (formerly known as mucic acid, 11) also inhibit β-glucuronidase fairly effectively (24), with K_i values of 10–100 μM. Since galactaric acid is a *meso* compound, its solutions will contain racemic mixtures of lactones, but no crystalline product has been isolated from such a solution. The inhibitory action of galactaric acid solutions became easier to understand when it was realized that β-glucuronidase also hydrolyzes β-galacturonides.

Glucaro-1:4-lactone remains one of the most powerful inhibitors of a glycosidase yet known and is the first example of this type of inhibitor to be recognized. It can be seen that if a trace of this lactone is present as an impurity in related sugar derivatives, it can give rise to spurious inhibitory effects. The introduction of an O-methyl group at C3 in glucaro-1:4-lactone reduces the inhibitory power more than 100-fold (24).

Glucarolactone has no action on other glycosidases (including α-glucuronidase), nor is β-glucuronidase inhibited by any other lactone yet examined, with the exception of galactarolactone. Both β-glucuronidase and α-glucuronidase are feebly inhibited by glucuronic acid, with K_i values of approximately 1 mM. The inhibition of β-glucuronidase has been dealt with in greater detail elsewhere (25,30).

B. DERIVATIVES OF GALACTOSE

The earlier work on aldonolactones suggested that similarity in ring size to the pyranosyl substrate was not an important feature of their inhibitory action. Thus β-glucuronidase was inhibited by glucaro-1:4-lactone. In the hydrolysis of a β-glucoside (12), no difference was detected in the inhibitory power of glucono-1:5-lactone (13) and glucono-1:4-lactone (14), and this was also true of inhibition of α-glucosidase (9). (12, 13, and 14 are the Haworth formulas for the compounds already shown in Fischer projection formulas in 5, 2, and 3, respectively.)

The first hint that ring size might be a factor arose during the study of inhibitors of β-N-acetylglucosaminidase (13). This enzyme hydrolyzes N-acetyl-β-D-glucosaminides (2-acetamido-2-deoxy-β-D-glucosides) and N-acetyl-β-D-galactosaminides (2-acetamido-2-deoxy-β-D-galactosides), that is, it is not specific for C4 (47). It was found to be very powerfully inhibited by 2-acetamido-2-deoxygluconolactone and 2-acetamido-2-deoxygalactonolactone (13), with a K_i value for a rat-epididymal preparation of less than 1 μM in each case. (K_m for the glucosaminide substrate was about 0.5 mM.)

Whereas the substituted gluconolactone employed was an impure amorphous solid, the corresponding galactose derivative was the crystalline 1:4-lactone. In aqueous solutions of the latter, changes in composition were observed that led to an increase in inhibitory power of more than 10-fold. Inhibition by the glycosaminic acid having been excluded, the increase in inhibition was attributed to the 1:5-lactone. No such changes in inhibitory power were detected in the solutions of 2-acetamido-2-deoxygluconolactone.

This effect of ring size was further investigated with inhibitors of the hydrolysis of β-D-galactosides (15). (See Ref. 28.) It can be seen that in passing from galactono-1:5-lactone (**16**)—or galactose—to galactono-1:4-lactone (**17**), there is a reversal in the configuration of the ring

CH₂OH   CH₂OH   CH₂OH
                                    HOCH

(12)   (13)   (14)

CH₂OH   CH₂OH   CH₂OH
                                    HOCH

(15)   (16)   (17)

that is not seen in the glucose series. The hydroxyl groups at C4 and C5 are *trans* in galactose (cf. galactaric acid, **11**). This change in configuration is reflected in the complete reversal in optical rotation that is observed when galactose is converted into galactono-1:4-lactone. Unlike the 1:4-lactone, galactono-1:5-lactone has not been isolated in the solid state.

The inhibitory power of crystalline galactono-1:4-lactone varied from specimen to specimen, but it could be recrystallized to give a constant low value. Sodium galactonate was prepared from the lactone by treating an aqueous solution with sodium hydroxide. With rat-epididymal β-galactosidase, assayed at pH 3, solutions of sodium galactonate increased rapidly in inhibitory power due to ring closure under the conditions of assay.

Figure 1 shows the changes that occurred in the inhibitory power of galactonic acid, compared with its 1:4-lactone, during incubation with the enzyme and substrate. Since incubation was at the relatively low pH of 3, ring closure of the acid was rapid, but the lactone was fairly stable. It is clear from the falling rate of enzyme reaction that galactonic acid itself was not an inhibitor, and that its apparent inhibitory power was caused by another entity formed during incubation with enzyme and substrate. The usual values quoted for relative inhibitory power are derived from the 1-hr intercepts on velocity curves of the type shown in Figure 1.

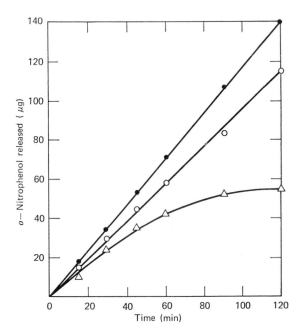

Fig. 1. Hydrolysis of 2.5 mM *o*-nitrophenyl β-galactoside by rat-epididymal β-galactosidase at pH 3 and 38° (●), and the effects of 1 mM sodium galactonate (△) and 1 mM galactono-1:4-lactone (○). From Levvy, McAllan, and Hay (28).

A solution with maximum inhibitory power toward β-galactosidase was obtained from sodium galactonate by bringing the solution to pH 2 and boiling it for 7.5 min. It then had $[\alpha]_D^{20} - 20°$, compared with $-76°$ for the 1:4-lactone. With the rat-epididymal enzyme, 50% inhibition was observed at pH 3 in the usual 1-hr assay with this solution at a concentration of 0.6 mM in terms of galactonate ion. Sodium galactonate itself and galactono-1:4-lactone caused 50% inhibition at 1.4 and 13 mM, respectively. It was evident that the enhanced inhibitory power of galactonate ion in acid solution could not be accounted for by the formation of galactono-1:4-lactone, and it was concluded that the 1:5-lactone was responsible. It was also possible to demonstrate direct conversion of the 1:4- into the 1:5-lactone below neutrality. As a means of obtaining the more inhibitory entity, direct interconversion of the lactones was less effective than ring closure of the galactonate ion.

C. MANNONOLACTONE

The next step was taken with D-mannonolactone (26). Like glucono-lactone, mannonolactone is unusual in two respects. Both mannono-1:5-lactone (**18**) and mannono-1:4-lactone (**19**) have been isolated as crystalline solids, and solutions of mannonolactone inhibit both α- and β-mannosidase. As in glucose (cf. gluconic acid, **1**), the hydroxyl groups at C4 and C5 are *cis* in mannose.

(18)   (19)

The 1:5-lactone was found to be much more powerful as an inhibitor for α- and β-mannosidase than the 1:4-lactone, the difference ap-proaching 500-fold in the case of rat-epididymal α-mannosidase. The K_i values for the 1:5- and 1:4-lactone were 0.071 and 32 mM respec-tively, compared with a K_m of 12 mM for p-nitrophenyl α-mannoside, the substrate employed. These two inhibitors were therefore convenient for following changes in the inhibitory power of lactone pairs in aqueous solution arising from interconversion and ring opening at different pH values. Each crystalline lactone was examined beforehand for the absence of the other by infrared spectroscopy. Determinations of the melting point and optical rotation did not suffice for this purpose.

As the pH of a lactone solution approaches alkalinity, there is opening of the lactone ring, but with a 1:4-lactone there is first of all conversion into the 1:5-lactone. Figure 2 deals with the stability of mannono-1:4-lactone, gauged by its inhibitory power, over varying periods at 38° in buffered solution. After incubation, the solutions were tested against mammalian α-mannosidase in the usual way. At pH 4, the 1:4-lactone was quite stable in aqueous solution at 38°. At pH 6 there was formation of the 1:5-lactone, as demonstrated by a marked increase in inhibition. This process reached a peak in 3 hr, after which the inhibitory power fell, due no doubt to opening of the lactone ring. These changes became more rapid and more pronounced at pH 7, and after 18 hr opening of the lactone ring appeared to be complete. Under the conditions employed for enzyme assay, mannono-1:4-lactone and mannonate ion had practically the same inhibitory power.

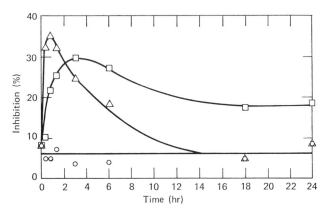

Fig. 2. The effect on the inhibitory power of mannono-1:4-lactone of incubation at 38° for varying periods in buffered solution at pH 4 (○), pH 6 (□), and pH 7 (△). After exposure to the buffer for the required time, the inhibitor solution was rapidly diluted and tested at 1-mM concentration against rat-epididymal α-mannosidase, assayed for 1 hr at 38° and pH 5, with 6 mM p-nitrophenyl α-mannoside as substrate. From Levvy, Hay, and Conchie (26).

Figure 3 shows the changes that take place in solutions of mannono-1:5-lactone at 38°. At pH 6, there was a progressive fall in the inhibitory power of the solution, accompanied by a fall in optical rotation, that indicated opening of the lactone ring. This process was more rapid at pH 7. The 1:5-lactone was remarkably stable in aqueous solution for short periods at pH values around 4. The fall in inhibitory power on prolonged incubation at pH 4 may have been a result of 1:4-lactone formation, as well as ring opening.

There is quite a big difference in the optical rotation of the two mannonolactones: $[\alpha]_D^{20}$ was $+122°$ for the 1:5-lactone and $+50°$ for the 1:4-lactone. The value for mannonic acid was virtually zero. A solution with maximum inhibitory power was prepared from sodium mannonate solution by boiling it at pH 2 for 10 min. Its inhibitory power corresponded to a mannono-1:5-lactone content of 36%, ignoring the negligible contribution of the 1:4-lactone. The solution had $[\alpha]_D^{21}$ $+56°$, from which it follows that only 25% of the mannonate ion was in the form of the 1:4-lactone. This solution was even more stable at acid pH than solutions of the pure 1:5-lactone, at least as regards inhibition of the enzyme.

There was no obvious explanation on structural grounds of the great difference in inhibitory power between the two mannonolactones (**18**

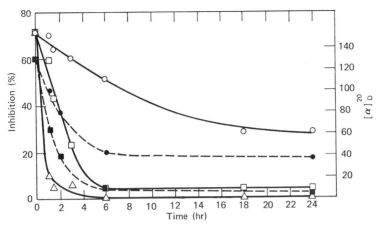

Fig. 3. The effect on the inhibitory power of mannono-1:5-lactone of incubation at 38° for varying periods in buffered solution at pH 4 (○), pH 6 (□), and pH 7 (△). Subsequent test of the lactone, at 0.25-mM concentration, was done as in Figure 2. Changes in the rotation of the buffered lactone solution (c 0.25) at pH 4 (●) and pH 6 (■) are also shown. From Levvy, Hay, and Conchie (26).

and **19**), such as applied to the analogous behavior of the galactono-lactones (**16** and **17**). On the other hand, this is equally true of the difference in optical rotation between the two mannonolactones. However, the gluconolactones (**13** and **14**) are very similar to one another in optical rotation. On reinvestigation, there was found to be a small, but measurable, difference in the inhibitory power of the two gluconolactones for α- and β-glucosidase, in favor of the 1:5-lactone (26).

D. CATALYTIC OXIDATION OF PYRANOSE SUGARS

It has already been mentioned that only two aldono-1:5-lactones, as opposed to the 1:4-lactones, were known in the solid state. A new development, however, made it possible to examine a number of other 1:5-lactones in an almost pure condition (8). It was found that, in aqueous solution at a slightly acid pH, a pyranose could be oxidized at C1 by gaseous oxygen in the presence of a platinum catalyst to give the 1:5-lactone, without ring opening. Unfortunately, evaporation of the product to dryness usually led to extensive transformation into the 1:4-lactone, and perhaps also to ring opening. The solid product was examined by infrared spectroscopy for 1:4- and 1:5-lactone peaks. Evaporation was carried out *in vacuo* at 40°, and it seems probable that decomposition of the 1:5-lactone might have been largely averted by the use of modern freeze-drying equipment. However, the aqueous filtrates from catalytic oxidation were suitable for immediate test for glycosidase inhibition. In general, the results bore out the conclusion that the 1:5-lactones are more powerful inhibitors than the corresponding 1:4-lactones. Direct interconversion of the two lactone forms in aqueous solution at acid pH, as well as ring opening at alkaline pH, could be observed by following changes in the degree of enzyme inhibition.

The sugars selected for catalytic oxidation were D-mannose, N-acetyl-D-glucosamine, N-acetyl-D-galactosamine, D-glucuronic acid, D-galacturonic acid, D-galactose, L-arabinose, and D-fucose. They were chosen on two grounds: as representative of different types of aldopyranose and as likely to yield products of particular interest in connection with the inhibition of glycosidases. Only in the case of mannose was the product isolated as the crystalline 1:5-lactone.

For purposes of illustration, it is proposed, first of all, to return to the inhibition of β-N-acetylglucosaminidase by 2-acetamido-2-deoxygalactonolactone. Catalytic oxidation of N-acetylgalactosamine was quantitative and the aqueous solution thus obtained caused 50% inhibition of a preparation of β-N-acetylglucosaminidase from pig epididymis at 0.19 μM compared with a figure of 8 μM for crystalline 2-acetamido-2-deoxygalactono-1:4-lactone. This was therefore not only the most powerful inhibitor of β-N-acetylglucosaminidase yet encountered, but it may prove to be the most potent of all the aldonolactones in their inhibition of the corresponding glycosidases. The optical rotation of the oxidized N-acetylgalactosamine, $[\alpha]_D^{20}$, was +66° compared with $[\alpha]_D^{22}$ −24° for the crystalline 2-acetamido-2-deoxygalactono-1:4-lactone. Evaporation of the oxidized solution caused relatively little decomposition, and the solid product caused 50% inhibition at 0.26 μM, while the infrared spectrum of the solid indicated the presence of the 1:5-lactone only.

Figures 4 and 5 illustrate the interconversion of the two lactones of 2-acetamido-2-deoxygalactonic acid, as well as ring opening, at different pH values. The 1:4-lactone (Fig. 4) appeared to undergo immediate transformation into the more inhibitory 1:5-lactone at pH 7. However, at this pH the 1:5-lactone was itself unstable (Fig. 5). Both processes were slower at lower pH values. After incubation at 38°, transformation of the 1:4- into the 1:5-lactone was seen at pH 5-6, and there was ring opening above pH 7 (Fig. 4). The only pH at which the 1:5-lactone

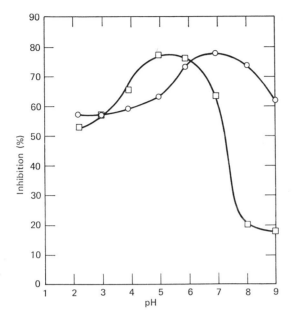

Fig. 4. The effect of exposure to buffers of varying pH on the inhibitory power of 2-acetamido-2-deoxygalactono-1:4-lactone, measured immediately (○) and after 1 hr at 38° (□). The inhibitor solution was rapidly diluted and tested at 9 μM concentration against pig-epididymal β-N-acetylglucosaminidase, assayed for 1 hr at 38° and pH 4.4, with 5 mM p-nitrophenyl N-acetyl-β-glucosaminide as substrate. From Conchie et al. (8).

was at all stable at 38° was 3 (Fig. 5). At higher pH, ring opening presumably occurred, and at lower pH, 1:4-lactone formation.

Figure 6 gives velocity curves for the inhibition of β-N-acetyl-glucosaminidase by 2-acetamido-2-deoxygalactonic acid and its two lactones. Decomposition of the 1:5-lactone was demonstrated by the increasing enzyme velocity. It can be seen that the value obtained from the usual 1-hr assay underestimates the affinity of the lactone for the enzyme. Conversion of the 1:4-lactone into the 1:5-lactone was evident from the progressive decrease in enzyme velocity, similar to that produced by ring closure of the free acid.

In the usual 1-hr assay, the same preparation of β-N-acetylglucosa-minidase was inhibited to the extent of 50% by the impure preparation of 2-acetamido-2-deoxygluconolactone mentioned in Section II.B at a

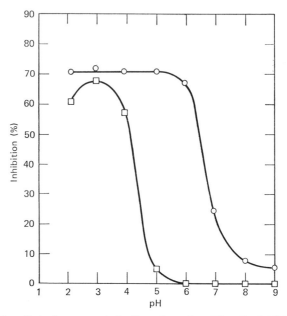

Fig. 5. The effect of exposure to buffers of varying pH on the inhibitory power of 2-acetamido-2-deoxygalactono-1:5-lactone, measured immediately (○) and after 1 hr at 38° (□). The lactone was tested, at 0.44 μM concentration, as in Fig. 4. From Conchie et al. (8).

concentration of 2.3 μM. The compound was not made by catalytic oxidation of the sugar. A specimen that had been subjected to intensive purification to yield the crystalline 1:5-lactone (11) caused 50% inhibition at 0.6 μM. An impure specimen of 2-acetamido-2-deoxy-glucono-1:4-lactone that was free from the 1:5-lactone, as shown by its infrared spectrum, caused 50% inhibition at 8 μM. Catalytic oxidation of N-acetylglucosamine was not quite quantitative, and the aqueous product caused 50% inhibition at 1.2 μM. The solid obtained from the evaporated solution caused 50% inhibition at 1.6 μM.

Of particular interest were the oxidation products from glucuronic acid and galacturonic acid as inhibitors of β-glucuronidase. Both were more powerful inhibitors than glucaro-1:4-lactone. The solution of glucaro-1:5-lactone caused 50% inhibition at 1.3 μM, compared with a figure of 4.5 μM for a solution freshly prepared from the crystalline 1:4-lactone. Figure 7 shows interconversion of the two glucarolactones

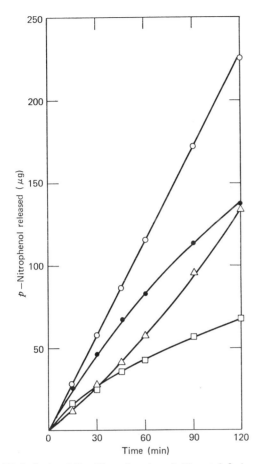

Fig. 6. Hydrolysis of 5 mM p-nitrophenyl N-acetyl-β-glucosaminide by pig-epididymal β-N-acetylglucosaminidase at pH 4.4 and 38° (○), and the effects of 0.36 μM 2-acetamido-2-deoxygalactono-1:5-lactone (△), 14 μM 2-acetamido-2-deoxygalactono-1:4-lactone (□), and 14 μM sodium 2-acetamido-2-deoxygalactonate (●). From Conchie et al. (8).

on incubation at pH 5. There was an initial rapid rise with the 1:4-lactone and fall with the 1:5-lactone, after which the inhibitory power of both solutions stayed fairly constant for several hours.

Oxidation of galacturonic acid was nearly quantitative and the aqueous product caused 50% inhibition of β-glucuronidase at 1.6 μM, compared with the figure of 4.5 for glucaro-1:4-lactone. When it was

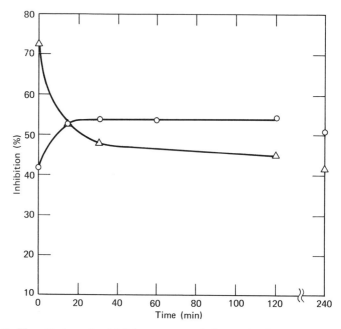

Fig. 7. The effect on the inhibitory power of glucaro-1:4-lactone (○) and of glucaro-1:5-lactone (△) of incubation for varying periods in buffered solution at pH 5 and 38°. After exposure to the buffer for the required time, both types of solution were diluted and tested against rat preputial-gland β-glucuronidase at 2.7 μM concentration. The enzyme assay was for 1 hr at 38° and pH 4.5, with 0.63 mM phenolphthalein β-glucuronide as substrate. From Conchie et al. (8).

converted into a solution of galactaric acid by way of the dibasic salt, and boiled, the final inhibitory power of the oxidation product was only 1% of the original. A boiled solution of authentic galactaric acid gave an identical figure in the inhibition trials.

It seems likely that boiled solutions of galactaric acid contain the DL-1:4-lactone and the DL-1:5-lactone, the former predominating. If one accepts the argument that the configuration of the 1:4-lactone ring excludes inhibition in the galactose series (cf. **16** and **17**), then inhibition by these solutions must be entirely due to D-galactaro-1:5-lactone. It can be seen that no more than 1% need be present in the equilibrium mixture to account for the inhibition observed.

The fresh oxidation product from galacturonic acid was very sensitive to changes in pH (Fig. 8). After incubation for 1 hr at 38°, a major

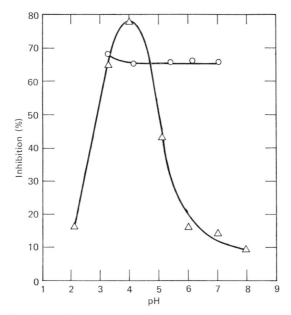

Fig. 8. The effect of exposure to buffers of varying pH for 1 hr at 38° on the inhibitory power of D-galactaro-1:5-lactone (△) and of a boiled solution of galactaric acid (○). The 1:5-lactone solution was tested at 6.7 μM concentration and the galactaric acid solution at 0.5 mM. Both were tested as in Figure 7; the untreated solutions caused 87% and 66% inhibition, respectively. From Conchie et al. (8).

proportion of the inhibitory power was retained only at pH 4. On the other hand, the much feebler inhibitory power of a boiled solution of galactaric acid remained unchanged over a wide range of pH.

Before leaving the subject of catalytic oxidation of sugars, it may be noted that the aqueous product from galactose fell 25-fold in its inhibitory power toward β-galactosidase after converting it into the free acid and boiling the solution for 7.5 min (see Section II.B). From this, one may conclude that when such solutions are prepared from galactono-1:4-lactone, 4% is transformed into the 1:5-lactone.

E. COMPARISON OF ALDONO-1:4- AND 1:5-LACTONES AS INHIBITORS

We have already seen that partial (4%) conversion of galactono-1:4-lactone into the 1:5-lactone led to a 20-fold increase in the inhibitory power of the aqueous solution for β-galactosidase, and it was pointed out that there is a reversal in ring configuration relative to the parent sugar on passing from the 1:5- (**16**) to the 1:4-lactone (**17**). It was postulated that only the 1:5-lactone caused inhibition, and that inhibition by the 1:4-lactone was due to traces of the former arising in the aqueous solution. This explanation could also apply to other pairs of isomeric lactones related to galactose. 2-Acetamido-2-deoxygalac-tono-1:5-lactone was 40 times more powerful an inhibitor of β-N-acetylglucosaminidase than the corresponding 1:4-lactone, and D-galactaro-1:5-lactone was a much more powerful inhibitor of β-glucuronidase than one might have anticipated from the action of a lactonized solution of galactaric acid. Enzyme inhibition by the isomers of D-fuconolactone (6-deoxy-D-galactonolactone) also conformed to this pattern.

This explanation clearly could not apply to the mannonolactones (**18** and **19**), although the crystalline 1:5-lactone was 500 times more powerful than the 1:4-lactone as an inhibitor of mammalian α-mannosidase. In the case of the gluconolactones, there was a small difference in favor of the 1:5-lactone as an enzyme inhibitor, and glucaro-1:5-lactone was more than three times as powerful as the 1:4-lactone toward the appropriate enzyme. There is no doubt that 2-acetamido-2-deoxyglucono-1:5-lactone inhibited β-N-acetylglucosaminidase more strongly than the 1:4-lactone, although an exact figure cannot be put upon inhibition by the latter.

As an all-embracing hypothesis, it was suggested (8) that in every instance inhibition is caused entirely by the 1:5-lactone, and that the apparent inhibition by a 1:4-lactone is a measure of its transformation into the former. The smaller the difference in the inhibitory power of a pair of isomeric lactones, therefore, the greater is their ease of interconversion under the conditions employed for enzyme assay. To take two extreme examples, in which the enzyme assays were done under identical conditions, at pH 5, the mannonolactones must be regarded as relatively stable, and the gluconolactones as unstable.

It is true that interconversion of lactone pairs can be very rapid without ring opening. However, the hypothesis requires, that in certain instances, at least, interconversion can proceed more rapidly than ring closure at acid pH. Glucaro-1:4-lactone was a much more powerful inhibitor of β-glucuronidase at pH 5.2 than glucaric acid, the difference being about 300-fold. The question as to whether a 1:4-lactone as such can ever inhibit a glycosidase, or whether conversion into the 1:5-lactone is always obligatory, could perhaps be answered by infrared spectrophotometry of the aqueous solution. The mannonolactones would provide a particularly good test system for this purpose.

F. CHEMICAL CHANGES IN LACTONE SOLUTIONS

Very little reliable information on the interconversion in aqueous solution of an aldonic acid and its two lactones has been obtained by physical–chemical measurements. Enzyme inhibitory studies have probably provided as much information, at least so far as the formation and stability of the 1:5-lactone are concerned.

In the past, attempts have been made to follow the transformations in solutions of aldonic acids and their lactones by titration and by polarimetry. These two methods are not specific, but they do indicate that, in general, aldono-1:4-lactones are more stable in aqueous solution than aldono-1:5-lactones. In the case of the gluconolactones, however, it would now appear doubtful whether either method can entirely distinguish between the two isomers because interconversion is so rapid in aqueous solution. The early work has been summarized in Pigman (14), and the following mechanism (eq. 2) given for the interconversion of 1:4- and 1:5-lactones:

$$1:4\text{-lactone} \rightleftharpoons \text{aldonic acid} \rightleftharpoons 1:5\text{-lactone} \tag{2}$$

From a study of the transformation of the gluconolactones in aqueous solution, using the hydroxylamine color reaction to measure residual lactones, Jermyn (20) concluded that below pH 7 the route of conversion is as follows (eq. 3):

$$1:4\text{-lactone} \rightarrow 1:5\text{-lactone} \rightarrow \text{aldonic acid} \tag{3}$$

This supports the general conclusion arrived at from enzyme inhibition studies that direct conversion of an aldono-1:4-lactone to an aldono-1:5-lactone is possible without intervention of the free acid.

The most intensive study of the transformations in an aldonic acid solution has been made by Takahashi and his co-workers, who again employed the gluconolactones. Takahashi and Mitsumoto (45), by means of paper-chromatographic separation of the two lactones, found that between pH 4.6 and 7.5 glucono-1:4-lactone is formed from the 1:5-lactone. Between pH 6.5 and 7.5, there was some formation of glucono-1:5-lactone from the 1:4-lactone. Under the conditions of the experiments, the transformations seemed to be virtually instantaneous, and there was no evidence of lactone formation from the aldonic acid.

Spontaneous hydrolysis of the two lactones in buffers of varying pH was followed by the hydroxylamine method. The rate of hydrolysis of the 1:4-lactone increased steadily as the pH was raised. The stability of the 1:5-lactone went through a maximum between pH 3.5 and 4.5, very similar to those that we have observed for other 1:5-lactones by measurements of enzyme inhibition (see Figs. 5 and 8). Takahashi and Mitsumoto concluded that interconversion in aqueous solution follows the scheme shown (eq. 4):

$$1:4\text{-lactone} \rightleftharpoons 1:5\text{-lactone}$$

aldonic acid   (4)

An infrared spectrophotometric study on the interconversion and hydrolysis of glucono-1:4- and 1:5-lactones in deuterium oxide at p^2H 6.6 was made by Shimahara and Takahashi (44). The two lactones and the aldonic acid can be distinguished from each other in the infrared, but high intensity equipment is essential for work in solution. Starting with the 1:4-lactone, there was a lag in the appearance of gluconic acid, accompanied by a temporary accumulation of the 1:5-lactone in the first 10 min. In the decomposition of the 1:5-lactone, both 1:4-lactone and gluconic acid were formed at once. It was concluded that the interconversion of 1:4-lactone and 1:5-lactone proceeds in either direction without involving gluconic acid as an intermediate, and that only the 1:5-lactone is hydrolyzed directly, in accordance with the amended scheme (eq. 5):

$$1:4\text{-lactone} \rightleftharpoons 1:5\text{-lactone} \rightleftharpoons \text{aldonic acid} \qquad (5)$$

There is thus support from physical evidence for the view that the relatively small difference between the two gluconolactones in their inhibitory power toward the glucosidases is a result of their rapid interconversion in aqueous solution. It seems unlikely, however, that many other aldono-1:4-lactones are as unstable as glucono-1:4-lactone in aqueous solution.

Couling and Goodey (10) studied the composition of the original 2-acetamido-2-deoxygluconolactone preparation employed as an inhibitor of β-N-acetylglucosaminidase by Findlay et al. (13) (see Sections II.B and II.D). They were able to separate the solid into three components, N-acetylglucosaminic acid, the 1:4-lactone and the 1:5-lactone, by paper chromatography. The 1:5-lactone could be distinguished from the 1:4-lactone by its instability at pH 7. Specific colorimetric determination of the 1:4-lactone was possible by the use of an alkaline-hydroxylamine reagent that destroyed the 1:5-lactone. The hydrolysis of the individual lactones in an aqueous solution was measured in parallel with its declining inhibitory power for β-N-acetylglucosaminidase. It was concluded that inhibition was caused entirely by the 1:5-lactone.

III. The Specificity of Inhibition

The specificity of action of aldonolactones has already been touched upon. For inhibition, as we have seen, the lactone requires the same structure and configuration as the sugar residue in the substrate for the glycosidase. In the majority of cases, the lactone inhibits only the β-glycosidase; for example, α-glucuronidase and α-N-acetylglucosaminidase are only feebly inhibited, if at all, by the corresponding lactones. However, with gluconolactone and mannonolactone, inhibition extends to the α-glycosidase; that is, the lactone does not discriminate qualitatively between enzymes hydrolyzing anomeric glycosides. Lactone inhibition is always competitive in nature, and it appears to be independent of the type of molecule to which the glycosyl residue is attached, provided that the residue is in a terminal position (38,40). In general, therefore, the specificity of inhibition parallels the substrate specificity of a glycosidase, and considerable information as to enzyme specificity can be derived from the selective effects of different aldonolactones. Aldonolactones can thus be used to identify different glycosidases in a mixed preparation and to confirm the type of linkage that is hydrolyzed.

Aldonolactones are also of value in settling questions as to the identity or nonidentity of enzymes catalyzing different reactions. This is a familiar problem, which arises when two related types of enzyme activity have not been physically separated. One example we have already encountered is β-glucuronidase. This enzyme hydrolyzes β-galacturonides, as well as β-glucuronides; its effect on either type of substrate is inhibited both by glucarolactone and by galactarolactone. The K_i figure for either lactone is independent of the type of substrate employed (29,33).

Two recent papers illustrate the use of lactone inhibitors in the study of glycosidase specificity (6,7). These deal with the C4 and C6 specificity of β-glucosidase and β-galactosidase. It had been suggested that these two activities reside in a single enzyme in almond-emulsin preparations (16); that is, there is a lack of specificity for C4. On the other hand, it has been concluded that, in certain instances, the substrate specificity of β-galactosidase is not strict for C6 and the enzyme can also hydrolyze β-D-fucosides. In the limpet, however, physical separation of these two activities had been demonstrated (27).

Table I shows the concentrations of glucono-, D-fucono-, and galactonolactone required for 50 % inhibition of β-glucosidase, β-D-fucosidase, and β-galactosidase activity in almond emulsin and in the limpet. With emulsin, these were virtually the same for each lactone acting against all three enzyme activities. However, in the limpet, there was a sharp distinction between β-galactosidase on the one hand, and β-glucosidase and β-fucosidase on the other. Whereas β-galactosidase was inhibited only feebly by fuconolactone and not at all by gluconolactone, it was powerfully inhibited by galactonolactone. Glucono- and fuconolactone were both good inhibitors of glucosidase and fucosidase, but galactonolactone had practically no effect on these enzyme activities.

Although they can vary with the concentration and K_m value of the substrate employed, figures for 50 % inhibition reflect the K_i value for an inhibitor. Of course, K_i is independent of the concentration and affinity of the substrate chosen. Table I includes some figures for K_i for the different lactones. These figures bear out the conclusion that, in emulsin, the same enzyme hydrolyzes β-glucosides, β-D-fucosides, and β-galactosides; whereas, in the limpet, hydrolysis of β-glucosides and β-fucosides results from the action of a single enzyme that is distinct from β-galactosidase. For purposes of comparison, Table I also shows some typical K_m values.

The two types of glycosidase activity in a limpet preparation can be separated from each other by heat inactivation at suitable pH values (6). Incubation for 2 hr at 27° and pH 8.5 caused the β-galactosidase activity to fall to 6%, but the β-glucosidase and β-fucosidase only to 60%. A preparation that had been maintained at 50° and pH 3 for 10 min retained about one-third of its β-galactosidase activity and lost almost all of its β-glucosidase and β-fucosidase activity.

TABLE I

The Inhibition of β-Glucosidase, β-D-Fucosidase, and β-Galactosidase in Emulsin and in the Limpet, *Patella vulgata*

	Enzyme activity[a] (units/g starting material)	K_m[a] (mM)	Inhibitor concentration[a] (mM)					
			Gluconolactone		Fuconolactone		Galactonolactone	
			K_i	50% Inhibition	K_i	50% Inhibition	K_i	50% Inhibition
Emulsin								
β-Glucosidase	2.4×10^6	6.0	0.036	0.09	0.012	0.02	—	17
β-D-Fucosidase	1.8×10^6	1.8	0.042	0.09	0.013	0.03	—	>5
β-Galactosidase	0.4×10^6	23	—	0.07	0.013	0.02	—	12
Limpet								
β-Glucosidase	6.4×10^4	0.9	0.015	0.16	0.004	0.08	—	>20
β-D-Fucosidase	4.6×10^4	0.6	0.014	0.06	0.003	0.01	1.9	11
β-Galactosidase	6.4×10^4	3.0	—	No effect	0.34	0.92	0.019	0.15

For each inhibitor, values are shown for K_i and for the concentration causing 50% inhibition of the different enzyme activities. Concentrations of the inhibitor solutions are expressed in terms of the starting materials. Gluconolactone was the crystalline 1:5-lactone. Fuconolactone was a solution of D-fucose. Galactonolactone was a solution of maximum inhibitory power prepared from the 1:4-lactone, and it was therefore the inhibitor solution with the lowest proportion of 1:5-lactone. Nitrophenyl glycosides were employed as substrates.

[a] The values are quoted from refs. 6, 7, 27.

173

It was surprising to find that limpet β-glucosidase hydrolyzed β-fucosides but did not hydrolyze β-galactosides. To explain the coupled lack of specificity for C4 and C6, it was suggested that inversion of the configuration at C4 in glucose led to interference by the hydroxyl group in galactose, and that this group could only be accommodated by the enzyme when the hydroxyl group on C6 was removed, as in fucose.

In both limpet and emulsin, α-L-arabinosidase activity was associated with β-D-fucosidase activity, as one might expect from comparison of the structures of the substrates. In a barley preparation, this relationship did not hold, but the α-L-arabinosidase activity was very low and it was probably a result of multiple unspecific effects of other enzymes in the preparation (7).

The structural relationships between the different sugar residues discussed in this section are shown in the accompanying formulas.

β-D-Glucose β-D-Galactose β-D-Fucose α-L-Arabinose

IV. The Mechanism of Inhibition

Aldonolactones have been put to most of the same uses as other specific inhibitors of enzymes. Their instability in aqueous solution at slightly alkaline pH, however, limits their usefulness *in vivo*. This lends urgency to the quest for a modified, more stable, molecule that still retains the inhibitory properties of a lactone. Detailed knowledge of the specificity and mechanism of action of a lactone is essential to this end. Aldonolactones are not all foreign compounds in the body. Gluconolactone is an intermediary in carbohydrate metabolism. Glucarolactone is a metabolite of glucuronic acid, and small quantities are found in normal urine (31,32). Vastly greater quantities have been found in the urine of a mentally retarded patient (46).

A. ALDONO-1:5-LACTONE STRUCTURE IN RELATION TO GLYCOSIDE HYDROLYSIS

At the present time, considerable interest has been shown in the potential value of aldonolactones in elucidating the mechanism of glycoside hydrolysis by enzymes. Their possible importance in this connection was first pointed out in 1968 by Leaback (22). In a study of the inhibition of β-N-acetylglucosaminidase by 2-acetamido-2-deoxyglucono-1:5-lactone, Leaback drew attention to the conformational similarity between the lactone and the hypothetical transition state in an enzyme-catalyzed pyranoside hydrolysis, which he felt accounted for the high affinity of the lactone for the enzyme.

It is generally believed that acid-catalyzed hydrolysis of a pyranoside proceeds by way of a carbonium ion intermediate (eq. 6) (3,17).

$$(6)$$

An aldono-1:5-lactone can form a similar ion (eq. 7).

$$(7)$$

The glycosyl ion is expected to assume a half-chair conformation, and delocalization of electrons can occur. This is shown for a β-D-glycoside (eq. 8), starting with the favored chair conformation of the intact sugar ring.

$$(8)$$

On structural grounds, an aldono-1:5-lactone favors the half-chair conformation; this too can be stabilized by delocalization of electrons (39), as shown in equation 9.

$$(9)$$

Hydrolysis of a pyranoside by a glycosidase is believed to resemble acid-catalyzed hydrolysis in proceeding by way of the carbonium ion.

Since enzymatic hydrolysis does not lead to racemization, the configuration at C1 must somehow be preserved by binding the glycosyl group to the enzyme.

X-Ray studies of the structure of the endo-enzyme, lysozyme,* have cast further light on the mechanism of glycoside hydrolysis by enzymes. Phillips and his colleagues (4,37) have concluded that the glycosyl group at which cleavage occurs must first be distorted into the half-chair conformation in order to fit onto lysozyme (eq. 10). This distortion of the intact sugar ring in the transition state is considered to be an essential step in the catalytic formation of the glycosyl cation. An aldono-1:5-lactone is ideally designed, as Leaback (22) suggests, to act as a competitive inhibitor of such an enzymatic process.

$$(10)$$

Hackert and Jacobson (15) have reversed this argument. They proved by X-ray diffraction that crystalline glucono-1:5-lactone does in fact have the half-chair conformation illustrated. Therefore, they conclude, the high affinity of this lactone for glucosidases provides the best evidence that intermediates with a half-chair conformation are indeed involved in the enzymatic hydrolysis of the pyranosides.

Beddell et al. (2) have studied the binding of aldonolactones by lysozyme, but the results were far from straightforward. In comparing the action of a glycosidase with that of lysozyme, it must not be forgotten that the latter splits an internal glycosyl bond in a sugar chain, whereas the former acts only upon a terminal glycosyl group.

Secemski and Lienhard (42) have reported powerful inhibition (50% at 0.7 μM) of lysozyme by the substituted aldono-1:5-lactone obtained by oxidation of tetra-N-acetylchitotetraose. This supports the model put forward by Phillips and his colleagues (37) for the enzyme–substrate complex formed by lysozyme. In this model, six sugar residues are bound to the protein, and fission occurs between the fourth and fifth

* Lysozyme hydrolyzes β-1:4-linked polymers composed of alternate residues of N-acetylmuramic acid and N-acetylglucosamine or of N-acetylglucosamine only. It acts with retention of the configuration at C1 in the glycosyl group.

of these residues, counted from their nonreducing end. It is suggested that the substituted lactone replaces residues 1 to 4, so that the lactone group takes the place of the distorted glycosyl group, 4.

Provided that the glycosyl group is identical to that in the normal substrate, glycosyl fluorides are specifically hydrolyzed by glycosidases, and the same mechanism of hydrolysis appears to apply (1). Aldonolactone inhibition extends to the hydrolysis of the glycosyl fluorides. The fluorine atom is isoelectronic with the glycosyl oxygen in a sugar or glycoside. It could hardly, however, have been foreseen that the fluorides would behave as substrates for the enzymes.

No theory of the mechanism of inhibition by lactones, or for that matter of glycosidase action, is complete that does not take into account the fact that α-glycosidases are only feebly inhibited, if at all, with the notable exceptions of α-mannosidase and α-glucosidase. There may be a fundamental difference in the mechanism of action between different α-glycosidases that is reflected in their susceptibility to inhibition by the aldonolactones. Furthermore, the whole question as to whether furanosidases are inhibited by aldono-1:4- or 1:5-lactones has still to be explored. Any theory on the glycosidase–inhibitor reaction must ultimately consider the results of such experiments also.

B. RELATED COMPOUNDS

1. Glycals

Lee (23) found that β-D-galactosidase from a number of different sources was competitively inhibited by galactal, with K_i values varying between 20 and 100 μM. These are within the same range as K_i values for the solution of maximum inhibitory power prepared from galactono-1:4-lactone (see Section II.B and Table I). Other glycosidases examined were not inhibited by galactal.

(20)

Galactal (20) is believed to take up the half-chair conformation shown alongside; Lee suggests that its inhibitory action can be explained by its resemblance in shape to the glycosyl cation (see eq. 8). On closer examination, however, it can be seen that galactal and the glycosyl cation adopt quite different conformations. It is difficult to see how the configurational requirements that determine enzyme specificity can be met completely by galactal if it has to replace the galactosyl cation. Leaving aside conformational considerations, the inhibitory action of galactal may merely indicate that the hydroxyl group on C2 is not essential for the binding of substrate by β-galactosidase.

Reese et al. (39) studied glucal as an inhibitor of glucosidases. It was equally effective toward α- and β-glucosidase. Glucono-1:5-lactone was slightly more powerful than glucal as an inhibitor of α-glucosidase and much more powerful as an inhibitor of β-glucosidase. Glucal was much less effective than mannono-1:5-lactone as an inhibitor of α-mannosidase (41).

2. Nojirimycin and Its Derivatives

Nojirimycin (5-amino-5-deoxy-D-glucopyranose, **21**), an antibiotic from certain strains of *Streptomyces*, is a nitrogen analogue of glucose, with the same conformation (19,36). In aqueous solution it is in equilibrium with the imine (**22**), which may be expected to assume a half-chair conformation, as shown. Both compounds display acidic properties, due to protonation of the nitrogen. Hypoiodite oxidation of nojirimycin in alkaline solution gives D-glucono-1:5-lactam (**23**), which is known to prefer the same half-chair conformation as glucono-1:5-lactone. Reduction of nojirimycin yields 1-deoxynojirimycin (**24**).

Niwa et al. (35) found that nojirimycin was 10 times more powerful than glucono-1:5-lactone (**13**) as an inhibitor of apricot-emulsin β-glucosidase. They attributed the inhibitory power of nojirimycin to its similarity to glucose in configuration and conformation (see p. 153). The action of nojirimycin was shown to be specific for α- and β-glucosidase by Reese et al. (39), who considered it more likely that the imine (**22**) or the iminium ion was the inhibitory entity, because of the resemblance to the glycosyl cation. Niwa et al. (35) found that β-glucosidase was also inhibited by glucono-1:5-lactam (**23**), to the same extent as by glucono-1:5-lactone. This certainly supports the theory that the powerful inhibitory action of aldono-1:5-lactones arises from their conformational similarity to the carbonium ion intermediate in glycoside hydrolysis.

On the other hand, β-glucosidase was inhibited by L-idono-1:5-lactam (**25**), which had about one-quarter of the inhibitory power of glucono-1:5-lactone (35). This lactam is considered (19) to prefer the boat conformation shown. Although protonation can probably occur, the molecule resembles the glycosyl cation in shape only in so far as C2, C1, N, and C5 are in one plane.

1-Deoxynojirimycin (**24**) barely inhibited β-glucosidase, 50% inhibition requiring 25 mM. This adds interest to an earlier observation (6) that glycosidases are only very feebly inhibited, if at all, by the appropriate 1:5-anhydrohexitols. 1-Deoxynojirimycin is a nitrogen analogue of 1:5-anhydroglucitol ("1-deoxyglucose") and the latter did not inhibit β-glucosidase. [Glucose, incidentally, has a very low affinity for β-glucosidase (16), as one might expect for the product of a reaction that is essentially irreversible.] In this regard, a recent study of myoinositol as a specific inhibitor of α-galactosidase is also perhaps relevant: in spite of some similarity in configuration and conformation, the inhibitor had only about one-tenth of the affinity of the substrate for the enzyme (43).

These results strongly support the view that, to be really powerful, an inhibitor should resemble the glycosyl cation. One concludes that the efficiency of nojirimycin (**21**) as an enzyme inhibitor is largely accounted for by the formation of the imine (**22**), as proposed by Reese et al. (39).

Although β-glucosidase had a much higher affinity for nojirimycin than it had for glucono-1:5-lactone, it should perhaps be emphasized that the affinity of β-glucosidase for its lactone inhibitor is at least 50 times less than those of β-N-acetylglucosaminidase and β-glucuronidase for their corresponding inhibitors. As compared with the simpler aldonolactones, such as gluconolactone, the presence of a second "functional" group in the lactone molecule appears to enhance its action.

V. Summary

All β-glycosidases and certain α-glycosidases are powerfully inhibited by the aldono-1:5-lactones derived from the same sugars as the substrates. The status of aldono-1:4-lactones as glycosidase inhibitors is in doubt. Their inhibitory power may merely indicate the degree to which spontaneous conversion into the corresponding 1:5-lactone occurs in aqueous solution under the conditions of enzyme assay. Ring opening destroys the inhibitory power of lactones.

The specificity of glycosidase inhibition by an aldonolactone reflects the substrate specificity of the enzyme. The mechanism of inhibition by aldonolactones is discussed in relation to current views on the theory of glycoside hydrolysis by enzymes. The inhibitory effects upon glycosidases of certain other compounds that display a structural resemblance to aldono-1:5-lactones are also considered.

References

1. Barnett, J. E. G., Jarvis, W. T. S., and Munday, K. A., *Biochem. J.*, *105*, 669 (1967).
2. Beddell, C. R., Moult, J., and Phillips, D. C., *Ciba Found. Symp. Molecular Properties of Drug Receptors* p. 85 (1970).
3. BeMiller, J. N., *Advan. Carbohydrate Chem.*, *22*, 25 (1967).
4. Blow, D. M., and Steitz, T. A., *A. Rev. Biochem.*, *39*, 63 (1970).
5. Bose, R. J., Hullar, T. L., Lewis, B. A., and Smith, F., *J. Org. Chem.*, *26*, 1300 (1961).
6. Conchie, J., Gelman, A. L., and Levvy, G. A., *Biochem. J.*, *103*, 609 (1967).
7. Conchie, J., Gelman, A. L., and Levvy, G. A., *Biochem. J.*, *106*, 135 (1968).
8. Conchie, J., Hay, A. J., Strachan, I., and Levvy, G. A., *Biochem. J.*, *102*, 929 (1967).
9. Conchie, J., and Levvy, G. A., *Biochem. J.*, *65*, 389 (1957).
10. Couling, T. E., and Goodey, R., *Biochem. J.*, *119*, 303 (1970).

11. Cross, N. M., British Patent, 1,138,367 (1969).
12. Ezaki, S., *J. Biochem. (Tokyo)*, *32*, 107 (1940).
13. Findlay, J., Levvy, G. A., and Marsh, C. A., *Biochem. J.*, *69*, 467 (1958).
14. Green, J. W., in *The Carbohydrates*, W. Pigman, Ed., Academic Press, New York, 1957, p. 304.
15. Hackert, M. L., and Jacobson, R. A., *Chem. Commun.*, p. 1179 (1969).
16. Heyworth, R., and Walker, P. G., *Biochem. J.*, *83*, 331 (1962).
17. Hopkinson, S. M., *Quart. Rev. (London)*, *23*, 98 (1969).
18. Horikoshi, K., *J. Biochem. (Tokyo)*, *35*, 39 (1942).
19. Inouye, S., Tsuruoka, T., Ito, T., and Niida, T., *Tetrahedron*, *23*, 2125 (1968).
20. Jermyn, M. A., *Biochim. Biophys. Acta*, *37*, 78 (1960).
21. Karunairatnam, M. C., and Levvy, G. A., *Biochem. J.*, *44*, 599 (1949).
22. Leaback, D. H., *Biochem. Biophys. Res. Commun.*, *32*, 1025 (1968).
23. Lee, Y. C., *Biochem. Biophys. Res. Commun.*, *35*, 161 (1969).
24. Levvy, G. A., *Biochem. J.*, *52*, 464 (1952).
25. Levvy, G. A., and Conchie, J., in *Glucuronic Acid*, G. J. Dutton, Ed., Academic Press, New York, 1966, p. 301.
26. Levvy, G. A., Hay, A. J., and Conchie, J., *Biochem. J.*, *91*, 378 (1964).
27. Levvy, G. A., and McAllan, A., *Biochem. J.*, *87*, 206 (1963).
28. Levvy, G. A., McAllan, A., and Hay, A. J., *Biochem. J.*, *82*, 225 (1962).
29. Levvy, G. A., McAllan, A., and Marsh, C. A., *Biochem. J.*, *69*, 22 (1958).
30. Levvy, G. A., and Marsh, C. A., *Advan. Carbohydrate Chem.*, *14*, 381 (1959).
31. Marsh, C. A., *Biochem. J.*, *86*, 77 (1963).
32. Marsh, C. A., *Biochem. J.*, *99*, 22 (1966).
33. Marsh, C. A., and Levvy, G. A., *Biochem. J.*, *68*, 610 (1958).
34. Morita, Y., *J. Biochem. (Tokyo)*, *43*, 7 (1956).
35. Niwa, T., Inouye, S., Tsuruoka, T., Koaze, Y., and Niida, T., *Agr. Biol. Chem. (Tokyo)*, *34*, 966 (1970).
36. Paulsen, H., and Todt, K., *Advan. Carbohydrate Chem.*, *23*, 115 (1968).
37. Phillips, D. C., *Scientific American*, *215*, 78 (1966).
38. Reese, E. T., Maguire, A. H., and Parrish, F. W., *Can. J. Biochem.*, *46*, 25 (1968).
39. Reese, E. T., Parrish, F. W., and Ettlinger, M., *Carbohydrate Res.*, *18*, 381 (1971).
40. Reese, E. T., and Shibata, Y., *Can. J. Microbiol.*, *11*, 167 (1965).
41. Schwartz, J., Sloan, J., and Lee, Y. C., *Arch. Biochem. Biophys.*, *137*, 122 (1970).
42. Secemski, I. I., and Lienhard, G. E., *J. Amer. Chem. Soc.*, *93*, 3549 (1971).
43. Sharma, C. B., *Biochem. Biophys. Res. Commun.*, *43*, 572 (1971).
44. Shimahara, K., and Takahashi, T., *Biochim. Biophys. Acta*, *201*, 410 (1970).
45. Takahashi, T., and Mitsumoto, M., *Nature*, *199*, 765 (1963).
46. Westall, R. G., Cahill, R., and Sylvester, P. E., *J. Ment. Defic. Res.*, *14*, 347 (1970).
47. Woollen, J. W., Heyworth, R., and Walker, P. G., *Biochem. J.*, *78*, 111 (1961).

MECHANISM OF ACTION AND OTHER PROPERTIES OF SUCCINYL COENZYME A SYNTHETASE

By JONATHAN S. NISHIMURA, *San Antonio, Texas* and FREDERICK GRINNELL, *Dallas, Texas*

CONTENTS

I. Introduction

It has been several years since the subject of succinyl CoA synthetase* was last reviewed (1). In the ensuing period, many interesting facets regarding its mechanism of action and structure have been uncovered. The complexity of the reaction catalyzed by this enzyme,

* Succinate : CoA ligase (ADP), EC 6.2.1.5 and (GDP), EC 6.2.1.4, also known as the succinate-phosphorylating enzyme, P-enzyme, and succinic thiokinase.

as illustrated in equation 1, and its role in the citric acid cycle in the production of high energy phosphate at the substrate level have attracted the attention of many investigators.

$$\text{succinyl CoA} + \text{Pi} + \text{NDP} \overset{M^{2+}}{\rightleftharpoons} \text{NTP} + \text{succinate} + \text{CoA}*\dagger \qquad (1)$$

The reaction is freely reversible, as indicated by an equilibrium constant of 3.7 at 20° (2).

From a mechanistic point of view, succinyl CoA synthetase has been more or less resistant to solution. However, recent developments have provided a good basis for comprehending the reaction and achieving an understanding of the structure of the enzyme. It is primarily these considerations that will provide the focal point of this discussion, although the appropriate background material has been included at some risk of duplicating information from Hager's review (1).

II. General Considerations

A. FUNCTION OF THE ENZYME

An important physiological role played by succinyl CoA synthetase is its participation in the citric acid cycle in catalyzing the substrate level phosphorylation of nucleoside diphosphate (eq. 1). The reverse reaction can also be vital to anaerobic microorganisms that derive succinate from precursors other than α-ketoglutarate and require succinyl CoA for biosynthetic purposes (1). In like vein, the reverse reaction could be of value to organisms possessing a glyoxylate cycle (3) by which α-ketoglutarate can be bypassed as a precursor of succinate. In animals, some indication that succinate may be a precursor of succinyl CoA is derived from the report of Labbe et al. (4) that increased levels of both δ-aminolevulinate synthetase and succinyl CoA synthetase

---

* Abbreviations used: NTP, nucleoside triphosphate; ATP and ADP, adenosine tri- and diphosphates; GTP and GDP, guanosine tri- and diphosphates; ITP and IDP, inosine tri- and diphosphates; CoA, coenzyme A; Pi, inorganic orthophosphate; E, enzyme; E ~ P, phosphorylated enzyme; ~ ECoA, high energy nonphosphorylated enzyme; succinyl ~ P, succinyl phosphate; pHMB, p-hydroxymercuribenzoate.

† NTP can be ATP, GTP, or ITP, depending on the source of enzyme. It is assumed, though not proved, that all reactions catalyzed by succinyl CoA synthetase require a divalent metal ion, M^{2+}.

in mouse liver were induced by administration of drugs known to produce experimental porphyria. It appeared that citric acid cycle activity in both treated and untreated animals was the same, but that the succinate–succinyl CoA pool was higher in the former. The suggestion was made that the excess succinate arose from precursors other than α-ketoglutarate, probably fumarate. The induced succinyl CoA synthetase was found to have properties significantly different from those of the noninduced enzyme.

B. OCCURRENCE AND PURIFICATION OF SUCCINYL CoA SYNTHETASE

The occurrence of the enzyme is widespread, as it has been detected in animals, higher plants, and microorganisms. It is, therefore, not surprising that a number of sources have been suitable as starting materials for the purification of the enzyme. The catalytic activities of some of these preparations are listed in Table I. Since there has been considerable nonuniformity in the assay methods employed and, therefore, difficulty in comparing preparations with each other, the authors have taken the liberty of converting reported values to either the units defined by Kaufman et al. (5) or those of Ramaley et al. (6), or both. All of these values indicated are reflections of the enzyme samples to catalyze succinyl CoA formation (reverse direction of eq. 1). Hydroxylamine, used as a trapping agent of the activated succinyl group (7) in the Kaufman assay, provides a more convenient, albeit less accurate, method than that of succinyl CoA determination (8). Perhaps with the exception of the spinach enzyme, all other preparations listed were homogeneous or very nearly so. The broad variance with respect to the specific activities of the *E. coli* enzyme will be discussed later.

C. SUBSTRATE SPECIFICITY

1. Succinate

Robinson et al. (9) have reported that of 11 organic acids tested as possible substrates of the *E. coli* enzyme, only malate was effective. This compound was 3 % as active as succinate and only so if incubated at the relatively high concentration of 91 mM. Upper (10) had reported malate to be inactive with the same enzyme, but the concentration of malate used in his experiments was 1 mM. Strict specificity for

TABLE I

Specific Enzymatic Activities of Succinyl CoA Synthetase Preparations Purified
from Various Sources

Sources	Specific enzymatic activity		
	Reported	Kaufman et al. (5) units	Ramaley et al. (6) units
Spinach [Kaufman and Alivisatos (2)]	36 μmoles succinyl hydroxamate/(30 min)(mg)	36	1.6[a]
E. coli [Ramaley et al. (6)]	47 μmoles succinyl CoA/(min) (mg)	1080[a]	47
E. coli [Knight (26)]	4120 μmoles succinyl hydroxamate/(30 min)(mg)	874[a]	38[b]
E. coli [Gibson et al. (47)]	35 μmoles succinyl hydroxamate/(min)(mg)	449[a]	19.5[b]
E. coli [Grinnell and Nishimura (24,28)]	1000 μmoles succinyl hydroxamate/(30 min)(mg)	400–1000	17–43[a]
Jerusalem artichoke [Palmer and Wedding (18)]	18.7 μmoles succinyl CoA/(min)(mg)	Not estimated	Not estimated
Rhodopseudomonas spheroides [Burnham (11)]	36.2 μmoles succinyl hydroxamate/(min)(mg)	460[a]	20[b]
Porcine heart [Cha et al. (14)]	108 μmoles succinyl CoA/(min)(mg)	432	
Pigeon liver [Meshkova and Matveeva (50)]	40 μmoles succinyl CoA/(min)(mg)	160	

[a] Calculated on the basis that 23 Kaufman units are equal to 1 unit of Ramaley et al. (J. S. Nishimura and T. Mitchell, unpublished).

[b] Calculated by Ramaley et al. (6).

[c] Comparable data are lacking.

succinate has been reported for the enzymes from *R. spheroides* (11) and spinach (2).

2. Nucleoside Triphosphate and Diphosphate

Succinyl CoA synthetases from spinach (2) and *E. coli* (9) appear to be specific for ATP and ADP. The porcine kidney enzyme utilizes the nucleoside di- and triphosphates of inosine and guanosine (12). Cha and Parks (13) found that azaGTP and azaGDP were also functional as substrates and inhibitors of the porcine heart enzyme. In addition, it was observed that 6-thioguanosine triphosphate very effectively replaced GTP (14). Burnham's results (11) showed that ATP, GTP, and ITP, but not CTP and UTP, were active with the *R. spheroides* enzyme. This observation, coupled with the reports that a phosphoryl enzyme intermediate is formed with other synthetase preparations (see Section III.A), suggests that this enzyme may also function as a nucleoside diphosphate kinase with specificity for the nucleotides of adenosine, guanosine, and inosine.

3. Coenzyme A

Only dephospho CoA (14,15) appears to be able to replace CoA in the net catalytic reaction. However, pantetheine (16) and 4'-phospho-pantetheine (9) have been shown to stimulate partial reactions catalyzed by the *E. coli* enzyme.

4. Inorganic Phosphate

Replacement of Pi with arsenate in the forward direction (see eq. 1) results in the breakdown (arsenolysis) of succinyl CoA, which is independent of nucleoside diphosphate. For the porcine kidney (12), the porcine heart (13), and the *E. coli* enzymes (10), this reaction was found to proceed at a rate of about 20% that of the phosphorolysis reaction, which requires the presence of Pi and NDP.

5. Divalent Metal Ion

A divalent metal ion is an absolute requirement in the succinyl CoA synthetase reaction. Magnesium ion has been most commonly employed, but other divalent metal ions have also been found at least to partially replace magnesium. Mazumder et al. (17) showed that with the porcine kidney enzyme Mg^{2+} and Mn^{2+} were equally effective at $6 \times 10^{-3}\ M$. The *R. spheroides* synthetase (11) was about 75% as active with Mn^{2+} (at $5 \times 10^{-3}\ M$) as it was with Mg^{2+} (at $0.01\ M$). Upper (10) obtained maximum activation of the *E. coli* enzyme at $4 \times 10^{-4}\ M\ Mn^{2+}$ and at $0.01-0.02\ M\ Mg^{2+}$. In this case the relative rate with Mn^{2+} was 67% that with Mg^{2+}. Some reaction was also seen with Zn^{2+}, Co^{2+}, and Ca^{2+}. Succinyl CoA synthetase from the Jerusalem artichoke could be activated by Mg^{2+}, Mn^{2+}, and Co^{2+} in order of decreasing effectiveness, when these metal ions were tested at 5 mM (18).

TABLE II
Michaelis Constants of Substrates of Various Succinyl CoA Synthetase Preparations

Source and reference	Succinate $(M \times 10^3)$	CoA $(M \times 10^4)$	NTP $(M \times 10^5)$	Succinyl CoA $(M \times 10^5)$	Pi $(M \times 10^3)$	NDP $(M \times 10^4)$
Spinach (2)[a]	15	1	0.5 (ATP)			
Jerusalem artichoke (18)[a]	2	0.22	14 (ATP)	5.6	1.4	1.2 (ATP)
E. coli (19)	0.1	0.015	0.02 (ATP)			
E. coli (9)	1.5–4.3	0.076–0.29				
R. spheroides (11)[a]	1.9	0.4	1.4 (ATP)			
Porcine kidney (17)[a]	5.8	1	3.3 (ITP, ATP)			0.38 (IDP) 0.17 (GDP)
Porcine heart (13)	0.4–0.8	0.2	0.01 (GTP)	1.4		0.02–0.08 (GDP)

[a] Fixed substrate concentrations were those of the standard assays employed.

188

D. MICHAELIS CONSTANTS OF SUBSTRATES

A wide range of K_m values for the different substrates has been obtained in studies with succinyl CoA synthetases from various sources. This is illustrated in Table II. The differences in values seen for the enzyme from the same species can be at least partially ascribed to the conditions of the assays. The concentrations of the fixed substrate are quite important, as emphasized by Cha and Parks (13) and illustrated by the data of Moffet and Bridger (19). The complexity of the succinyl CoA synthetase reaction (three substrates, three products, and metal ion) is such that these studies are difficult to carry out and to evaluate.

III. Mechanism of the Succinyl CoA Synthetase Reaction

Progress in elucidating the mechanism of action of succinyl CoA synthetase has been facilitated by the use of exchange (or partial) reactions and by experiments in which relatively large concentrations of enzyme are employed. The dangers inherent in these approaches, for example, expression of contaminating enzyme activities and of minor catalytic activities of the enzyme itself, have been carefully considered. Thus S-phosphoryl CoA, which had been proposed as an intermediate on the basis of data from partial reactions (see ref. 1), has been ruled out by Grunau et al. (20).

A number of exchange reactions are catalyzed by succinyl CoA synthetase preparations and include: (1) ATP $\rightleftarrows$ Pi, (2) ^{18}O-Pi $\rightleftarrows$ succinate carboxyl; (3) succinate $\rightleftarrows$ succinyl CoA, and (4) ATP $\rightleftarrows$ ADP. These reactions have been quite adequately described by Hager (1), and will be referred to frequently in the discussion that follows.

There is convincing evidence for the participation of a phosphorylated enzyme intermediate (E $\sim$ P) in the succinyl CoA synthetase reaction. Considerable data have also been presented that implicate enzyme-bound succinyl phosphate as a second intermediate. At the same time it has been argued on kinetic grounds that reaction of succinate and CoA with E $\sim$ P, or of succinyl CoA and Pi with E, may involve a concerted reaction without the transitory formation of succinyl phosphate.

A. THE PHOSPHORYLATED ENZYME

Succinyl CoA synthetase can be phosphorylated either by Pi in the presence of succinyl CoA or by ATP (10,21). That the site on the enzyme which is phosphorylated is the N-3 (ring) position of a histidine residue has been demonstrated by Hultquist et al. (22). Kinetic evidence that the formation of $E \sim P$ occurs at a rate which is compatible with its participation in the overall reaction has been presented for the *E. coli* enzyme by Bridger et al. (23). Thus it was found that in the forward direction (see eq. 1) the steady state level of $E \sim P$ was attained before the establishment of the steady state rate of ATP formation. In the reverse direction, the initial rate of $E \sim P$ formation was greater than the steady state rate of Pi synthesis. Kinetic evidence that $E \sim P$ breaks down to yield Pi at a rate faster than Pi evolution from ATP has been obtained by Grinnell and Nishimura (24), who employed $E \sim {}^{32}P$ and ATP-γ-${}^{33}P$ in their studies. The formation of a phosphorylated intermediate is in harmony with the ATP $\rightleftharpoons$ ADP exchange reaction, which was first demonstrated with the spinach enzyme by Kaufman (25) and with the *E. coli* enzyme by Knight (26).

A puzzling aspect of the *E. coli* enzyme is the degree to which it can be phosphorylated. Ramaley et al. (6) reported that reaction of succinyl CoA synthetase with ATP resulted in the transfer of close to 1 phosphoryl group per mole of enzyme, based on an enzyme molecular weight of 141,000. On the other hand, phosphorylation by Pi in the presence of succinyl CoA resulted in 1.38 moles of bound phosphoryl groups per mole of enzyme, with values of 1.6 and 1.8 also observed. These workers also presented evidence that the catalytic activity of the phosphorylated preparations was not directly related to the extent of phosphorylation. A later report from the same laboratory (27) confirmed the latter observation. It is of considerable interest that in both reports the specific activity of homogeneous enzyme preparations varied significantly. In contrast to these results, Grinnell and Nishimura (28) ascertained the capacity for phosphorylation of succinyl CoA synthetase preparations to be directly related to their catalytic activity. Moreover, in the preparations of highest specific activity, approximately 2 moles of phosphoryl groups were found per mole of enzyme with ATP as the phosphorylating agent and 2.1–2.3 moles of phosphoryl groups were detected per mole of enzyme upon phosphorylation by Pi. Explanations for these discrepancies are not yet at hand. It is conceivable, however, that the enzyme is extremely sensitive to the methods of purification employed; while catalytic activity is retained, capacity for phosphorylation may be altered.

Cha et al. (29) confirmed formation of $E \sim P$ in highly purified porcine heart preparations of succinyl CoA synthetase. These workers (30) estimated that 2-4 phosphoryl groups could be bound per mole enzyme, based on an estimated molecular weight of 70,000 (14). Data concerning the relationships of catalytic activity to the capacity for phosphorylation of the mammalian enzyme have not been published. However, it is a practical certainty that the enzyme-bound phosphoryl group functions as an intermediate in the mammalian enzyme.

Intermediary protein-bound phosphohistidine formation has been observed in the heat-stable bacterial phosphoryl transfer protein (31), in the citrate cleavage enzyme (32), and in nucleoside diphosphate kinase isolated from the Jerusalem artichoke (33). Walinder (34) has reported isolation of 3-phosphohistidine, 1-phosphohistidine, and N-ε-phospholysine from digests of phosphorylated erythrocytic nucleoside diphosphate kinase.

Studies reported by Moffet and Bridger (19) did not yield evidence of "ping-pong" kinetics with succinyl CoA synthetase that might be anticipated from intermediary $E \sim P$ formation. Their results indicate that during the net catalytic reaction, for example, the reverse of equation 1, NDP is not released prior to the reaction of succinate and CoA with $E \sim P$. "Ping-pong" kinetics have been observed with nucleoside diphosphate kinase (35,36), another known phosphoenzyme. The findings with succinyl CoA synthetase emphasize the hazards involved in deducing mechanism from kinetic data alone.

B. HIGH ENERGY NONPHOSPHORYLATED FORM OF THE ENZYME. EVIDENCE FOR ITS NONINVOLVEMENT IN THE NET CATALYTIC REACTION

The proposal of this intermediate came about through elaboration of the original observations of Upper (10). In its simplest form (without the assumed binary and possible ternary complexes) the reaction sequence involved can be summarized as shown in equations 2a, 2b, and 2c.

$$E + \text{succinyl CoA} \rightleftarrows \sim E\text{–CoA} + \text{succinate} \qquad (2a)$$

$$\sim E\text{–CoA} + \text{Pi} \rightleftarrows E \sim P + \text{CoA} \qquad (2b)$$

$$E \sim P + \text{NDP} \rightleftarrows E + \text{NTP} \qquad (2c)$$

Evidence for the high energy nonphosphorylated mechanism was based largely on the following observations with both $E.\ coli$ and mammalian heart preparations: (a) incubation of $E \sim ^{32}P$ with CoA resulted in release of ^{32}Pi (10,16,29) with the apparent concomitant and covalent binding to the enzyme of an equivalent amount of CoA (16,30); (b) CoA stimulated $ATP \rightleftarrows Pi$ (10,16) and $E \sim P \rightleftarrows pi$ (16) exchange reactions catalyzed by the *E. coli* enzyme; (c) incubation of succinate with ^3H–CoA-treated $E \sim P$ or with ^{32}P–dephospho CoA-treated $E \sim P$ resulted in the appearance of labeled succinyl CoA (30) or labeled succinyl-dephospho CoA (15); (d) succinate $\rightleftarrows$ succinyl CoA exchange catalyzed by the enzyme proceeded without the addition of Pi (10) and, in fact, was observed to be inhibited by this anion (16).

The earlier work of Cohn (37) and Hager (38) showed that ^{18}O was transferred from Pi to a carboxyl group of succinate during the succinyl CoA synthetase reaction. The formation of a succinyl phosphate intermediate or a concerted mechanism involving all reactants would be reconcilable with these observations. However, verification of the high energy nonphosphorylated mechanism required that an oxygen atom be retained by the enzyme or by CoA (see eqs. 2a, 2b, and 2c) because, according to this hypothesis, succinyl CoA reaction with the enzyme to form $\sim E$ CoA and release succinate precedes reaction with Pi. Oxygen-18 experiments designed specifically to explore this possibility were carried out (39). No evidence for the participation of succinyl CoA synthetase or CoA oxygens was found in the CoA-stimulated $E \sim P \rightleftarrows$ Pi exchange reaction, the $ATP \rightleftarrows Pi$ reaction, or the succinate $\rightleftarrows$ succinyl CoA exchange reaction. Furthermore, it was found that the apparent release of ^{32}P from $E \sim ^{32}P$ by CoA could be explained in terms of an $E \sim P \rightleftarrows Pi$ exchange with unlabeled Pi present as a contaminant and that, in fact, $E \sim P$ actually did not break down significantly during this process (9). In addition, Kaufman (25), Cha et al. (30), and Grinnell and Nishimura (28) all observed significant stimulation of the succinate $\rightleftarrows$ succinyl CoA exchange reaction by Pi. These data provide convincing evidence that the mechanism of the succinyl CoA synthetase reaction does not involve a high energy enzyme-CoA intermediate.

The CoA-stimulated $ATP \rightleftarrows Pi$ and $E \sim P \rightleftarrows Pi$ exchange reactions catalyzed by succinyl CoA synthetase still remain to be explained in mechanistic and functional terms. It is quite likely that both reactions represent the same phenomenon and that they are related to a weak CoA and desulfo-CoA-stimulated ATPase activity of the enzyme (40) in which $E \sim P$ reacts with water. It is conceivable that $E \sim P$ is capable of phosphorylating compounds other than substrates of the enzyme and that water is only a poor reactant.

C. EVIDENCE FOR INTERMEDIARY FORMATION OF ENZYME-BOUND SUCCINYL PHOSPHATE

One of the most obvious possibilities indicated by the aforementioned ^{18}O studies of Cohn and Hager was that of enzyme-bound succinyl phosphate. This concept was entertained by Kaufman but tentatively considered unlikely on the basis of experiments with spinach succinyl CoA synthetase (25). However, Nishimura and Meister (41) isolated succinyl phosphate by paper electrophoresis after incubation of relatively large amounts of *E. coli* succinyl CoA synthetase with succinate and ATP–γ–^{32}P. Incubation of enzyme with chemically synthesized succinyl phosphate, labeled with ^{14}C or with ^{32}P, under various conditions led to the following results: (*a*) succinyl phosphate reacted with ADP to form ATP; (*b*) succinyl phosphate reacted with CoA to yield succinyl CoA; and (*c*) in the absence of ADP and CoA, succinyl phosphate phosphorylated the enzyme to a substantial extent. It was shown later that incubation of E $\sim$ P with succinate yielded succinyl phosphate (42). It was also deduced from this experiment that in the absence of CoA, succinyl phosphate dissociated from the enzyme. The rate of succinyl phosphate formation under these conditions was slow, and the concentrations of succinate required to evoke maximal reaction were relatively high.

For obvious reasons, direct assessment of the importance of CoA in the reaction of succinate with E $\sim$ P would be difficult. However, it was reasoned that an analogue of CoA which was incapable of acting as a succinyl acceptor would be useful for this purpose. Desulfo CoA, which possesses a hydrogen atom in place of the —SH group present in CoA, appeared to be a convenient analogue. The preparation of this compound by reduction of CoA with Raney nickel had already been described by Chase, Middleton, and Tubbs (43).

The effects of desulfo CoA on succinyl phosphate formation from the reaction of ATP, succinate, and succinyl CoA synthetase were rather striking (40). For example, an 80-fold stimulation of the rate of succinyl phosphate formation in the presence of enzyme, ATP, and succinate could be seen under the appropriate conditions. A K_m value for desulfo CoA in this reaction of 6.2×10^{-5} M was determined. The rate of succinyl phosphate formation was proportional to succinate concentrations up to 20 mM, indicating a high K_m value for succinate. The reaction rate was also proportional to enzyme concentration. In these experiments the quantity of enzyme employed was significantly less than that used earlier (41).

A reaction sequence describing the essentials of a mechanism involving enzyme-bound succinyl phosphate is given in equations $3a$, $3b$, and $3c$.

$$\text{E} + \text{succinyl CoA} + \text{Pi} \rightleftharpoons \text{E} \cdot \text{succinyl} \sim \text{P} + \text{CoA} \tag{3a}$$

$$\text{E} \cdot \text{succinyl} \sim \text{P} \rightleftharpoons \text{E} \sim \text{P} + \text{succinate} \tag{3b}$$

$$\text{E} \sim \text{P} + \text{NDP} \rightleftharpoons \text{E} + \text{NTP} \tag{3c}$$

Strong support for the involvement of enzyme-bound succinyl phosphate as an intermediate has come from Spector's laboratory (44,45). Hildebrand and Spector (44) have been able to demonstrate a dramatic synthesis of ATP upon incubation of highly pure, chemically synthesized succinyl phosphate with ADP and Mg^{2+} in the presence of *E. coli* succinyl CoA synthetase. In the presence of desulfo CoA, ATP synthesis occurred at one-fifth the rate seen with succinyl CoA and Pi. The pronounced stimulatory effect of desulfo CoA in this reaction was illustrated by a 75% reduction in rate when this CoA analogue was omitted. The addition of desulfo CoA raised the apparent K_m value for succinyl phosphate from 6×10^{-5} M to 5×10^{-4} M, while raising V_{max} approximately 10-fold. Coenzyme A also stimulated ATP synthesis from succinyl phosphate. It is interesting that oxy CoA, in which the sulfhydryl function of CoA is replaced by a hydroxyl group (46), was without effect.

Thirteen other acyl phosphates were synthesized and tested in the enzymatic synthesis of ATP (44). During a 90-min incubation at 37°, acetyl, carbamyl, and propionyl phosphates reacted with ADP to yield conversions to ATP of 95, 90, and 74%, respectively, compared with 49% with succinyl phosphate. The latter conversion was comparatively low because of the relatively short half-life (39 min under conditions of the assay). The homologous dicarboxylic compounds, malonyl, glutaryl, and adipyl phosphates, gave conversions to ATP of 14, 22, and 9%, respectively. The surprising result was that of the seven 4-carbon chain analogues tested (including malyl, maleyl, and fumaryl phosphates), only itaconyl and α-methylsuccinyl phosphates were active. The conversions in these cases were 41 and 25%, respectively. Thus it appeared that the short chain monocarboxyl phosphates had relatively easy access to the phosphorylating site (presumably, a histidinyl residue), while selectivity was exercised by the enzyme, in so far as the other compounds were concerned. It is of considerable interest that the reactions of acetyl phosphate and carbamyl phosphate were not affected by the omission of desulfo CoA, while those of the others were.

Dephosphorylation of E ∼ P by organic acids in the presence of CoA has been studied by Robinson et al. (9), who reported that in addition to succinate (42), malate and, to a lesser extent, fumarate and α-ketoglutarate were effective. If it is assumed that succinate contaminants were not present in the malate and fumarate substrates employed, these results offer a striking contrast to those of Hildebrand and Spector described above. It is of further interest that acetate and malonate, the phosphates of which were active in ATP synthesis, were both inactive, within experimental error, in dephosphorylation.

With regard to enzymatic synthesis of succinyl CoA from succinyl phosphate and CoA, Hildebrand and Spector (44) observed that synthesis proceeded at the same rate, with or without the enzyme. Superficially, these results diverge from those reported by Nishimura and Meister (41), who observed an enzyme-dependent synthesis of succinyl CoA. It should be pointed out, however, that in the latter work relatively high concentrations of enzyme were employed and the assay for succinyl CoA was more sensitive than that used by Hildebrand and Spector.

Walsh et al. (45) have done elegant experiments describing unique features of succinyl phosphate in relation to other biologically important acyl phosphates, which may be of great importance in its reactivity as an enzyme-bound intermediate. In brief, they have proposed that reaction with the sulfhydryl group of CoA involves a cyclic form of succinyl phosphate, a γ-disubstituted butyrolactone. The proposal was based on experiments using a variety of analogues of succinyl phosphate, from which it was concluded that orientation of the free carboxyl group in juxtaposition (cis) to the acyl phosphate group favored reaction with the thiol. Hildebrand and Spector (44) have suggested that the reaction between succinyl phosphate and coenzyme A on the enzyme surface may occur nonenzymatically. However, the data could just as conveniently suggest that groups in the enzyme molecule may assist in expediting proper orientation of the free carboxyl group with respect to the acyl phosphate function. In any event, this cogent hypothesis, though not proved, adds to the attractiveness of a succinyl phosphate intermediate.

In succinyl phosphate formation from ATP and succinate, one possible role of desulfo CoA would be to facilitate binding of succinate and/or reaction of succinate and the phosphoryl enzyme. It is interesting to note that desulfo CoA had little effect on succinyl phosphate formation from succinate and phosphoryl enzyme (24) in contrast to its substantial effect on succinyl phosphate synthesis from ATP and succinate (40). In addition, it was discovered (24) that desulfo CoA had little effect when phosphorylation of succinate was carried out at low ATP concentrations (0.05–0.01 mM). However, at higher concentrations of ATP, omission of desulfo CoA from the reaction mixtures led to marked inhibition of succinyl phosphate synthesis, while addition of the analogue resulted in substantial stimulation. It would appear that desulfo CoA (and perhaps CoA) exerts its effect in this particular case by counteracting a negative effect of ATP (or ADP). This observation could be related to those phenomena manifested by "substrate synergism," which has been discussed by Bridger et al. (23). According to this concept, partial reactions catalyzed by a multisubstrate enzyme may be accelerated by the addition of a substrate (or substrates) not directly involved in the partial reaction. For example, this has been evidenced in the stimulation by ATP of succinate $\rightleftarrows$ succinyl CoA exchange (23) and acceleration of NTP $\rightleftarrows$ NDP exchange by CoA and succinate, added individually or together (10,23,28,30).

A kinetic test of the involvement of E $\sim$ P in the formation of succinyl phosphate was carried out in a dual-label experiment, employing ATP–γ–^{33}P and E $\sim$ ^{32}P (24). It was observed that, when these labeled materials were incubated with succinate, ^{32}P appeared at a faster rate than ^{32}P in the succinyl phosphate product. These results supported the contention that succinyl phosphate did not arise by direct phosphorylation of succinate by ATP but proceeded through reaction of succinate with the E $\sim$ P intermediate (42).

IV. Structural Studies of the Enzyme

Purification procedures describing homogeneous (or close to homogeneous) preparations have been described for both the *E. coli* enzyme (6,28,47,48) and the mammalian enzyme (14,49,50). These procedures

yield quantities of protein sufficient to perform a variety of physicochemical and structural studies. Thus, in more recent years the
molecular weight of the *E. coli* enzyme has been estimated by gel
filtration and sedimentation equilibrium analysis, and reported at
141,000 (6) and 146,000 (27). A complete amino acid analysis has also
been reported (27). The molecular weight of mammalian succinyl CoA
synthetase was estimated at 70,000 by gel filtration (14).

A possible approach to the determination of quaternary structure
and conformation of succinyl CoA synthetase is that of using antibody
to native enzyme and antibody to dissociated structures. Grinnell
et al. (51) prepared antibody to native succinyl CoA synthetase and,
on a qualitative basis, were able to discern a loss of enzymatic activity
associated with the decreased formation of the typical immunoprecipitin
line, when the enzyme was treated with the mercurial Merthiolate. The
appearance of more diffuse precipitin lines, corresponding to faster
diffusing antigen, indicated dissociation of the native structure into
subunits. Subsequently, it was found (52) that dissociation of the
enzyme with either Merthiolate or with pHMB gave products that
reacted in immunodiffusion with antiserum to carboxymethylated
succinyl CoA synthetase. Reaction of the enzyme with ^{14}C-pHMB
showed that inactivation (and presumably dissociation) of the enzyme
could be correlated with the titration of 4 sulfhydryl groups/mole
enzyme. In addition, gel filtration of such solutions on Sephadex G-75
gave two major peaks of radioactivity, each containing approximately
half of the bound radioactivity. One peak corresponded to substances
of molecular weight 44,000 and the other was much larger, as evidenced
by its appearance in the excluded volume of the column. When the
same experiment was carried out with ^{32}P–phosphoryl enzyme,
practically all of the ^{32}P was eluted in the 44,000 molecular weight
region. Thus these results indicated at least two types of subunits, in
harmony with the then unpublished data of Bridger. [See footnote 5
in Ramaley et al. (6)].

Leitzmann et al. (27) have carried out dissociation studies, using a
number of methods to effect dissociation by employing the techniques
of acrylamide gel electrophoresis, gel filtration, sucrose density centrifugation, and sedimentation equilibrium for detection of this process.
The results obtained by these investigators can be summarized as
follows: (*a* pHMB treatment of succinyl CoA synthetase gave rise to a
decreased sedimentation coefficient (from 6.2 to 2.3 s) of the protein; (*b*)

treatment of the enzyme with succinic anhydride in the presence of urea resulted in what appeared to be the equivalents of quarter-molecules (about molecular weight 35,000); (c) incubation of the enzyme with urea and iodoacetamide gave what appeared to be half-molecules (about 70,000 mol. wt.); and (d) treatment with iodoacetamide and succinic anhydride in urea gave rise to a population of molecules with a weight average molecular weight of 32,000 and a number average molecular weight of 25,000. Aggregation of the treated enzyme posed difficulties. However, these investigators have presented arguments to the effect that the composition of the aggregates indicates that the observations listed above describe major alterations in enzyme structure. As of the writing of this review, the most recent experiments pertaining to E. coli succinyl CoA synthetase have been published by Bridger (53). On the basis of molecular weight determinations by polyacrylamide gel electrophoresis in the presence of sodium dodecyl sulfate under reducing conditions, Bridger has reported that succinyl CoA synthetase contains two different kinds of subunits with the markedly different molecular weights of 38,500 and 29,500. He has also determined that the phosphorylation site resides in the smaller polypeptide chain. If it is assumed that there are two of each subunit, the molecular weight of a tetramer so constituted would be 136,000, which is very close to the molecular weights reported for the native enzyme.

An interesting aspect of the mammalian enzyme relates to the possible existence of isozymes (4,14). Recently, Baccanari and Cha (54) have observed several forms of the porcine heart enzyme by using the technique of electrofocusing. Meshkova and Matveeva (50) have also made mention of multiple forms of succinyl CoA synthetase from pigeon breast muscle. It is not clear at this time to what extent aggregation phenomena may be involved in these observations. If, as appears to be the case with the E. coli enzyme, there are subunits for phosphorylation and other subunits essential for still undefined function, it will be of great interest to learn how different forms of these subunits can be assembled as hybrid macromolecules.

V. Conclusion

The quest for definitive answers about the mechanism of action of succinyl CoA synthetase has spanned two decades. It now appears that some reasonable conclusions have been reached. The succinyl phosphate hypothesis was derived logically from oxygen exchange studies and has received support by analogy from studies of related enzyme reactions. For example, ^{18}O exchange between the γ-carboxyl of glutamate and Pi (55,56) had been demonstrated in the glutamine synthetase reaction and served as a basis for later work in which it was shown that enzymatic synthesis of glutamine proceeds through an enzyme-bound γ-glutamyl phosphate intermediate (57,58). Similarly, Strumeyer (59) showed ^{18}O exchange between the carboxyl group of γ-glutamylcysteine and Pi in the enzymatic synthesis of glutathione, prior to the demonstration that enzyme-bound γ-glutamylcysteinyl phosphate was almost certainly involved in formation of the tripeptide (60,61). The evidence documented in this review strongly supports this hypothesis.

It would appear that kinetic considerations constitute the chief basis upon which the case for a concerted mechanism lies (9). Although the effects of desulfo CoA, the analogue of the natural substrate CoA, are quite pronounced, the rates of reaction are less than what one might expect ideally. It should be kept in mind, however, that free succinyl phosphate as an intermediate is not the major point of concern here. It is not known how fast succinyl phosphate attaches to the enzyme or dissociates from it. Therefore, these rates could be quite slow in comparison to those involved for the actual synthesis of succinyl phosphate on the enzyme, its conversion to either ATP or succinyl CoA and the release of either of those products from the enzyme. It may be remarkable, indeed, that the observed rates are as high as seen. Stimulation by desulfo CoA of the reactions involving succinyl phosphate appears to be consistent with what would be expected with CoA but at reduced rates. In a mechanistic sense, how this analogue could affect the catalytic activity of the enzyme is not known, and there is no basis upon which one can assess the relative importance of structural features of this molecule relative to those of CoA. Thus the lack of knowledge and/or data prevents quantitative appraisal of the necessary factors to correct for the lack of the sulfhydryl group in desulfo CoA. Data in support of the concept of substrate synergism would also appear to favor the less than objective notion, with regard to the reaction of succinyl phosphate with succinyl CoA synthetase, that all substrates with their essential functional groups are required for the expression of maximum catalytic activity.

It is highly likely that solutions to unanswered questions will be found when more is known regarding the structure of succinyl CoA synthetase. The recent findings regarding nonidentity of subunits should quickly pave the way for an attack on the nature of the polypeptide chain containing the phosphorylated histidine site. Present-day methodology is of such sophistication that the actual amino acid sequencing of each polypeptide chain should not be too far off. This knowledge would facilitate the understanding of the relationship of the various binding and catalytic sites, as they are determined, to each other.

Acknowledgment

Work done by the authors, which was cited in this review, was generously supported by grants from the National Institutes of Health, United States Public Health Service.

References

1. Hager, L. P., in *The Enzymes*, Vol. 6, P. D. Boyer, H. Lardy, and K. Myrback, Eds., Academic Press, New York, 1962, p. 387.
2. Kaufman, S., and Alivisatos, S. G. A., *J. Biol. Chem.*, *216*, 141 (1955).
3. Lowenstein, J. M., in *Metabolic Pathways*, 3rd ed., Vol. 1, D. M. Greenberg Ed., Academic Press, New York, 1967, p. 146.
4. Labbe, R. F., Kurumada, T., and Onisawa, J., *Biochim. Biophys. Acta*, *111*, 403 (1965).
5. Kaufman, S., Gilvarg, C., Cori, O., and Ochoa, S., *J. Biol. Chem.*, *203*, 869 (1953).
6. Ramaley, R. F., Bridger, W. A., Moyer, R. W., and Boyer, P. D., *J. Biol. Chem.*, *242*, 4287 (1967).
7. Lipmann, F., and Tuttle, L. C., *J. Biol. Chem.*, *159*, 21 (1945).
8. Stadtman, E. R., in *Methods in Enzymology*, Vol. 3, S. P. Colowick and N. O. Kaplan, Eds., Academic Press, New York, 1957, p. 931.
9. Robinson, J. L., Benson, R. W., and Boyer, P. D., *Biochemistry*, *8*, 2503 (1969).
10. Upper, C. D., Doctoral Dissertation, University of Illinois, Urbana, 1964.
11. Burnham, B., *Acta Chem. Scand.*, *17*, S 123 (1963).
12. Sanadi, D. R., Gibson, D. M., Ayengar, P., and Jacob, M., *J. Biol. Chem.*, *218*, 505 (1956).
13. Cha, S., and Parks, R. E., Jr., *J. Biol. Chem.*, *239*, 1968 (1964).
14. Cha, S., Cha, C-J. M., and Parks, R. E., Jr., *J. Biol. Chem.*, *242*, 2577 (1967).
15. Moyer, R. H., and Smith, R. A., *Biochem. Biophys. Res. Commun.*, *22*, 603 (1966).
16. Moyer, R. W., Ramaley, R. F., Butler, L. G., and Boyer, P. D., *J. Biol. Chem.*, *242*, 4299 (1967).

17. Mazumder, R., Sanadi, D. R., and Rodwell, V. W., *J. Biol. Chem.*, *235*, 2546 (1960).
18. Palmer, J. M., and Wedding, R. T., *Biochim. Biophys. Acta*, *113*, 167 (1966).
19. Moffet, F. J., and Bridger, W. A., *J. Biol. Chem.*, *245*, 2758 (1970).
20. Grunau, J. A., Knight, E., Hart, E. S., and Gunsalus, I. C., *J. Biol. Chem.*, *242*, 3531 (1967).
21. Kreil, G., and Boyer, P. D., *Biochem. Biophys. Res. Commun.*, *16*, 551 (1964).
22. Hultquist, D. E., Moyer, R. W., and Boyer, P. D., *Biochemistry*, *5*, 322 (1966).
23. Bridger, W. A., Millen, W. A., and Boyer, P. D., *Biochemistry*, *7*, 3608 (1968).
24. Grinnell, F. and Nishimura, J. S., *Biochemistry*, *8*, 4126 (1969).
25. Kaufman, S., *J. Biol. Chem.*, *216*, 153 (1955).
26. Knight, E., Jr., Doctoral Dissertation, University of Illinois, Urbana, 1961.
27. Leitzmann, C., Wu, J-Y., and Boyer, P. D., *Biochemistry*, *9*, 2338 (1970).
28. Grinnell, F. L., and Nishimura, J. S., *Biochemistry*, *8*, 562 (1969).
29. Cha, S., Cha, C-J. M., and Parks, R. E., Jr., *J. Biol. Chem.*, *240*, PC3700 (1965).
30. Cha, S., Cha, C-J. M., and Parks, R. E., Jr., *J. Biol. Chem.*, *242*, 2582 (1967).
31. Kundig, W., Ghosh, S., and Roseman, S., *Proc. Nat. Acad. Sci. U.S.*, *52*, 1067 (1964).
32. Cottam, G. L., and Srere, P. A., *Biochem. Biophys. Res. Commun.*, *35*, 895 (1969).
33. Norman, A. W., Wedding, R. T., and Black, M. F., *Biochem. Biophys. Res. Commun.*, *20*, 703 (1965).
34. Walinder, O., *J. Biol. Chem.*, *243*, 3947 (1968).
35. Mourad, N., and Parks, R. E., Jr., *Biochem. Biophys. Res. Commun.*, *19*, 312 (1965).
36. Mourad, N., and Parks, R. E., Jr., *J. Biol. Chem.*, *241*, 271 (1966).
37. Cohn, M., in *Phosphorus Metabolism*, Vol. 1, W. D. McElroy and B. Glass, Eds., Johns Hopkins, University Press, Baltimore, 1951, p. 374.
38. Hager, L. P., *J. Amer. Chem. Soc.*, *79*, 4864 (1957).
39. Benson, R. W., Robinson, J. L., and Boyer, P. D., *Biochemistry*, *8*, 2496 (1969).
40. Grinnell, F. L., and Nishimura, J. S., *Biochemistry*, *8*, 568 (1969).
41. Nishimura, J. S., and Meister, A., *Biochemistry*, *4*, 1457 (1965).
42. Nishimura, J. S., *Biochemistry*, *6*, 1094 (1967).
43. Chase, J. F. A., Middleton, B., and Tubbs, P. K., *Biochem. Biophys. Res. Commun.*, *23*, 208 (1966).
44. Hildebrand, J. G., and Spector, L. B., *J. Biol. Chem.*, *244*, 2606 (1969).
45. Walsh, C. T., Jr., Hildebrand, J. G., and Spector, L. B., *J. Biol. Chem.*, *245*, 5699 (1970).
46. Miller, T. L., Rowley, G. L., and Stewart, C. J., *J. Amer. Chem. Soc.*, *88*, 2299 (1966).
47. Gibson, J., Upper, C. D., and Gunsalus, I. C., *J. Biol. Chem.*, *242*, 2474 (1967).
48. Bridger, W. A., Ramaley, R. F., and Boyer, P. D., in *Methods in Enzymology*, Vol. 12, J. M. Lowenstein, Ed., Academic Press, New York, 1969, p. 70.
49. Cha, S., in *Methods in Enzymology*, Vol. 13, J. M. Lowenstein, Ed., Academic Press, New York, 1969, p. 62.

50. Meshkova, N. P., and Matveeva, L. N., *Biokhimiya*, *35*, 374 (1970).
51. Grinnell, F. L., Stollar, B. D., and Nishimura, J. S., *Biochim. Biophys. Acta* *185*, 471 (1969).
52. Grinnell, F., and Nishimura, J. S., *Biochim. Biophys. Acta*, *212*, 150 (1970).
53. Bridger, W. A., *Biochem. Biophys. Res. Commun.*, *42*, 948 (1971).
54. Baccanari, D. P., and Cha, S., *Fed. Proc.*, *30*, 1172 Abstr. (1971).
55. Boyer, P. D., Koeppe, O. J., and Luchsinger, W. W., *J. Amer. Chem. Soc.*, *78*, 356 (1956).
56. Kowalsky, A., Wyttenbach, C., Langer, L., and Koshland, D. E., Jr., *J. Biol. Chem.*, *219*, 719 (1956).
57. Krishnaswamy, P. R., Pamiljans, V., and Meister, A., *J. Biol. Chem.*, *237*, 2932 (1962).
58. Khedouri, E., Wellner, V. P., and Meister, A., *Biochemistry*, *3*, 824 (1964).
59. Strumeyer, D. H., Doctoral Dissertation, Harvard University, Cambridge, Mass., 1959.
60. Nishimura, J. S., Dodd, E. A., and Meister, A., *J. Biol. Chem.*, *238*, PC1179 (1963).
61. Nishimura, J. S., Dodd, E. A., and Meister, A., *J. Biol. Chem.*, *239*, 2553 (1964).

BIOSYNTHESIS AND METABOLISM OF 1,4-DIAMINOBUTANE, SPERMIDINE, SPERMINE, AND RELATED AMINES

By HERBERT TABOR and CELIA WHITE TABOR, *Bethesda, Maryland*

CONTENTS

I. Introduction

The principal diamines and polyamines in biological materials are:

$NH_2(CH_2)_3NH_2$ 1,3-diaminopropane

$NH_2(CH_2)_4NH_2$ 1,4-diaminobutane (putrescine)

$NH_2(CH_2)_5NH_2$ 1,5-diaminopentane (cadaverine)

$NH_2(CH_2)_3NH(CH_2)_4NH_2$ spermidine

$NH_2(CH_2)_3NH(CH_2)_4NH(CH_2)_3NH_2$ spermine

In this review we will summarize the recent developments in the synthesis and metabolism of these amines *in vitro* and *in vivo*. We have previously reviewed the earlier work in this field (421,428), and the reader is referred to these publications for discussion and references on the earlier material.*

* Certain aspects of the polyamine field have been reviewed recently by Bachrach ["Metabolism and Function of Spermine and Related Amines" in *Ann. Rev. Microbiol*, 1970 (ref. 18)], Cohen ["Introduction to the Polyamines" (ref. 76c)], Stevens ["The Biochemical Role of Naturally Occurring Polyamines in Nucleic Acid Synthesis" in *Biol. Rev.*, 1970 (ref. 392)], and Smith ["The Occurrence, Metabolism, and Functions of Amines in Plants" in *Biol. Rev.*, 1970 (ref. 380)] and ["The Physiology of the Polyamines and Related Compounds" in *Endeavour*, 1972 (ref. 380)]. Two symposia in this area have also been published recently, one in *Fed. Proc.*, *29*, 1560–1588 (1970), and the other in *Ann. N.Y. Acad. Sci.*, *171*, 691–1009 (1970). For the older literature, in addition to the bibliography cited in refs. 421 and 428, the reader is directed to M. Guggenheim, "Die biogenen Amine," 4th ed., S. Karger, Basel, Switzerland, 1951.

For convenience in this review, the term "polyamines," is often used to include 1,4-diaminobutane, as well as spermidine and spermine.

II. Biosynthesis and Metabolism of 1,4-Diaminobutane, Spermidine, and Spermine in Microorganisms

A. CONCENTRATION AND DISTRIBUTION OF DIAMINES AND POLYAMINES

Diamines and polyamines have been reported in a variety of microorganisms (18,33,66,77,97,147,168,215,262,308,316,331,420,421,423,443, 452,453). The concentration of the amines* and the relative ratios of the different amines vary markedly in different organisms. These concentrations vary with such factors as pH, conditions of growth, and composition of the medium.

1. Bacteria and Fungi

The highest concentrations of polyamines are found in gram-negative bacteria (168,420). *Escherichia coli*, for example, when grown in a minimal glucose–salts medium at 37°, contains 0.06 μM of 1,4-diaminobutane and 0.017 μM of spermidine per mg dry weight (2.5 $\times$ 10^9 cells) but no spermine; the calculated intracellular content in the cell water is therefore 0.019 M for 1,4-diaminobutane and 0.006 M for spermidine (423,424). Such cells do not contain either acetylated amines (405,406) or glutathionylspermidine (406,411,412).† Much higher concentrations

---

* Until very recently, there were no convenient sensitive and specific analytical procedures for the assay of all of the amines present in a bacterial or tissue extract. The older techniques used in our laboratory depended on ion-exchange chromatography (420) and were relatively insensitive and time-consuming. More recently, several investigators have developed more sensitive techniques, including condensation with o-phthalaldehyde (105,224,257–259) or with dansyl chloride (79a,80b,93,353a,b,c,354–356) to give fluorescent derivatives, elution after paper or thin layer electrophoresis or chromatography (39,51,151,180,272,307,321,377), gas chromatography (39,204,377), or measurement of the products of enzymatic degradation by spermidine dehydrogenase (29,33) or serum amine oxidase (33,438). Although these methods offer considerable promise, most do not assay all of the amines or are not entirely specific.

Recently, automated ion-exchange methods have been described (51,160a,266, 270,440); these offer the most promising approach to this analytical problem at this time.

† Conditions for the occurrence of acetyl-1,4-diaminobutane, acetylspermidine and glutathionylspermidine are presented in Section II.D.

of polyamines have been reported for *E. coli* grown on an enriched medium (168); presumably, these elevated levels result from the amines that are taken up from the medium, or from the decarboxylation of any accumulated ornithine or arginine. Similarly, cultures that accumulate ornithine because of a genetic block (as *E. coli* 15 TAU⁻; see refs. 77,79b,308,315), or arginine because of inhibition of protein synthesis (452) may increase their amine levels by decarboxylation of these amino acids.

In contrast, most gram-positive bacteria have lower concentrations of total polyamines than gram-negative organisms (147,168). *Bacillus subtilis*, for example, has a much lower polyamine concentration than *E. coli* (168), although the spermidine concentrations are comparable (22). Guirard and Snell (147) showed that a series of *Lactobacillus* strains contain lower concentrations of polyamines than *E. coli*, even when grown on a relatively complex medium. Only traces of either amine are present in *Staphylococcus aureus* (168,331). The reason for this difference between gram-positive and gram-negative organisms is not known.

Although few strains have been studied exhaustively for other amines, 1,3-diaminopropane has been described in some strains (421, 443), and (+)hydroxyputrescine (1,4-diaminobutan-2-ol) has been found in a *Pseudomonas* species (227,330,436). In addition to other amines, spermine has been found in *Bacillus stearothermophilus* and various strains of *Lactobacillus* (147,394). Fungi, such as *Aspergillus nidulans*, *Saccharomyces cerevisiae*, and wheat rust, contain substantial quantities of spermidine and spermine (66,215,420,421). 1,5-Diaminopentane and an analog of spermidine (*N*-3-aminopropyl-1,5-diaminopentane) have been found in an *E. coli* mutant that is deficient in agmatine ureohydrolase (see Section II.B.4) (91b).

The localization of the polyamines within the bacterial cell is not known. They are found in the trichloroacetic acid-soluble fractions, and thus do not appear to be covalently linked to any macromolecular structures. The distribution of the amines in various isolated cell components does not necessarily reflect the distribution of polyamines *in vivo*. After a cell is broken the polyamines, being basic substances, are likely to redistribute themselves by forming salt linkages to the various acid macromolecules. It has been shown that further redistribution can occur during isolation procedures such as sucrose gradient centrifugation, particularly if Mg^{2+} is present (407). Thus, even though isolated ribosomes or cell membranes may contain high concentrations of amines (22,70,71,78,84,184,216,284,306,330,391,393, 407), one cannot conclude that the amines were actually associated with these structures *in vivo*. Indirect evidence for some type of localization of 1,4-diaminobutane in the bacterial cell has been presented by Morris and Koffron (269), who, on the basis of kinetic studies, concluded that there are two separate intracellular pools of 1,4-diaminobutane in *E. coli*.

2. Viruses

The high concentration of polyamines in the T-even bacteriophages of *E. coli* has been of particular interest (12,13). These polyamines account for about 40% of the phage cations and are derived from the pool of polyamines of the infected bacteria. These observations represent the only unequivocal demonstration of the physical associ- ation of polyamines and nucleic acids *in vivo*. As opposed to the limitations discussed above for the relationship between polyamines and ribosomes, the association of polyamines and nucleic acids in T-even bacteriophages is not complicated by secondary uptake during isolation, since the T-even bacteriophages do not take up polyamines from the medium.

Spermidine and a small amount of spermine have been isolated from purified turnip yellow mosaic virus; the amounts that are present are enough to neutralize approximately 20% of the viral ribonucleic acid (39). This study did not confirm the previous report (204) that the triamine, bis(3-aminopropyl)amine, was present in this virus.

B. BIOSYNTHESIS OF DIAMINES AND POLYAMINES IN *E. COLI*

Most of the studies on the biosynthesis of 1,4-diaminobutane and spermidine have been carried out in *E. coli*. 1,4-Diaminobutane is formed either by the decarboxylation of L-ornithine or by the decarboxy- lation of L-arginine, followed by the hydrolysis of the resultant agmatine (Fig. 1). As discussed below, separate constitutive ("biosynthetic") and induced enzymes have been described for the decarboxylation of ornithine and arginine.

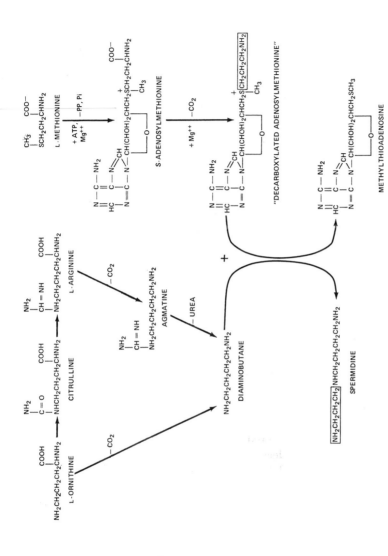

Fig. 1. Biosynthesis of 1,4-Diaminobutane and Spermidine in *E. coli.*

1. Enzymatic Formation of 1,4-Diaminobutane, 1,5-Diaminopentane, and Agmatine

The early studies on the formation of these amines in cultures of *E. coli* and other bacteria were summarized by Gale (133) in a review in *Advances in Enzymology* in 1946. The decarboxylases for ornithine, lysine, and arginine studied in these experiments were induced enzymes, since they were not formed in *E. coli* growing on a glucose–salts medium unless the specific amino acid was present and the growth medium was slightly acid (pH ≃ 5); optimum formation of arginine decarboxylase required, in addition, a crude medium containing casein digest. Gale (132,133) was also able to demonstrate specific arginine, ornithine, and lysine decarboxylases in cell-free extracts of *E. coli* and to show that the reaction required pyridoxal phosphate. The pH optima for the activity of these decarboxylases are pH 4.5 to 6.0. Each amino acid decarboxylase is specific for a single L-amino acid, permitting the use of these enzymes as analytical tools for the quantitative assay of these amino acids.

Following these early studies, it was generally assumed that these decarboxylases were responsible for the synthesis of the diamines in bacteria, although the acidic pH optimum for both the synthesis of these enzymes and for their activity made the physiological significance of the enzymes questionable. It was also hard to understand why, if no decarboxylases are formed when the organism is grown at neutral pH in the absence of the inducing amino acid, the bacteria still contain 1,4-diaminobutane. This problem was resolved by Morris and Pardee (271,272) who showed that these bacteria do indeed contain ornithine and arginine decarboxylases, but that these are completely different enzymes with different characteristics and pH optima (about pH 8.0) from the induced enzymes discussed above (pH optima ≃ 5.0). These constitutive enzymes appear to be the decarboxylases used for the biosynthesis of the amines and shall be referred to as "biosynthetic" enzymes in the discussion below. The occurrence of a biosynthetic arginine decarboxylase is supported by the recent reports of mutants lacking this activity (see Section II.B.4). Preliminary mapping experiments (240) show that the genetic locus for the biosynthetic arginine decarboxylase is at a different location from the gene for the induced arginine decarboxylase.

The discovery of these biosynthetic decarboxylases, plus the more recent studies on mutants lacking the biosynthetic arginine decarboxylase, brings up the question of the function of the induced decarboxylases. There have been no recent developments that clarify this question, and we are still left only with the older viewpoint that the induced decarboxylases are synthesized in acid cultures in order to produce amines to help buffer the bacteria against the noxious effects of the acid medium (133).

a. **Ornithine Decarboxylases.** When *E. coli* is grown in a minimal salts-glucose medium at neutral pH, only the *biosynthetic ornithine decarboxylase* is formed (271,272). The specific activity, measured in the crude extract, is 0.035 μmoles/min/mg protein; the pH optimum is 7.5.

Under the best conditions for induction (i.e., acid pH, poor aeration, added ornithine), the specific activity of the biosynthetic ornithine decarboxylase is decreased to 0.009 μmoles/min/mg protein, but the extract also contains the inducible ornithine decarboxylase [specific activity: 0.019 μmoles/min/mg protein]. The pH optimum of the induced enzyme is 5.3; the induced enzyme is much more labile to heating than the biosynthetic enzyme. Both enzymes require pyridoxal phosphate.

Morris et al. (273) have reported a 30-fold purification of the biosynthetic ornithine decarboxylase. This preparation has an absolute requirement for pyridoxal phosphate, an optimal activity at pH 8.0, and is inhibited by 1,4-diaminobutane and by spermidine.

b. **Arginine Decarboxylases.** When *E. coli* is grown at neutral pH, only the *biosynthetic arginine decarboxylase* is formed (272,454). The specific activity in crude extracts is 0.012 μmoles/min/mg protein, (or about 30% of the value for the biosynthetic ornithine decarboxylase). Preliminary results have been reported on a 200-fold purification of the biosynthetic arginine decarboxylase (454). The purified enzyme has an absolute requirement for pyridoxal phosphate and magnesium ion, and shows optimal activity at pH 8.4. The enzyme is inhibited by 1,4-diaminobutane and spermidine; this inhibition is consistent with the feedback inhibition of arginine decarboxylation found when 1,4-diaminobutane or spermidine was added to whole cells (see p. 216).

The *induced arginine decarboxylase* is present in high concentrations when *E. coli* is grown under inducing conditions (43,133,252,363); the specific activity of the induced enzyme in the crude extracts is 50- to 1,000-fold greater than that listed above for the biosynthetic enzyme. The induced arginine decarboxylase has been purified to homogeneity from induced *E. coli*, and the characteristics of the crystalline enzyme have been extensively studied by Snell and his group (43,45–48). The homogeneous enzyme is 30-fold purified over the crude extract and has a specific activity of 410 μM CO_2 released per minute per milligram of protein. L-Arginine is the best substrate, but canavanine is also decarboxylated at 40% of the rate for arginine. L-Ornithine is a very poor substrate (about 1% of the rate with L-arginine) with a high K_m. The pH optimum is 5.2.

The enzyme has a molecular weight of 820,000. It is easily dissociated into five subunits of equal size (mol. wt. = 165,000) when stored at pH > 6.5 in dilute salts in the absence of divalent cations. The subunits can be fully reassociated with a recovery of about 90% of the original activity of the enzyme by dialysis against a buffer containing divalent cations or a high concentration of monovalent cations at a pH less than 6.0. Electron micrographs indicate that the undissociated arginine decarboxylase has a regular pentagonal structure, apparently containing a planar arrangement of five subunits.

After reduction, carboxymethylation, and treatment with guanidine, it is possible to show that each of the 165,000 molecular weight subunits can be further dissociated into two subunits of 82,000 molecular weight. These results have been confirmed by acrylamide gel electrophoresis, sulfhydryl analyses, end group determination with carboxypeptidase or hydrazinolysis, and peptide maps after tryptic digestion. Thus each of the five major subunits seen in the electron micrographs must itself be a dimer.

The isolated arginine decarboxylase has an absorption maximum at 420 nm. The holoenzyme can be resolved by dialysis against cysteine and reconstituted by the addition of excess pyridoxal phosphate. The pyridoxal phosphate was shown to be bound to the ϵ-amino group of a lysine residue. The enzyme binds 10 moles of pyridoxal phosphate per mole of enzyme; that is, 1 pyridoxal phosphate for each 85,000 molecular weight subunit. Recently (46) the sequence around the site binding pyridoxal phosphate has been determined after fixation of the cofactor by $NaBH_4$ reduction. This sequence has been shown to be highly homologous to the peptides containing pyridoxal phosphate in three other *E. coli* decarboxylases:

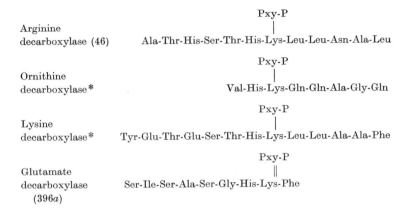

Arginine
decarboxylase (46)

Pxy-P
|
Ala-Thr-His-Ser-Thr-His-Lys-Leu-Leu-Asn-Ala-Leu

Ornithine
decarboxylase*

Pxy-P
|
Val-His-Lys-Gln-Gln-Ala-Gly-Gln

Lysine
decarboxylase*

Pxy-P
|
Tyr-Glu-Thr-Glu-Ser-Thr-His-Lys-Leu-Leu-Ala-Ala-Phe

Glutamate
decarboxylase
(396a)

Pxy-P
‖
Ser-Ile-Ser-Ala-Ser-Gly-His-Lys-Phe

c. Agmatine Ureohydrolase.   Agmatine ureohydrolase is present in crude extracts of *E. coli*. The activity of this enzyme is easily measured by assay of the urea produced (268,269) because *E. coli* contains no other enzymes that form urea, nor any urease. No purification of agmatine ureohydrolase has been reported yet.

2. Enzymatic Formation of Spermidine

The steps involved in the biosynthesis of spermidine are shown in Figure 1. The first evidence for this pathway was obtained in 1958 (420), when it was shown that [2-^{14}C]ornithine and [1-^{14}C]-1,4-diaminobutane were incorporated into spermidine by intact *E. coli*. Other experiments showed that [^{14}C-^{15}N]-1,4-diaminobutane was incorporated directly into spermidine, since the spermidine that was synthesized contained the same ratio of ^{14}C:^{15}N as that in the added 1,4-diaminobutane.

The propylamine moiety of spermidine is derived from methionine (143); the intermediates in this conversion are adenosylmethionine and decarboxylated adenosylmethionine (419,420).

---

* D. Sabo, unpublished data (personal communication from Dr. E. H. Fischer).

a. Synthesis and Decarboxylation of S-Adenosylmethionine

L-methionine $+$ ATP $\rightarrow$ S-adenosyl-L-methionine $+$ PPi $+$ Pi

S-adenosyl-L-methionine $\rightarrow$
decarboxylated S-adenosylmethionine $+$ CO_2

Enzymes for the synthesis of S-adenosylmethionine have been partially purified from liver, yeast, and *E. coli* (63,64,276,277,361,425) and can be considered as catalyzing the first step in the biosynthesis of spermidine.

The second step in this pathway is the decarboxylation of S-adenosylmethionine (403,420). The enzyme catalyzing this reaction has been purified to homogeneity from *E. coli* (448,449). The purified enzyme has a molecular weight of 113,000 and contains approximately eight identical subunits of 15,000 molecular weight. The enzyme requires Mg^{2+} for activity (420) and is specific for S-(-)-adenosyl-L-methionine (469).

The most striking characteristic of this enzyme is the presence of pyruvate, rather than pyridoxal phosphate, as a prosthetic group (448). One mole of pyruvate is present per mole of enzyme; this was shown by measurement of the spectrum obtained upon treatment of the enzyme with phenylhydrazine and by isolation of radioactive lactate after reduction of the enzyme with NaB^3H_4 and subsequent acid hydrolysis. The importance of the pyruvate for the activity of the decarboxylase was shown by the complete loss of activity upon exposure to such carbonyl reagents as phenylhydrazine, cyanide, and borotritide. Thus adenosylmethionine decarboxylase is one of a new group of enzymes with unusual carbonyl-containing cofactors; these include proline reductase (178,179) and histidine decarboxylase (320), which have pyruvate; urocanase, which has α-ketobutyrate (134,135); and phenylalanine ammonia-lyase (153,154) and histidine ammonia-lyase (138,447), which have the closely related cofactor dehydroalanine.

The S-adenosylmethionine decarboxylase of *E. coli* differs considerably from the mammalian enzyme (see Section III.B.3). The enzyme from *E. coli* is not significantly stimulated by 1,4-diaminobutane, in contrast to the marked dependence of the mammalian and yeast enzymes on the addition of this amine (116,201,299). The bacterial enzyme requires Mg^{2+}, while no Mg^{2+} requirement has been found with the mammalian and yeast enzymes (201,299). A pyridoxal phosphate inhibitor, NSD-1055, inhibits the mammalian enzymes but not the *E. coli* enzyme (116,299).

b. Propylamine Transferase. Propylamine transferase carries out the last step in the biosynthesis of spermidine, that is, the transfer of the propylamine group to 1,4-diaminobutane (49,404,420). In this reaction the high energy sulfonium ion of decarboxylated adenosylmethionine is converted to the thioether of methylthioadenosine. The propylamine transfer reaction is analogous to the well-known methylation reactions, in that in each case there is an alkyl group transfer from a sulfonium compound to a receptor.

This enzyme was purified 2000-fold from extracts of *E. coli* W (49). The enzyme has a molecular weight of 73,000 and is composed of two identical subunits of 36,000 molecular weight. The purified enzyme has no demonstrable requirements or cofactors; it is inhibited by sulfhydryl binding reagents and by the products of the reaction.

The optimal pH for enzymatic activity is about 10.3, although considerable activity (about 60%) is still present at pH 8.0. The activity is greatest with 1,4-diaminobutane as the substrate, but significant activity can also be shown at pH 10.3 with either 1,5-diaminopentane or with spermidine as the substrate; with the latter substrate spermine is formed.

The other product (49) in the biosynthesis of spermidine is methylthioadenosine, a compound normally found in *E. coli* (72). A nucleosidase that hydrolyzes methylthioadenosine to adenine and methylthioribose has been purified from *E. coli* by Duerre (100). The deamination of methylthioadenosine to form methylthioinosine has not been described in *E. coli*, but this conversion has been observed with the "nonspecific" adenosine deaminase of *Aspergillus oryzae* (350a,350b). The resultant spectral shift can be used as a convenient quantitative assay for methylthioadenosine.

3. Control Mechanisms Involved in the Biosynthesis of 1,4-Diaminobutane and Spermidine in E. coli

The pathway for the synthesis of ornithine, arginine, and the amines in *E. coli* has several sites at which control mechanisms may operate. Particularly noteworthy are the two branch points for the biosynthesis of 1,4-diaminobutane; that is 1,4-diaminobutane may be formed either by decarboxylation of ornithine or by decarboxylation of arginine to agmatine, which is then hydrolyzed to 1,4-diaminobutane and urea (Fig. 1).

Until recently, there have been rather few studies on the control of the specific enzymes involved in the biosynthesis of the amines in *E. coli* or on any effects of 1,4-diaminobutane or spermidine on the overall biosynthetic pathway for arginine. This is a little surprising, since the biosynthesis of the amines represents a significant part of the overall pathway; in fact, in *E. coli* (423) grown on a mineral salts–glucose medium, approximately 24% of the total arginine biosynthetic pathway is used for the biosynthesis of polyamines (0.077 μM of 1,4-diaminobutane plus spermidine per milligram of dry weight, compared with a total arginine content of 0.25 μM/mg).

a. Ornithine and Arginine Decarboxylases. As mentioned above, the occurrence of two branch points (at ornithine and arginine) is a very striking aspect of the biosynthetic pathway for 1,4-diamino-butane. Morris and Koffron (269) and Morris et al. (273) have reported that, in *E. coli*, about 75–98% of the 1,4-diaminobutane is synthesized by ornithine decarboxylation rather than by arginine decarboxylation, but this percentage varies with the strain and with conditions of growth. The reason for the double pathway to 1,4-diaminobutane is not at all clear, although one might speculate that decarboxylation at the ornithine level avoids the wasteful requirement for ATP in the ornithine → citrulline → arginine → agmatine → 1,4-diaminobutane path; contrariwise, if exogenous arginine is available, it is more economical to decarboxylate this exogenous arginine than to make ornithine for this purpose. However, it is unclear, if this were the case, why any constitutive arginine decarboxylase should be present in *E. coli*. One would expect, rather, that no such activity would be present in *E. coli* in the absence of arginine, and that all of the arginine decarboxylase activity would be inducible.

The biosynthetic ornithine and arginine decarboxylases appear to be controlled by both feedback inhibition and by repression (273,423). The addition of 1,4-diaminobutane or spermidine to a culture of an ornithine or arginine auxotroph of *E. coli* in a chemostat results in a marked decrease in the levels of ornithine and arginine decarboxylation, and consequently of amine synthesis. This effect was shown to be the

result of a decrease in the level of ornithine and arginine decarboxylases, as well as inhibition of preexisting enzyme. Indirect evidence
for comparable repression of ornithine decarboxylase has also been
presented by Morris and his associates (267,273) in their studies on
mutants partially deficient in arginine decarboxylase and agmatine
ureohydrolase.

An additional factor that has to be considered in interpreting studies
on arginine decarboxylation is the partial separation of arginine taken
up by the bacteria from the medium from that formed endogenously
from citrulline (424). In chemostat experiments with *E. coli* orn⁻,
exogenous arginine was more readily decarboxylated to form amines
than endogenous arginine, indicating that the exogenously derived
material had been exposed to arginine decarboxylase before mixing
completely with the total intracellular arginine pool. Previous to this
work, Sercarz and Gorini (357) had presented evidence for two pools
of arginine to explain a differential effect of exogenous and endogenous
arginine on repression of the arginine biosynthetic enzymes, and Davis
(86) had reported that exogenous and endogenous ornithine are
handled differently by *N. crassa*. Morris and Koffron (269) have also
considered such compartmentalization as a partial explanation for the
shift from the ornithine decarboxylase pathway to the arginine decarboxylase pathway upon addition of arginine to the culture medium.

b. Spermidine Biosynthesis. Although relatively little work has
been done on the specific control of spermidine biosynthesis, the
available data indicate that *E. coli* maintains its intracellular level of
spermidine more rigidly than its level of 1,4-diaminobutane. We have
found (423) that cells grown under a moderate limitation of arginine
in a chemostat showed a decrease of 30–40% in the concentration of
1,4-diaminobutane but essentially no change in the concentration
of spermidine. With a very severe limitation of arginine, the level of
1,4-diaminobutane decreased by 97%, but the spermidine level was
still 65% of the normal level (i.e., the amount found in cells grown
with excess arginine). It is noteworthy that, even with severe limitation, 4.5% of the added arginine is used for spermidine biosynthesis.
Thus, if one wishes to express these results teleologically, it appears
as if the spermidine is sufficiently important to *E. coli* so that it will
use some of the very scarce arginine for spermidine biosynthesis
instead of for protein synthesis. Spermidine was also maintained,

despite a $>99\%$ decrease in the level of 1,4-diaminobutane, in a mutant-lacking arginine decarboxylase (267).

In view of the large amount of polyamines in the T-even bacteriophages (see Section II.A.2), one would have expected considerable interest in the influence of phage infection on the control of amine biosynthesis; however, little has been published on this subject. Amine production continues after infection, paralleling the increase in DNA production (76a), but there is no evidence that the phage infection stimulates an increased amine production over that found in uninfected cells. In fact, Dion and Cohen (91a) showed that the same amount of polyamine synthesis occurred, even when the infecting bacteriophage was defective in the synthesis of DNA. Bachrach and Ben-Joseph (21), on the other hand, have reported a 50% increase in ornithine decarboxylase after phage infection.

4. Mutants for Arginine Decarboxylase and Agmatine Ureohydrolase

The growth of a number of organisms has been shown to require, or to be stimulated by, polyamines. In some cases, the requirement has been shown to be related to the stabilization of osmotically fragile organisms by the polyamines (243,244,421). The latter effect was studied most extensively by Mager with *Neisseria perflava*, *Pasteurella tularensis*, and such halophilic organisms as *Achromobacter fisheri* (243, 244). On the other hand, definitive growth requirements for 1,4-diaminobutane have been shown for *H. parainfluenzae* (166,167) and for *A. nidulans* (383), although the requirement of *H. parainfluenzae* could also be satisfied by several other amines (421). Until recently, no mutants requiring 1,4-diaminobutane had been found for *E. coli*, presumably because the alternative pathways for 1,4-diaminobutane synthesis in *E. coli* made the selection of a 1,4-diaminobutane—requiring mutant difficult.

Recently, this problem has been resolved in two laboratories (177, 240,267,273) by taking advantage of the effect of arginine in repressing ornithine biosynthesis. When arginine is added to the growth medium, ornithine is not synthesized, and therefore all of the 1,4-diaminobutane has to be formed via arginine decarboxylase and agmatine ureohydrolase. By growing *E. coli* after mutagenesis in such an arginine-containing medium, Morris and Jorstad (267) were able to isolate a mutant that was markedly deficient in arginine decarboxylase and one that was markedly deficient in agmatine ureohydrolase. When grown on arginine, these mutants are partially deficient in the synthesis of 1,4-diaminobutane because (a) the pathway from arginine is blocked by the mutation and (b) the pathway from ornithine is inoperative because ornithine biosynthesis is repressed by the arginine in the medium. These mutants contained much less 1,4-diaminobutane than wild-type cells (<1–20%), but their spermidine content was affected much less (50–140% of the normal value). However, even in the mutant with a 99% decrease in the 1,4-diaminobutane level, the growth rate was decreased by only 10%.

Similar mutants for arginine decarboxylase and agmatine ureohydrolase but with a more complete block have been isolated by Maas and his co-workers (177,240). They found that these mutations were detected more readily in a particular canavanine resistant mutant; it is completely unclear why this should be the case. The arginine decarboxylase and agmatine ureohydrolase mutants map close to, but definitely separate from, the canavanine resistance locus and from the gene for arginine permease; all these genes map near the Ser A gene. In these studies Maas et al. eliminated synthesis via the ornithine decarboxylase pathway, either by completely preventing the biosynthesis of ornithine by a genetic block before ornithine or by repressing the synthesis of ornithine in a prototroph by the addition of exogenous arginine. Dion and Cohen (91b) have also studied the mutant lacking agmatine ureohydrolase and have demonstrated a polyamine depletion and an inhibition of growth. These depleted bacteria did not support bacteriophage growth. Addition of 1,4-diaminobutane stimulated the growth rate and permitted bacteriophage production.

When grown on arginine, the mutant deficient in agmatine ureohydrolase grew much more slowly than normal, as measured by optical density. A good growth response was obtained by the addition of 1,4-diaminobutane or spermidine but not of 1,5-diaminopentane. Spermine, iminobispropylamine, and 1,3-diaminopropane were also effective, even though these amines are not normally found in E. coli.

Electron microscopic studies showed that the mutants which could not make amines developed into long filamentous forms, which did not form septa or divide until after 1,4-diaminobutane was added (240). On the basis of these results, Maas et al. (240) have postulated that the amines may have a role in cell division. Further evaluation of these most interesting findings will have to await a more extensive study.

C. BIOSYNTHESIS OF 1,4-DIAMINOBUTANE AND SPERMIDINE IN MICROORGANISMS OTHER THAN *E. COLI*

1. Biosynthesis of 1,4-Diaminobutane

The presence of ornithine and arginine decarboxylases in bacteria other than *E. coli* was studied by Gale (132,133), but no purification of these enzymes has been reported. The preparation of a crystalline L-lysine decarboxylase from *Bacterium cadaveris* has been described by Soda and Moriguchi (387).

In *Neurospora crassa*, all of the 1,4-diaminobutane is synthesized by ornithine decarboxylase (87); that is, none is formed by decarboxylation of arginine and hydrolysis of agmatine. Davis et al. (87) took advantage of this fact to isolate strains having a partial requirement for 1,4-diaminobutane. In a strain defective in ornithine biosynthesis (or in which ornithine biosynthesis is repressed by added arginine), the ornithine for 1,4-diaminobutane biosynthesis has to come from arginine by the action of arginase (which is normally present in *Neurospora*). If the strain also lacks arginase, no 1,4-diaminobutane is made, since the arginase-less strain cannot convert arginine to ornithine. Such a mutant was obtained by Davis. This mutant grows very slowly, unless 1,4-diaminobutane is added. The hyphae from the starved cultures are short, contorted, and swollen to about two times the normal diameter.

In *N. crassa* no mutants have been found which lack the enzymes responsible for the synthesis of spermidine from 1,4-diaminobutane.

2. Biosynthesis of Spermidine

Isotopic evidence has been reported (147,420) for the conversion of 1,4-diaminobutane to spermidine and spermine in intact *A. nidulans*, *S. cerevisiae*, *Azotobacter vinelandii*, *Azotobacter chroococcum*, and *Lactobacillus casei*. With *N. crassa*, Greene (143) was the first to find that methionine is the precursor of the aminopropyl moiety of spermidine; the incorporation of [14C]methionine into spermidine has also been reported for *B. subtilis* (22), *N. crassa* (22), *P. aeruginosa* (22), and *A. nidulans* (22).

Recently, Jänne et al. (201) have studied the enzymes involved in spermidine biosynthesis from *S. cerevisiae*. S-Adenosylmethionine decarboxylase was purified 400-fold, and partially separated from the propylamine transferase. The enzyme behaves more like the enzyme from rat prostate than that from *E. coli*, since it has an absolute requirement for 1,4-diaminobutane, does not require Mg^{2+}, and is strongly inhibited by decarboxylated adenosylmethionine. No evidence has been found for either a pyridoxal phosphate or pyruvate cofactor. The purified preparation also contained propylamine transferase activity; the specific activity of the propylamine transferase in this preparation was 50-fold greater than in the crude extract.

D. METABOLISM OF 1,4-DIAMINOBUTANE AND SPERMIDINE IN *E. COLI*

Growing logarithmically in a glucose–salts medium, *E. coli* does not metabolize either 1,4-diaminobutane or spermidine, except for the conversion of the 1,4-diaminobutane to spermidine. No other derivatives are normally detected (406).

1. Conversion of Spermidine to Glutathionylspermidine (96,411,412,422)

When *E. coli* is at the end of its growth period (i.e., when the optical density is relatively high, the pH slightly acid, and the culture is effectively anaerobic) all of the spermidine and 50% of the bacterial glutathione are converted to glutathionylspermidine. Glutathionylspermidine is present in the bacteria at a concentration of 3–4 $\mu M/g$ wet weight; this represents about 50% of the total sulfur content of the trichloroacetic acid-soluble fraction and 7–10% of the total sulfur in the bacteria. Glutathionylspermidine has been isolated from *E. coli*, and evidence was obtained for the following structure: γ-glutamylcysteinylglycylspermidine. No data are available on whether this compound is present in the sulfhydryl or disulfide form *in vivo*.

When a dense culture of *E. coli*, containing large amounts of glutathionylspermidine, is diluted into fresh medium at pH 7.0, the glutathionylspermidine disappears very rapidly (<10 min) with the appearance of free spermidine.

Crude extracts of *E. coli* can synthesize glutathionylspermidine when incubated with high concentrations of glutathione, Mg^{2+}, ATP, and spermidine at pH 8.0 (422,426). Crude extracts also contain another enzyme that degrades glutathionylspermidine to spermidine (411,412). Both of these enzymes have only been partially purified, and little is known of their characteristics.

Spermine can also be converted to a glutathionyl derivative when added to a culture of *E. coli*, but the details of this conversion have not been studied yet (96).

The function of glutathionylspermidine is intriguing, especially since it represents a combination of two compounds, each of unknown function, which are present in large amounts in *E. coli*. The rapid synthesis and breakdown might lead one to speculate that glutathionylspermidine is being formed and metabolized rapidly, even in cells in the rapid growth period, but that only under the special conditions mentioned above does it accumulate. It would be attractive to postulate that glutathionylspermidine is involved in cell division, thereby correlating the recent work of Maas et al. (240) and Inouye and Pardee (189,190) on the possible relation of polyamines to cell division (see Section V.A) with the older speculations of Rapkine, Mazia, and others (251) on the relationship of sulfhydryl groups to cell division in sea urchin eggs. However, at this moment there are no definitive data indicating the function of this interesting compound.

2. Acetylation

Earlier work (77,97,410) described the occurrence of monoacetyl-1,4-diaminobutane and of monoacetylspermidine in *E. coli*.* However, more recent studies have shown that *E. coli*, growing in a glucose–salts medium, does not contain any acetylated polyamines (405,406). The discrepancy between these results was explained by the finding that only cells that are cooled (as during harvesting) contain acetylated

---

* Two forms of acetylspermidine were described by Dubin and Rosenthal (97). They used the terms "acetylspermidine A" for the material acetylated on the 1,4-diaminobutane moiety and "acetylspermidine B" for the material acetylated on the 1,3-diaminopropane end.

More recently (427), we have been using the following terminology, suggested by Drs. R. S. Tipson and W. E. Cohn; this is based on a modified aza-nomenclature (IUPAC Rule 814.6):

$$\overset{1}{N}H_2\overset{2}{C}H_2\overset{3}{C}H_2\overset{}{C}H_2\overset{4}{N}H\overset{5}{C}H_2\overset{6}{C}H_2\overset{7}{C}H_2\overset{8}{C}H_2NH_2$$

The order of the numbering is fixed by the requirement that the secondary NH be numbered as N-4, rather than N-5 (i.e., the lower number).

$$NH_2CH_2CH_2CH_2NHCH_2CH_2CH_2CH_2NHCOCH_3$$

N^8-acetylspermidine (acetylspermidine A)

$$CH_3CONHCH_2CH_2CH_2NHCH_2CH_2CH_2CH_2NH_2$$

N^1-acetylspermidine (acetylspermidine B)

amines (405,406). If a culture containing no acetylated amines is stored at 6° for 2 hr, as much as 10% of the 1,4-diaminobutane and 60% of the spermidine are acetylated.

The physiological significance of this acetylation that occurs in the cold is completely unclear. It is noteworthy that, at 6°, two other striking events are observed in *E. coli:* (1) the ribosomes dissociate into subunits within 3 hr after chilling (122); (2) although protein, DNA, and RNA syntheses continue at a slow rate for several days, cell division does not occur and long filaments form (362).

Acetylation of polyamines can also be observed without cooling. if excess 1,4-diaminobutane, spermidine, or spermine is added to the culture medium (97,406). It seems likely that the acetylation of amines reported for organisms that have a genetic block in the ornithine → arginine conversion (such as *E. coli* 15 TAU⁻; see refs. 77,314) is a reflection of a high internal pool of 1,4-diaminobutane, resulting from the decarboxylation of the accumulated ornithine.

The enzymatic acetylation of 1,4-diaminobutane and of spermidine *in vitro* has not been described.

3. Transamination

Kim and Tchen (212–214) have shown that a mutant of *E. coli* that is able to grow on 1,4-diaminobutane as the sole source of carbon and nitrogen contains a pyridoxal phosphate-requiring enzyme (diamine: α-ketoglutarate aminotransferase) for the transamination of diamines. In this reaction 1,4-diaminobutane is converted to γ-aminobutyralde-hyde.* The enzyme has been purified 80-fold and shows maximum activity at pH 9.0 to 10.0. The relative activity with different amines is 1,4-diaminobutane (100%); 1,5-diaminopentane (107%); 1,6-diaminohexane (65%); 1,7-diaminoheptane (30%); and (+)1,4-diaminobutan-2-ol (30%). 1,3-Diaminopropane, as well as a variety of other amines and amino acids, are inactive.

---

* The product, γ-aminobutyraldehyde, spontaneously cyclizes to Δ^1-pyrroline:

$$NH_2CH_2CH_2CH_2CHO \longrightarrow$$

Δ^1-Pyrroline is easily followed in a reaction mixture, since it reacts with *o*-aminobenzaldehyde to form a yellow product (2,3-trimethylene-1,2-dihydroxy-quinazolinium) with an absorption at 435 nm.

E. METABOLISM OF 1,4-DIAMINOBUTANE AND SPERMIDINE IN MICROORGANISMS OTHER THAN *E. COLI*

1. Diamine Metabolism

a. Diamine Oxidation. Diamine oxidation has been described in a number of organisms (253,347,421), although the enzymes have not been purified. Preliminary evidence indicates that in some organisms the first step is not an oxidation but a transamination. An apparent "diamine oxidase" activity, for example, would result from a primary transamination reaction (212–214,253), followed by oxidation of the γ-aminobutyraldehyde to γ-aminobutyric acid, succinic semialdehyde, and succinic acid (15,192).

Aspergillus niger contains a monoamine oxidase that also oxidizes diamines at a slow rate. The enzyme has been purified 350-fold and crystallized. The purified enzyme has an absorption maximum at 480 nm; it contains copper in the cupric form and is inhibited by chelating agents and carbonyl reagents (397,456–459).

b. Putrescine Oxidase. *Micrococcus rubens* contains an enzyme that oxidizes 1,4-diaminobutane to γ-aminobutyraldehyde (3,455,460, 463,464). 1,5-Diaminopentane, but not other diamines, is also oxidized at a slower rate. Spermidine is slowly oxidized to γ-aminobutyraldehyde and 1,3-diaminopropane; no ammonia is formed. This enzyme has been purified to homogeneity, has a molecular weight of 80,000, and contains 1 mole of FAD/mole of enzyme.

c. Carbamylation of 1,4-Diaminobutane. Carbamylation of 1,4-diaminobutane* has been reported in extracts of *Streptococcus faecalis* (329), baker's yeast (124), and *A. nidulans* (124).

1,4-diaminobutane + carbamylphosphate $\rightleftarrows$
    N-carbamyl-1,4-diaminobutane + Pi
    (*N*-carbamylputrescine)

The enzyme has been purified 60-fold from *S. faecalis* grown on agmatine and has been shown to be distinct from ornithine transcarbamylase.

Agmatine can substitute for arginine as an energy donor for *S. faecalis*, and presumably supplies energy by the reactions:

agmatine $\longrightarrow$ *N*-carbamyl-1,4-diaminobutane $\rightleftarrows$
    carbamylphosphate $\xrightleftharpoons{+ADP}$ ATP + CO_2

* See Section III.B.2 for a discussion of carbamyl-1,4-diaminobutane in plants.

The substrate specificity of the *S. faecalis* enzyme has not been reported, but the extracts from yeast and *A. nidulans* have been shown to be specific for 1,4-diaminobutane, since 1,5-diaminopentane and a variety of other amines were not substrates for the enzyme.

d. Hydroxylation of 1,4-Diaminobutane. (+)1,4-Diamino-butan-2-ol (i.e., hydroxyputrescine) has been isolated from *Pseudomonas* and has been shown to be synthesized from 1,4-diaminobutane (227, 256,330,436). No enzymatic studies have been reported yet on this oxidation.

e. Acetylation. Acetylation of added 1,4-diaminobutane, spermidine, and spermine has been described in cultures of *S. aureus* (331). No enzymatic experiments have been carried out.

2. Spermidine Metabolism

a. Spermidine Dehydrogenase. *Serratia marcescens* contains an enzyme that oxidizes spermidine at the secondary nitrogen (15,16,29, 62,408);*

$$\text{spermidine} \rightarrow \text{1,3-diaminopropane} + \Delta^1\text{-pyrroline}$$

The enzyme has been purified 5000-fold to homogeneity (408). The enzyme contains approximately 1 mole of iron-protoporphyrin IX and 1 mole of FAD/mole of enzyme (mol. wt. 76,000). The enzyme does not react directly with molecular oxygen but requires an electron acceptor, such as ferricyanide, dichlorophenolindophenol, or cytochrome c. The presence of both heme and a flavin in a single enzyme has also been reported for yeast cytochrome b_2, sulfite reductase, and nitrate reductase. The absorption spectrum of the enzyme is characteristic of a heme with a maximum at 414 nm. When substrate is added in the absence of an electron acceptor, the absorption spectrum resembles that of a reduced heme, with a maximum at 427 nm, and two small peaks at 530 and 562 nm.

In addition to spermidine, N^8-acetylspermidine, N,N'-bis(3-amino-propyl)-1,3-propanediamine, and N-(3-aminopropyl)-1,3-propanedia-mine are oxidized, although the K_m is higher and the rate of oxidation

* The product, γ-aminobutyraldehyde, spontaneously cyclizes to Δ^1-pyrroline. See footnote on page 223.

slower. Spermine is also oxidized slowly. A variety of other amines are not oxidized.

Spermidine is also oxidized to 1,3-diaminopropane and Δ^1-pyrroline by the putrescine oxidase of *Micrococcus rubens* (see p. 224). When spermidine is oxidized by *Pseudomonas* (289), the oxidation takes place at the other side of the secondary nitrogen:

$$\text{spermidine} \rightarrow 1,4\text{-diaminobutane} + 3\text{-aminopropionaldehyde}$$

3-Aminopropionaldehyde has been described in human serum (305), but it is not known whether it is formed by analogous reaction.

b. Edeine. Two complex antibiotic peptides have been isolated from the medium of *Bacillus brevis* cultures. One (edeine "A") contains spermidine, while the other (edeine "B") contains guanylspermidine [*N*-guanyl-*N'*-(3-aminopropyl)-1,4-diaminobutane]. Both have unusual amino acids linked to the amine through glycine (169–172). (See refs. 170 and 172 for additional articles on edeine.)

The biosynthesis of edeine has been shown in cell-free extracts of *B. brevis*. The synthesis does not require ribosomes and is not inhibited by the usual inhibitors of protein synthesis (229b).

F. UPTAKE OF 1,4-DIAMINOBUTANE AND SPERMIDINE BY *E. COLI*

The uptake of polyamines from the medium has been shown in a variety of organisms (18,123,147,260,331,421). This is even true for those organisms that normally do not contain certain amines, and thus, particularly if the growth is in crude media, the amines found in the bacteria might not represent the amines normally synthesized by the organism.

"Active uptake" has been shown in *E. coli* for 1,4-diaminobutane, spermidine, and spermine (410). The K_m values for spermidine and 1,4-diaminobutane are very low (less than $8 \times 10^{-8} M$ and $2 \times 10^{-7} M$, respectively), while that for spermine is higher. The rates of uptake for spermidine and spermine are higher at pH 8.0 than at 7.0, while the optimum pH for the uptake of 1,4-diaminobutane is pH 7.0. These experiments are complicated because of the tendency for these amines, particularly spermine, to adsorb to cell surfaces, millipore filters, and glass and suitable washing procedures with unlabeled amines are necessary to permit reliable measurement of the actual active uptake.

III. Biosynthesis and Metabolism of 1,4-Diaminobutane, Spermidine, and Spermine in Animals and Plants

A. CONCENTRATION

1. Animals

The content of polyamines has been determined in many animal and insect tissues (14a,20,33,59,92,93,98,99,150,182,193,224,225a,225b,226, 235,258,259,282,283,291,301,333,343,354,364b,364c,366,451); for a complete review of the early literature see refs. 421 and 428). Various tissues differ markedly, both in the concentration of total amines as well as in the relative concentrations of spermidine and spermine. The highest concentration of amines is found in human semen, where 12–14 μmoles/ml of spermine are found (333,421,450). Semen does not have any spermidine, but most mammalian tissues have more spermidine than spermine. Particularly high concentrations have been found in pancreas (e.g., in the rat, pancreas has 9 μmoles of spermidine and 1 μmole of spermine per gram of wet weight), and in the prostate (e.g., in the rat, the prostate has 8 μmoles of spermidine and 5 μmoles of spermine per gram wet weight) (295,333). Moderately high amine concentrations have also been found in liver, spleen, and kidneys.

1,4-Diaminobutane is present in low concentrations in many tissues, 10–20% that of spermidine or spermine in most instances (58,93,225b, 314,336a,337,339,340). 1,5-Diaminopentane (cadaverine) has also been described in chick embryo (55) and in cystinuric urine (51,221,421).

Diguanido derivatives of spermidine (hirudinone*) and 1,4-diaminobutane (arcaine), as well as the monoguanido derivatives of spermidine and 1,4-diaminobutane (agmatine), have been described in the leech (14b,322,323).

A large number of papers have been published on a phospholipid, malignolipin, which has been reported to contain spermidine and to be present in cancer tissue and the sera of cancer patients (421). However, there is still too much doubt as to the nature and existence of the compound, or whether it actually contains spermidine, to warrant further discussion at this time (37c). Bachrach and Robinson (34a) have

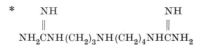

$$* \quad \overset{NH}{\underset{\|}{}} \qquad \qquad \overset{NH}{\underset{\|}{}}$$
$$NH_2CNH(CH_2)_3NH(CH_2)_4NHCNH_2$$

reported that they could not confirm previous reports on elevated spermine levels in the serum of cancer patients. However, recently, high concentrations of 1,4-diaminobutane (3–72 mg/day), spermidine (10–84 mg/day), and of spermine (<2.5–64 mg/day) have been reported in the urine of cancer patients, contrasting with the trace quantities present in normal urine (336c).

2. Plants

1,4-Diaminobutane, spermidine, and/or spermine have been found in various plants (39,74,166,167,327,376,379,380,467), seeds (35), pollen (236), and plant viruses (39). In addition, N-carbamylputrescine, agmatine, and tetramethylputrescine are present in significant amounts in certain species (249,379–381). Certain alkaloids contain spermidine (lunarine and palustrine), spermine (homaline), and hydroxyspermine (pithecolobine), linked covalently to other moieties (304b,380,421,433). 3,4-Dihydroxycinnamoylputrescine, feruloylputrescine, coumaroylagmatine, and 2-hydroxyputrescine amides of ferulic and p-coumaric acids have also been found in plants (446a,261,380). The spermidine analogue, homospermidine, [1,9-diamino-5-azanonane, $NH_2(CH_2)_4NH-(CH_2)_4NH_2$], and its tetramethylacylated derivatives are found in Santalum album L and Solanum tripartitum (229a,230). Steroidal diamines have been isolated from tropical African plants (246,369). For a summary of the recent studies on the distribution and concentration of these amines in plants, see refs. 376, 379, and 380.

Special mention should be made of the dramatic increase found in the amine content of plants grown under conditions of extreme potassium deficiency. In barley, for example, the 1,4-diaminobutane level increases by 50-fold (accounting for 10% of the total nitrogen), while the agmatine concentration increases 10-fold. In some plants carbamylputrescine accumulates in potassium deficiency (376,380).

B. BIOSYNTHESIS OF 1,4-DIAMINOBUTANE, SPERMIDINE, AND SPERMINE IN ANIMALS AND PLANTS

Studies of the enzymes involved in the biosynthesis of the amines have shown that, in general, the pathways in animal tissues resemble those described for bacteria (193,225b,292,293,295,297,299,300,307,340, 367,421,451). The biosynthetic pathways in plants have not yet been completely defined, but certain differences have been described in the biosynthesis of 1,4-diaminobutane (see Section III.B.2).

1. Biosynthesis of 1,4-Diaminobutane in Animals,

Ornithine decarboxylase appears to be the only enzyme that synthesizes 1,4-diaminobutane in animals, since there is still no evidence for the arginine decarboxylase-agmatine ureohydrolase pathway in animal cells (303,313). Ornithine decarboxylase has been found widely distributed in animal tissues, although it is usually present in very low concentration   (75,193,194,198,225b,295,297,311,341,342,343,367,385, 389,451). The levels of ornithine decarboxylase in various stages of synchronously-growing D on C cells have recently been reported (125a). (See also Section V.B.5.)

a. Purification of Ornithine Decarboxylase from Rat Prostate. Jänne and Williams-Ashman have obtained a 300-fold purification of the ornithine decarboxylase from the rat ventral prostate (198). The enzyme acts on L-ornithine and forms equimolar quantities of CO_2 and 1,4-diaminobutane. There is an absolute requirement for pyridoxal phosphate with the purified enzyme. The enzyme appears homogeneous by acrylamide gel electrophoresis, but no physical-chemical studies have been reported.

This enzyme has a sulfhydryl requirement; the activity in a crude extract is increased 10-fold by the addition of dithiothreitol during homogenization and in the assay (198,199).

b. Purification of Ornithine Decarboxylase from Regenerating Rat Liver. Friedman et al. described the 175-fold purification of the ornithine decarboxylase of regenerating rat liver (125b,c). Regenerating liver was used as the source material, since the concentration of ornithine decarboxylase is increased manyfold in regenerating liver. (See Section V.B.5). The enzyme is nearly pure on acrylamide electrophoresis and has an absolute requirement for pyridoxal phosphate. The enzyme is stimulated by mercaptoethanol and inhibited by N-ethylmaleimide.

The observation that sulfhydryl compounds cause marked increases of both the rat-liver and rat-prostate enzymes is particularly important in evaluating ornithine decarboxylase activities. Numerous studies have been concerned with the increase in ornithine decarboxylase activity in regenerating liver or after a variety of stimuli, but most of the assays have not been carried out with added reducing agents. (See Section V.B.5.)

2. Biosynthesis of 1,4-Diaminobutane in Plants

Many studies have reported on the overall pathway of the biosynthesis of 1,4-diaminobutane in plants,* but little work has been done on purification of the enzymes involved. 1,4-Diaminobutane is usually formed from arginine via agmatine and carbamylputrescine. Only indirect evidence exists for ornithine decarboxylase (158,467); its importance in the biosynthesis of 1,4-diaminobutane in plants is not known (380).

a. Carbamylputrescine Pathway. The major precursor for the biosynthesis of 1,4-diaminobutane in plants appears to be arginine (249,379,380). The postulated pathway is shown in Figure 2; it seems likely that ornithine is not an intermediate. The arginine decarboxylase of plants cleaves arginine to form agmatine and CO_2 (373,380). In contrast with the pathway in *E. coli* in which agmatine is hydrolyzed directly to 1,4-diaminobutane (putrescine) by a ureohydrolase, in plants agmatine iminohydrolase hydrolyzes agmatine to NH_3 and N-carbamylputrescine (375,380,381). The enzyme is specific for agmatine and has a pH optimum of about 7.8. No purification has been reported for this enzyme.

Two different types of enzymatic reaction have been described for the conversion of carbamylputrescine to putrescine. In pea seedlings an enzyme has been purified 1000-fold, which carries out reaction II, Figure 2 (219,380). This reaction is analogous to the ornithine

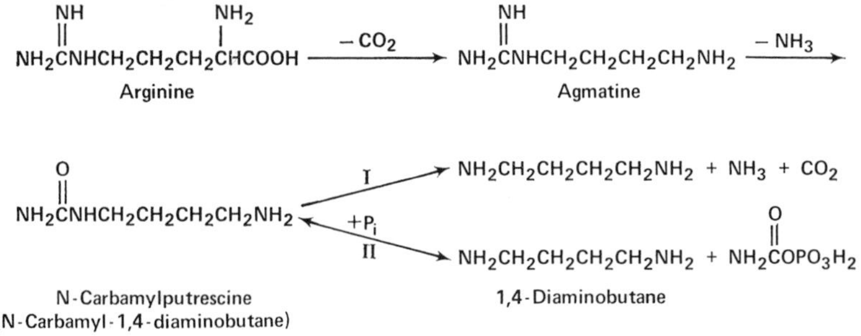

Fig. 2. Biosynthesis of 1,4-diaminobutane in plants (380).

* An extensive review of the polyamines in plants has recently been published by Smith (380).

transcarbamylase reaction and is reversible. The purified enzyme also carries out the ornithine transcarbamylase reaction, but it is not certain whether this activity represents a contaminating activity. (The usual animal ornithine transcarbamylase does not react with putrescine.)

A second possible pathway for the conversion of carbamylputrescine to putrescine is a direct hydrolysis by N-carbamylputrescine amido-hydrolase, reaction I, Figure 2 (374,380). This activity has been found in extracts of a number of different plants, but the enzyme has not been purified. The reaction observed could be the result of the reaction discussed in the previous paragraph, plus a hydrolysis of carbamyl-phosphate to CO_2 and NH_3 in the crude extracts used. The only evidence against this possibility is that the overall reaction took place in the absence of added phosphate, but this cannot be considered definitive proof with crude preparations.

b. Effect of Potassium Deficiency. As indicated above, in potassium-deficient plants there is a marked stimulation of the formation and accumulation of 1,4-diaminobutane and agmatine. Consistent with this finding is the increase in arginine decarboxylase (10-fold) and N-carbamylputrescine amidohydrolase (2- to 3-fold) in potassium-deficient barley, and the appearance of citrulline decarboxylase in potassium-deficient *Sesamum indicum* L. (373,374,379,380). Since the levels of the enzymes concerned in 1,4-diaminobutane formation are also increased after acid feeding, Smith has suggested that the increased synthesis of these organic bases may be a mechanism to help maintain the plant's pH at a time of K^+ shortage (379,382).

c. Diamine Transamination. Diamine transamination has been described in a number of plant tissues; the reactions appear to be reversible, and Smith suggests that the enzyme may provide alternative biosynthetic pathways for the amines (159,380).

3. Biosynthesis of Spermidine and Spermine in Animals

a. Purification of S-Adenosylmethionine Decarboxylase from Rat Prostate and Rat Liver. In earlier studies of the enzyme from rat prostate, it had appeared as if both the decarboxylation of adenosyl-methionine and the transfer of the propylamine moiety were carried out by one protein (299,451). With increased purification, it became clear that the adenosylmethionine decarboxylase and the propylamine transferase of rat ventral prostate are separate proteins (197,200,201).

The S-adenosylmethionine decarboxylase from rat prostate has been purified over 500-fold (200). In contrast with the enzyme from $E. coli$, this enzyme does not require Mg^{2+} but is stimulated by 1,4-diaminobutane and, to a lesser extent, by spermidine (296,299). The enzyme is strongly inhibited by decarboxylated S-adenosylmethionine.

S-Adenosylmethionine decarboxylase has also been purified from rat liver (115,116,152a). The specific activity of a preparation that is 350-fold purified is only 1/100 that of the purified enzyme from rat prostate. The molecular weight is about 45,000. This enzyme, like the one from prostate, is stimulated by 1,4-diaminobutane, and to a lesser extent, by spermidine. Hannonen et al. have recently shown that the adenosylmethionine decarboxylase and the propylamine transferase from rat liver are two separate proteins (152a). A partially purified S-adenosylmethionine decarboxylase has also been obtained from rat brain (310).

b. Purification of Propylamine Transferase from Rat Prostate and Rat Brain. A preliminary report has described an 80-fold purification of the propylamine transferase from rat ventral prostate (197). The partially purified enzyme carries out the transfer of the propylamine group from decarboxylated S-adenosylmethionine to 1,4-diaminobutane to form spermidine; no synthesis of spermine is detectable (197). Methylthioadenosine is formed in this reaction, and can be enzymatically degraded by an enzyme from rat prostate, which requires phosphate (298).

Propylamine transferase was also purified from rat brain and separated into two fractions (152b,310). One is active in the synthesis of spermidine from 1,4-diaminobutane and decarboxylated adenosylmethionine; this was purified 150-fold. The other is active in the synthesis of spermine from spermidine and decarboxylated adenosylmethionine; this was purified 15-fold. Two propylamine transferases have also been detected in regenerating rat liver (152a,309).

C. METABOLISM OF 1,4-DIAMINOBUTANE, SPERMIDINE, AND
SPERMINE IN ANIMALS AND PLANTS

1. Amine Oxidases

Although the terms "monoamine oxidase" and "diamine oxidase" are still widely used, it is now evident that there is a marked overlap in the substrate specificities of the various enzymes. For example, the beef plasma enzyme described below oxidizes polyamines such as spermidine and spermine, monoamines, and long chain diamines, but not 1,4-diaminobutane or histamine. Kidney diamine oxidase oxidizes histamine and short chain aliphatic diamines, but not polyamines, and has minimal activity against monoamines. The pig plasma enzyme does not oxidize polyamines or aliphatic diamines but does oxidize both histamine and monoamines, such as benzylamine. Mitochondrial mono-amine oxidase normally oxidizes monoamines, such as benzylamine, but does not act on diamines or polyamines. Furthermore, even the sub-strate and inhibitor specificity of mitochondrial monoamine oxidase has been shown to depend on the oxidation state of its sulfhydryl groups (6).

The discussion below will cover the enzymes particularly concerned with the oxidation of 1,4-diaminobutane, spermidine, and spermine in animal tissues.

a. **Diamine Oxidase (Animal).** Diamine oxidase* is present in many animal tissues and has been studied particularly extensively in hog kidney. The enzyme has been purified to homogeneity from this source (142a,142b,263a,264a,264b,462). The pure enzyme acts both on short chain diamines and on histamine (38,54,305,415,462,470). The substrate most commonly used for the assay of diamine oxidase is 1,4-diaminobutane. This is converted to γ-aminobutyraldehyde, which cyclizes to Δ^1-pyrroline.

$$NH_2(CH_2)_4NH_2 + O_2 \longrightarrow NH_2(CH_2)_3CHO + H_2O_2 + NH_3$$

Rotilio et al. (334) have presented evidence that superoxide is formed during the reaction; superoxide is converted to H_2O_2 at neutral pH. The kinetics of the reaction have been studied (80a,118).

The purified enzyme has 1 g atom of copper/87,000 g and an electron paramagnetic resonance spectrum that is typical of complexes of cupric copper (142b,142c,263c,264b,c,265,462). The enzyme activity is inhibited by such copper-binding agents as diethyldithiocarbamate (264b). However, the role of copper in the mechanism of the enzyme oxidation is unclear, since only a portion of the copper is reduced by

* For a summary of the earlier literature on diamine oxidase and histaminase, see the reviews by Buffoni (54), Kapeller-Adler (207), Tabor (415), and Zeller (470).

the addition of substrate under anaerobic conditions, and the rate of this reduction is too slow for it to be enzymatically significant (264b, 264c,265). This aspect of the diamine oxidase studies has recently been reviewed by Mondovi et al. (264c).

The enzyme has absorption maxima at 410 and 500 nm, which have been attributed to copper and pyridoxal phosphate (263a,263b,263c, 264b). The assumption that pyridoxal phosphate is involved has been based on several indirect observations (142b,263a,263b,263c). These include the effect of carbonyl binding agents on the spectrum of the enzyme, and the presence of 1 mole of phosphorus/90,000 g. Additional evidence for the presence of pyridoxal phosphate in the purified enzyme is based on the isolation of a semicarbazide derivative with chromatographic properties similar to that of the comparable pyridoxal phosphate derivative (263c), and chromatographic evidence for the formation of pyridoxylhistamine by treatment of the enzyme with borohydride in the presence of histamine (228). Material isolated from diamine oxidase preparations showed a little pyridoxal phosphate-like activity when tested with apo-aspartate aminotransferase. In some experiments there was also some stimulation of diamine oxidase by added pyridoxal phosphate. However, despite these observations, there is still no definitive proof of either the presence of pyridoxal phosphate in the purified enzyme or of a definitive requirement of the enzyme for this cofactor.

There is some disagreement on the molecular weight of the enzyme. The minimum molecular weight, based on the amount of copper present, is 87,000. Yamada et al. reported a molecular weight for the intact enzyme of 185,000 (462). Mondovi et al. (264b) obtained values of approximately 120,000 but indicated that the sedimentation data were complicated by association-dissociation interactions.

Diamine oxidase has been of particular interest to physiologists and pharmacologists because of its action as "histaminase" in oxidizing histamine to imidazoleacetaldehyde (54,207,415,470). This enzyme is strikingly elevated in the plasma in the last half of human pregnancy, reflecting the high amount of this enzyme in the maternal placenta (372,388). Diamine oxidase has, in fact, been highly purified from this source (54,207,372) and oxidizes both histamine and the aliphatic diamines (54,372). Because of its clinical implications there is a very large body of literature on histaminase. For a review of this literature see refs. 54, 207, 415, 470.

The findings that purified preparations from hog kidney and placenta oxidize both short chain diamines and histamine apparently resolves a long-standing controversy over the identity of diamine oxidase and histaminase in favor of Zeller's original suggestion that diamine oxidase and histaminase are really the same enzyme (470). However, this identity has not been shown definitively for the enzyme from all other sources; in fact, as mentioned above, the pig plasma enzyme oxidizes benzylamine and histamine but does not oxidize the diamines (54).

b. Diamine Oxidase (Plant). Diamine oxidase is present in high concentration in pea seedlings but not in the adult plant. The enzyme that has been purified 200-fold from this source contains divalent copper and has an absorption maximum at 505 nm. Upon removal of the copper, the activity is lost, and the absorption maximum shifts to 480 nm. There is no evidence for pyridoxal phosphate, although the enzyme is strongly inhibited by carbonyl-binding reagents (174).

The enzyme is most active with 1,4-diaminobutane and 1,5-diaminopentane but also has significant activity with other diamines, histamine, spermidine, agmatine, and tyramine. The kinetics of the pea seedling amine oxidase have been studied by Yamasaki et al. (465). Oxidation of spermidine by the diamine oxidase from peas (380), and of spermine by a crude barley extract (378), yield 1-(3-aminopropyl)-pyrroline.

c. Beef Serum Amine Oxidase. Spermine and spermidine are oxidized by a soluble enzyme in beef serum, which was first described by Hirsch in 1953 (421). This enzyme has been isolated as a homogeneous protein; the molecular weight is 170,000 (466a,466b). It has 1–2 atoms of copper per mole and an incompletely defined cofactor that may be pyridoxal· phosphate (466a,466b); it contains no flavin cofactor (281). The enzyme is inhibited by carbonyl reagents (421). The enzyme oxidizes spermine and spermidine as follows (413):

$$NH_2(CH_2)_4NH(CH_2)_3NH_2 \xrightarrow{O_2} NH_2(CH_2)_4NH(CH_2)_2CHO + NH_3 + H_2O_2$$

spermidine $\qquad$ N-(4-aminobutyl)-3-aminopropylaldehyde

$$NH_2(CH_2)_3NH(CH_2)_4NH(CH_2)_3NH_2 \xrightarrow{2O_2}$$

spermine

$$OHC(CH_2)_2NH(CH_2)_4NH(CH_2)_2CHO + 2NH_3 + 2H_2O_2$$

N,N'-bis(3-propionaldehyde)-1,4-diaminobutane

Various long chain monoamines and diamines, and benzylamine are also substrates for this enzyme.

As first shown by Hirsch and Dubos for tubercle bacilli (421), the products formed after oxidation of spermine and spermidine are toxic to a variety of cells, including tumor cells, bacteria, trypanosomes, viruses, and bacteriophages (7,17,19,23,27,28,32,34b,129,208,218,226, 286,287,409,421,461). The mechanism of the toxicity appears to be the binding of the aminoaldehyde to the nucleic acid of the cell or phage; evidence for an irreversible, base-specific binding, as well as a non-specific reversible ionic binding to DNA, has been presented (24–26,30, 31,104). Recently, the toxicity of the spermine oxidation product has been confirmed with a synthetic N,N'-bis(3-aminopropionaldehyde)-1,4-diaminobutane (129).

The mechanism of the toxic effects observed above is not entirely clear. The aldehydes produced are rather unstable, forming polymers by condensation, and forming acrolein by a nonenzymatic β-elimination reaction (217,414). It is very possible that some of the toxic effects found with the oxidation products of spermidine and spermine may be caused by the acrolein formed by this secondary breakdown of the primary product (7–10,217,218,414,421).

Spermine oxidase has been found in the serum of ruminants, and an extensive study of the distribution of this enzymatic activity in a variety of animals has been carried out by Blaschko and his associates (42). The level of this enzyme increases strikingly with age (42,437).

An amine oxidase similar to that found in beef serum has been purified 170-fold from the connective tissue of bovine aorta. This enzyme also requires copper for activity and is inhibited by carbonyl reagents. The substrate specificity is very similar to that of the beef serum enzyme; however, the connective tissue enzyme also oxidizes the peptidyl lysine of porcine vasopressin, while the serum enzyme does not (335).

Recently, a new amino acid, N(4-aminobutyl)-3-aminopropionic acid, putreanine, has been found in mammalian brain and nervous tissues (approximately one-tenth the spermidine concentration) (206, 279a,364a). The most likely biosynthetic pathway of putreanine is the further oxidation of N(4-aminobutyl)-3-aminopropionaldehyde, formed by a postulated tissue amine oxidase similar to the beef serum enzyme.

2. Metabolism in vivo in Animals

When 1,4-diamino-[^{14}C]butane is administered to a rat, one-third of the injected dose is converted to CO_2 within 2 hr (193). A small amount of the administered 1,4-diaminobutane is also converted to spermidine. Previous studies by Rosenthal and Tabor reported that when spermine is administered parenterally to animals, up to 20% of the dose appears in the urine unchanged, while 4–8% is excreted as spermidine (333). These experiments have now been extended by Siimes (365), with the use of ^{14}C-polyamines, which permitted the administration of smaller doses. After the administration of either labeled spermidine or spermine, 4–8% was recovered as radioactive CO_2 and 18–25% was present in the urine as an unidentified acid-hydrolyzable polyamine derivative. Evidence was found for the conversion of spermine to spermidine, and of spermidine to spermine and 1,4-diaminobutane (365,384).

In vivo acetylation is indicated by the occurrence, in normal urine, of the monoacetyl derivatives of 1,4-diaminobutane, 1,5-diaminopentane, and spermidine (279b,301).

IV. Miscellaneous Effects of Polyamines

A. EFFECTS ON BACTERIOPHAGES

Mutant bacteriophage strain T4rII is ordinarily unable to multiply in *E. coli* K12 (λ), but the production of the phage is markedly increased by the addition of Mg^{2+}, 1,4-diaminobutane, spermidine, or spermine (53,117). The mechanism of these effects is not understood, although it is of interest that, after an infection with the rII mutant, there is a striking loss of 1,4-diaminobutane from the bacteria into the medium. Whether or not this is a specific effect of the rII infection, or merely reflects a more general change in permeability or ATP levels (53,76a, 117), is not clear.

Polyamines have been shown to protect *E. coli* bacteriophage T5 against inactivation by metal-binding agents as citrate or EDTA (416) or urea-shocked bacteriophage (π) against inactivation by dilution or heating (121). Protection was observed with as little as $2 \times 10^{-8} M$ spermine.

Spermine has been reported as inhibiting the production of some bacteriophages (12,107,117,144,348), although in most of the studies a toxic effect of spermine on the bacterial cell was not excluded. In some instances spermine will inhibit lysis of bacteria by mature phage

(107,144). An inhibitory effect of 1,4-diaminobutane on the synthesis of T4-induced late enzymes, such as lysozyme, has been reported (358,359).

B. EFFECTS ON GROWTH

Several miscellaneous effects of polyamines have been reported, including reduction in the lag time of washed yeast (66), prevention of lysis of starved *S. lactis* (435), and stimulation of the growth of cultures of Jerusalem artichoke (41), Chinese hamster cells (149), human fibroblasts (304a), and a soil amoeba (5), as well as the growth of the larvae of a grain beetle (85).

C. TOXICITY

Spermine, when administered to animals, even in small doses, causes death after 5 days as a result of severe damage to the proximal convoluted tubules of the kidney (332,409). This work by Rosenthal et al. has been summarized in our previous review (421). No further studies have been carried out on the specific mechanisms of this effect.

Spermine and spermidine have been reported to be toxic to mammalian cells grown in cell culture (140,173,288,421), but some of these studies are difficult to interpret, since calf serum is present in many culture media; this serum contains the amine oxidase that converts spermine to a toxic aldehyde (see Section III.C.1.C).

Earlier work on the toxicity of spermine in bacteria was carried out by Silverman and Evans and by Bichowsky-Slomnitzki, Gurevitch, Razin, Rozansky, Grossowicz, and others, and was summarized in our previous review (421). In general, gram-positive organisms are more sensitive than gram-negative organisms, the toxicity of spermine is higher at a more alkaline pH, and relatively large concentrations of spermine are required for inhibition. Although the early work indicated that spermine might be the antibacterial factor in human semen, this has been questioned more recently (110). The toxicity of spermine to bacteria has been studied again recently by several laboratories (73,108, 123,260). When toxic doses of spermine are added to a culture, an inhibition of protein synthesis precedes the loss in viability (260). Ezekiel and Brockman (108) also showed an increased breakdown of bacterial protein in the presence of toxic doses of spermine.

Spermine has been reported to suppress the mutagenic action of an *E. coli* mutator gene (203).

Two classes of polyamine derivatives, steroidal diamines and N-acyltriamines, have been shown to be bacteriocidal or bacteriostatic, and to affect cell permeability (369). The interaction of the steroidal diamines with DNA and with polynucleotides has been studied by Mahler and Green (246).

D. TRANSGLUTAMINASE

Waelsch and his co-workers have demonstrated that transglutaminase, an enzyme present in a variety of animal tissues, can catalyze the replacement of the amide group of protein by primary amines (278, 421). Several workers have used the incorporation of radioactive 1,4-diaminobutane and 1,5-diaminopentane as assays for this enzyme and have shown that the amines can be incorporated into a variety of proteins, including actin, tropomyosin, and fibrinogen (90,101,119,231). Few studies, however, have been carried out on the effect of this substitution on the biological activity of the proteins studied.

E. REVERSAL OF LEVORPHANOL TOXICITY

Levorphanol has been shown to increase cell permeability, to decrease RNA synthesis, to cause a marked fall in the ATP level, and to completely inhibit the growth of $E.$ $coli$; addition of Mg^{2+} was able to reverse these changes. Recently, Simon and his associates have found that there is also a rapid loss of 1,4-diaminobutane from the cell, and that spermidine, as well as magnesium, can reverse the toxicity (370,371). It is of interest that the loss of 1,4-diaminobutane and the fall in ATP, caused by levorphanol, and the reversal by Mg^{2+} or spermidine, are similar to the effects noted above for the bacteriophage T4rII infection. (See Section IV.A.)

F. MISCELLANEOUS

Polyamines stimulate glucose metabolism and inhibit lipolysis in isolated fat cells of the rat (239), stimulate the metabolism of estradiol by rat liver microsomes (202a,202b), decrease the concentration of follicle-stimulating hormone in the rat pituitary (446b), and increase the synthesis of nucleotides in chick embryos (56,59).

The effects of polyamines on several miscellaneous enzymes were described in our previous review (421). More recently, spermine was found to stimulate the activity of yeast uridine diphosphate galactose-4-epimerase and to promote the formation of an 11S aggregate from a

6S moiety (83). The polyamines also activate phosphorylase b in the presence of AMP (441).

DeMeis has shown that spermine and spermidine produce relaxation of smooth and skeletal muscle made to contract by a variety of chemical stimuli (88). Polyamines also activate the uptake of Ca^{2+} by muscle microsomes; spermine inhibits the ATPase activity of the microsomes (89).

Fabro et al. have shown that diamines and polyamines react with thalidomide *in vitro*. At 37° thalidomide acylates diamines and polyamines rapidly, and the suggestion was made that this reaction might be a factor in the toxicity of the drug (109).

V. Postulated Function for the Polyamines

Although no definitive functions for the action of the polyamines *in vivo* have been established with certainty, a variety of biological effects has been described. Most of these seem to be related to the polycationic nature of these amines,* and their strong affinity for polyanions, such as phospholipids and nucleic acids; however, it is not known whether these effects represent the actual biological function of the amines *in vivo*. The recent work on mutants of *E. coli* (177,240, 267), (see Section II.B.4), together with the older demonstration of requirements for polyamines in *H. influenzae* (166,167) and for 1,4-dia-minobutane in *A. nidulans* (383), strongly suggests that these amines have some essential function and are not just a nonspecific form of intracellular cation.

A. CELL SURFACES AND CELL DIVISION

One of the earliest observations of the effects of polyamines on bacteria was their ability to prevent lysis of fragile bacteria in media of low ionic strength (e.g., *P. tularensis* and *S. aureus*); this effect is particularly striking in halophilic organisms (243–245,319,421). Subsequent studies showed that the polyamines also prevented the

---

* The dissociation constants (pK) for 1,4-diaminobutane (328) are 10.4 and 9.3, and for spermine (176) 8.19, 9.05, 10.21, and 10.99 at 25°. Somewhat lower values have been reported for spermine at a lower ionic strength in a less extensive study by Rogers et al. (326). The effect of temperature on the pK values for spermine has been reported by Hirschman, Leng, and Felsenfeld (176).

Several X-ray diffraction studies of the crystal structure of spermine phosphate hexahydrate (186,187), spermine tetrahydrochloride (82,136,238), and spermidine trihydrochloride (81) have been published.

lysis of bacterial spheroplasts and protoplasts (145,146,157,245,318, 390,401,402,421,435). These results, together with other data on the binding of spermine to phospholipids, cell wall components, and cell membrane (22,111,204,306,421,428) and stabilization of a membrane-ATPase complex (2), are consistent with the postulation that the polyamines stabilize the cell membrane (421); this stabilization could result from the binding of the polybasic amines to acid groups in the cell surface, with the consequent reduction of the repulsive effect of these groups without diminishing such cohesive factors as hydrogen bonding and van der Waal's forces. Most of these effects, however, can be duplicated with larger amounts of Mg^{2+} or Ca^{2+}, and there is no evidence that the effects observed necessarily reflect the *in vivo* function.

More recently attention has been directed to cell surfaces by the findings reported by Maas et al. (240) and summarized in Section II.B.4. In this work, mutants of *E. coli* that are limited in the synthesis of polyamines formed long filaments and did not septate until after the addition of polyamines. On the basis of other studies, Inouye and Pardee (189,190) also postulated a relationship between the polyamines and cell division, although they did not specifically imply a direct effect of the polyamines on the cell surface. Using certain amino acid auxotrophs in which synchronous growth was initiated by feeding the required amino acid after a period of starvation, they found that addition of 1,4-diaminobutane after starvation overcame the synchrony. Spermidine, on the other hand, counteracted the effect of the diaminobutane and maintained the synchrony. They proposed that changes in the spermidine to 1,4-diaminobutane ratio within the cell are a critical factor in determining cell division (189).

Of interest in this connection are the studies on the bacterium, *Myxococcus xanthus*, which develops microcysts when nutrients are limited. 1,4-Diaminobutane is one of several agents that can also induce the formation of microcysts. In contrast to the results with 1,4-diaminobutane, spermidine markedly inhibits microcyst formation, regardless of the inducing agent (453).

B. RELATIONSHIP OF POLYAMINES TO NUCLEIC ACID AND PROTEIN SYNTHESIS*

A number of earlier observations by Bichowsky-Slomnitzki, Razin and Rozansky, Mahler, Tabor, and others (summarized in our 1964

* Only a brief summary of this literature will be included here, since a complete survey has been published recently by Stevens in *Biological Reviews* (392).

review, ref. 421) pointed out that polyamines have a high affinity for nucleic acids. Particularly significant was the finding by Ames, Dubin, and Rosenthal (12,13) that polyamines are packaged together with nucleic acid in T-even bacteriophages. Subsequently, Herbst, Raina, Snyder, S. S. Cohen, and others have suggested that in some way polyamines have a role in controlling RNA synthesis. In evaluating the various papers concerned with the interrelations of polyamines and nucleic acids, the following factors must be kept in mind:

1. In most systems the effects of the polyamines are not qualitatively different from that of divalent cations, particularly Mg^{2+}. The polyamines are usually effective at a much lower concentration than needed for the divalent cations, probably as a result of their higher affinity for polyacids (see Section V.B.1).

2. In enzymatic experiments, in which a substrate, product, or primer is a polyanion, such as a nucleic acid, it is very difficult to distinguish between an effect of the polyamine on the polyanion and on the enzyme.

3. As indicated earlier in this review, the strongly basic polyamines can redistribute among the polyacids in the cell upon cellular disruption; therefore, the location of the amines in isolated cell components may not be an adequate indication of their location *in vivo*.

4. In experiments on the effects of polyamines added *in vivo*, either in animals or in microorganisms, it is difficult to evaluate whether the changes observed are direct effects of the polyamines or secondary to changes caused by the polyamines in some other systems. It is also necessary to distinguish between physiological and toxic effects, especially if high doses of the polyamines are used.

5. When parallel changes in the levels of polyamines and of some other cellular components result from experimental stimuli, it is necessary to evaluate whether a direct cause and effect relationship is involved or whether the changes measured are only distantly interrelated.

1. The Binding of Polyamines by Nucleic Acids

It is well established that the diamines and polyamines bind to nucleic acids and polynucleotides; spermine is the most tightly bound. This interaction of polyamines and nucleic acids has been shown by the many studies on the effects of polyamines on the susceptibility of nucleic acids to heat denaturation (139,141,176,183,232,238,246–248, 250,390,392,398–400,417,418,421) and to shearing (205), as well as studies on sedimentation coefficients (442) and solubility (5,161,442). The physical interaction of the polyamines and nucleic acids has been investigated with the use of x-ray diffraction (238,396b), optical rotatory dispersion (242,326,400), electron microscopy (69), nuclear magnetic resonance (130), circular dichroism (130), and miscellaneous other techniques (65,67,188,237,238,324,360). In considering the physical state of nucleic acids in the cell, one should note that the solubility studies indicate that if all of the polyamines in the cell were complexed with nucleic acids, the nucleic acids would not be in solution.

As discussed above, however, it has not been possible, with the use of conventional cell fractionation techniques, to test whether polyamines are associated with nucleic acids *in vivo*. A few attempts have been made, however, to answer this question by other techniques. For example, evidence against a preferential binding of polyamines to DNA was presented by Michaels and Tchen, who showed that the concentration of polyamines is similar in the DNA-free minicells and in the parent cell which contains the DNA (254,255). This study, however, is somewhat limited, since DNA is only a small fraction of the total bacterial nucleic acids, and, if the polyamines were bound equally to all nucleic acids, the differences obtained might have been too small to detect. Radioautography studies have been carried out with *Drosophila melanogaster* (92,93) and with *Tetrahymena pyriformis* GL (181) with added polyamines, but such techniques would not give any information on the localization of endogenous polyamines *in vivo*.

2. Effect of Polyamines on Enzymes That Synthesize and Degrade Nucleic Acid

The effect of polyamines on the synthesis and degradation of nucleic acids has been studied both with whole cells (37b,50,61,236,349,421,435) and with enzymatic preparations. The latter have included DNA-dependent RNA polymerase, (1,36,59,60,120,128,148a,148b,223,285, 302,314,338,386,392,395,471), DNA polymerase (52,285,352,392), deoxyribonuclease (26), and ribonuclease (106,131,210b,421). Stimulation, inhibition, or no effect has been observed, depending, in part, on the specific amine added and its concentration.

3. Effect of Polyamines on Ribosome Structure and Protein Synthesis

Polyamines have been found to stimulate protein or polypeptide synthesis in many *in vitro* systems. The amount of stimulation obtained depends on the type and purity of the ribosomes and protein-synthesizing enzymes used, as well as the relative ratio of spermidine: Mg^{2+} in the incubation mixture. In most experiments the stimulatory effects of polyamines were most pronounced in the presence of a low Mg^{2+} concentration (11,74,95,175,262,421,429); little or no effects were seen when polyamines were added to optimal Mg^{2+} levels. With certain other systems, however, little polypeptide was synthesized in the absence of spermidine, even in the presence of various concentrations of Mg^{2+}. The increase in polypeptide synthesized, when spermidine was added, was 2- to 10-fold in one experiment with extracts of dormant Jerusalem artichoke tubers (74), 10- to 40-fold in another with extracts from *B. thuringiensis* (68), and 5-fold in a cell-free system from bovine pituitary gland containing reconstituted ribosomes (4). Furano (personal communication) has also noted that spermidine was essential, in addition to magnesium, for optimal synthesis of protein in a system containing purified ribosomes and enzymes obtained from *E. coli*. Polyamines cannot completely replace the magnesium requirement in protein-synthesizing systems without a marked decrease in the activity (11,68,74,216,445).

A very important site of action of the polyamines in protein synthesis is the ribosome.* Cohen and Lichtenstein (78) first demonstrated that the polyamines can act like magnesium in maintaining ribosomal structure *in vitro*. Since then, many other investigators have confirmed these observations and have shown that polyamines have an effect in (1) stabilizing free 30 S and 50 S ribosomal subunits, (2) maintaining the association of these subunits as 70 S monomers, and (3) maintaining the polysomes (68,70,71,84,126,155,156,184,262,280,284,317,368,394,439, 445). However, Weiss and Morris (445) have shown that *E. coli* ribosomal subunit structure and function cannot be maintained by polyamines alone, but require a critical concentration of Mg^{2+} in addition to the polyamines. Part of the stabilizing action of the polyamines may be the result of the prevention by polyamines of the release of latent

---

* Although isolated ribosomes may contain high concentrations of amines, these data cannot be interpreted as indicating the association of ribosomes and polyamines *in vivo* because of the problems of redistribution of amines after cell lysis (see Section II.A.1).

ribosomal ribonuclease (106). A different effect of spermine on rat liver ribosomes was shown by Khawaja and Raina, who showed that the attachment of free ribosomes to the endoplasmic reticulum membrane at 0° was dependent on either spermine or magnesium (211a,b). Polyamines have been found associated with bacterial ribosomes (see Section II.A.1) and with mammalian microsomes (137,317). As we have previously discussed (Section II.A.1), because of the problem of redistribution after disruption of the cells, these findings do not necessarily indicate the localization of the polyamines in ribosomes or microsomes *in vivo*.

In addition to the effect of polyamines on the ribosomes, polyamines have been shown to have a stimulatory effect on many reactions which are important for protein biosynthesis. These include the renaturation of tRNA, the methylation of tRNA, the binding of aminoacyl tRNA and poly U to ribosomes, and the aminoacylation of tRNA (44,76b,94, 137,185,191,210a,233,234,294,325,346,430–432,434,468). In general, the stimulation appears to be the result of the substitution of the amines for the required magnesium. Cohen et al. (76b,79a) have reported that isolated tRNA contains 1–2 moles of spermidine per mole of tRNA; however, since this spermidine does not appear to be covalently linked, further studies are needed on the specificity of the binding and on the effect of the binding on the biological activity of the tRNA. Polyamines have also been shown to cause misreading of the code *in vitro* (127,444).

4. Studies in vivo on the Relationship of Polyamines and RNA in E. coli

Cohen and his associates (77,79b,308,316) have studied the changes in the concentrations of polyamines under various conditions in which there is a change in RNA synthesis *in vivo*. They have concluded that, in general, there is a correlation between the concentration of free spermidine and RNA synthesis; that is, with a decreased RNA synthesis there is a decreased spermidine accumulation. Their studies included the changes in RNA after chloramphenicol, streptomycin, and levorphanol treatment, and, in particular, the changes in stringent and relaxed strains upon deprivation of a required amino acid. In a stringent strain, RNA synthesis ceases upon removal of the amino acid, and Cohen and his associates (77,79b,308) showed that addition of large amounts of spermidine permitted RNA synthesis to proceed despite the absence of the amino acid. Although these results appeared to indicate a direct relationship between spermidine and RNA control, Ezekiel and Brockman (108) have presented evidence that the effect of spermidine is indirect. They showed that the large concentrations of spermidine cause an increase in protein degradation, as well as some inhibition of protein synthesis. The resultant increase in the amino acid pool would then relieve the amino acid starvation and permit RNA synthesis. In view of the latter findings, it is not possible at present to state that spermidine is directly involved in the control of RNA synthesis in stringent versus relaxed strains of E. coli. For a more detailed discussion of these results, the reader is referred to references 77,79b, 103,308, and 313, as well as the review of Stevens (392).

5. Changes in Ornithine Decarboxylase Activity in Rapidly Growing Tissues

In animal tissues, 1,4-diaminobutane appears to be synthesized by ornithine decarboxylase (297). Ornithine decarboxylase has a widespread distribution in animal tissues, although normal tissues contain only low levels of the enzyme. Ornithine decarboxylase has two very unusual characteristics:

1. Ornithine decarboxylase has an apparent half-life of only 11 min (342,344,384,385), which is considerably less than any other known mammalian enzyme. Such a short half-life would permit very rapid fluctuations of the quantity of enzyme present in response to physiological changes.

2. Ornithine decarboxylase activity is greatly increased in rapidly growing tissues. In regenerating liver, the ornithine decarboxylase activity is as much as 40- to 50-fold higher than in normal liver, the maximum usually occurring at 12–16 hr after hepatectomy (112,193, 194,307,311,312,339b,341,342,384). This large increase in the activity of ornithine decarboxylase appeared to be specific for this enzyme (341). Substantial increases in ornithine decarboxylase were also found in other rapidly growing tissues, such as embryonic tissues and some tumors (307,336b,337,341,343,384), and in target organs after administration of certain hormones (75,220,295,345,451). There have also been two reports of increases in ornithine decarboxylase activity during rapid growth of cells *in vitro:* (*a*) in human lymphocytes

stimulated by phytohemagglutinin (209); and (b) in epidermal cells and in mouse testes, which have a 10- to 20-fold increase in their ornithine decarboxylase activity after incubation with epidermal growth factor (389).

The increase in ornithine decarboxylase activity, however, is not always correlated with rapid growth. For example, some tumors do not exhibit a significant elevation of the activity (341). Also, the peak of ornithine decarboxylase activity occurs at 5 days in chick embryos and at 12 days in rat embryos, after the period of most rapid growth (385). In addition, a wide variety of stimuli that have no apparent effect on overall growth also cause a substantial increase in the ornithine decarboxylase activity of the liver, although the changes are smaller than that obseived after hepatectomy. These include such procedures as sham operation, intraperitoneal injections of 20% mannose or glucose, injections of histidine or thioacetamide, administration of casein hydrolysate or protein, intraperitoneal injections of celite, and administration of growth hormone to normal or pituitarectomized rats (112–114,160b,193,195,196,341,344), or of thyroxine, insulin, and glucagon to adrenalectomized rats (290,351). After almost all of these stimuli, the pattern of increase in the ornithine decarboxylase activity of the liver is similar: a rapid rise, often reaching the maximum at 4–12 hr after the stimulus, and a rapid fall approaching the normal levels over the next 24 hr (112,114,220,290,312,351,384,385,389).

Somewhat parallel to the increases noted in ornithine decarboxylase are changes in the levels of the amines and of RNA, which have been described in rapidly growing tissues or after various stimuli (55,57–59, 93,102,162,164,165,194,195,209,222,274,275,307,312,313,315,336a,336b, 339a–341,343,385). This correlation has been interpreted by many investigators as indicating a causal interrelationship between the synthesis of RNA and the increased activity of ornithine decarboxylase (and increased levels of amines). The correlation of amine levels and synthesis with RNA synthesis is very close in some systems, such as regenerating liver (112,162,195,307,312,315,385), developing chick embryo (55,307), embryos of Bufo bufo (59) and of Xenopus laevis (336a), proliferating endoneural cells of rats (353d), and after some hormones and drugs (58,209,293,312,354). The relationship is less clear in other systems, where the synthesis of RNA is not stimulated or where the peak of RNA synthesis and the peak of ornithine decarboxylase activity occur at different times (92,339b).

In interpreting the large amount of data that has been accumulated in this area, it is important to consider the following factors:

1. Berlin and Schimke have clearly shown that the rapid increase in the level of a given enzyme in response to an inducing agent does not mean that the stimulus is specific (40). In a situation where the synthesis of all proteins is stimulated equally, the level of enzymes with a rapid turnover (i.e., short half-life) will increase much more rapidly than those with a slow turnover. On the basis of these findings of Berlin and Schimke, one would expect that a general stimulation of all protein synthesis (as in regenerating liver) would lead to a rapid increase in ornithine decarboxylase levels because of its short half-life, without any specific stimulation of ornithine decarboxylase.

Such an increase in ornithine decarboxylase activity, though non-specific, may be physiologically important. Thus ornithine decarboxylase may have such a fast turnover in order to permit a rapid increase of this enzyme during periods of rapid growth without involving a more specific inducing mechanism.

2. For optimal activity of ornithine decarboxylase, dithiothreitol or mercaptoethanol has to be present both during the extraction and during the assay of ornithine decarboxylase (198,199,290,311). These conditions were not used for most of the assays reported.

3. The experiments presented do not unequivocally differentiate between increases in ornithine decarboxylase activity resulting from new enzyme synthesis and that caused by activation of preexisting protein. All of the studies showing an increase in ornithine decarboxylase have been based on activity measurements. The only data indicating that new protein is being formed to account for the increased ornithine decarboxylase activity was obtained in experiments with such compounds as cycloheximide, puromycin, and actinomycin D (112,114, 351). Interpretations of such data, however, are usually complicated by the toxic side reactions secondary to the high doses of inhibitor required for effects *in vivo*. Complications due to toxicity of the antibiotics, of course, would be particularly serious when one is studying a rapidly growing tissue, since the antibiotics would inhibit the growth of the cells, and thus would indirectly affect many systems.

Thus more data are needed to show that the increase in ornithine decarboxylase activity is entirely the result of new enzyme synthesis.

Immunologic determination of the specific enzyme protein increase would, of course, be very desirable.

4. Different tissues predominate at different stages of embryonic growth; since each tissue would be expected to have different amine levels, this factor complicates interpretations of the changes observed during development of the fetus. Similarly, in other rapidly growing tissues, one type of cell may predominate at a certain period and therefore influence the observed pattern of enzyme activity.

5. The increases in amine levels and in RNA synthesis may both be a reflection of rapid growth, without their necessarily having a causal relationship. Indeed, the maximum level of the RNA synthesis has been shown to precede, coincide with, or follow the peak of ornithine decarboxylase in different systems.

6. S-Adenosylmethionine Decarboxylase and Propylamine Transferase Activity in Rapidly Growing Tissues

The control of S-adenosylmethionine decarboxylase during growth has been investigated less extensively than the control of ornithine decarboxylase. However, increased adenosylmethionine decarboxylase activity has been found in embryos (336b,339a), in tumors (337), after the administration of growth hormone (339a), in the rat uterus after estrogen adeministration (345), and after hepatectomy in 2-month-old (but not older) rats (339a). The half-life of this enzyme was found to be 60 min (345). The amine concentration did not vary consistently with the enzyme levels in these systems.

Direct assays of the propylamine transferase activity in rapidly growing tissues have not been reported. Several authors have considered that the spermidine and spermine concentrations in tissues are an indication of the propylamine transferase activity. However, the concentration of these amines in a given tissue is influenced by many other factors and may not directly reflect changes in the propylamine transferase activity in the specific tissue under study. Some of these factors are (a) availability of 1,4-diaminobutane and of decarboxylated adenosylmethionine, which may vary as a result of changes in synthesis, degradation, or urinary excretion; and (b) changes in the capacity to take up exogenous amines, an example of which is the increase in the uptake of 1,4-diaminobutane in regenerating rat liver (102).

7. Relationship of Polyamines to Nucleoli and Nuclei

The demonstration of a physical association between polyamines and nuclei is difficult because of the problem of redistribution of polyamines *in vitro*, as mentioned several times earlier in this review. Stevens (392) has recently introduced the use of a nonaqueous solvent for the preparation of nuclei from animal tissues in an attempt to avoid this problem. His results did not support a close association, since equal concentrations of spermine and spermidine were found in the nuclei and the extranuclear fraction (392).

A physical association of ornithine decarboxylase and nuclei has been reported for chick embryo (384). About 25% of the enzyme is found in the nuclear fraction. Consistent with this finding is the simultaneous increase in ornithine decarboxylase activity and the nucleolar enlargement that results from the administration of thioacetamide, a compound known to stimulate nucleolar enlargement and RNA synthesis (112,113).

A relationship between ornithine decarboxylase and nucleoli was supported by work on embryos of an anucleolate mutant of *Xenopus laevis*. Embryos of this mutant, which cannot synthesize RNA, have a marked decrease in ornithine decarboxylase and S-adenosylmethionine decarboxylase as compared to a control embryo (336a,336b). However, interpretation of these experiments is difficult. Without ribosomal RNA synthesis, there is no increase in the number of ribosomes, and thus no increase in protein synthesis. As indicated in Section V.B.5, changes in the rate of protein synthesis are first reflected in the levels of enzymes with a rapid turnover (40), such as ornithine decarboxylase and adenosylmethionine decarboxylase.

The addition of polyamines to isolated nuclei or nucleoli has been shown to stimulate both RNA and DNA synthesis (37a,59,140,241). A very interesting stabilizing effect of spermine has been observed with isolated nuclei of embryos of *Rana pipiens*. The transplantation of nuclei of the late gastrula stage into enucleated eggs results in the normal development of only a low percentage of embryos. When spermine is added to the donor cells and nuclei, the number of embryos that undergo normal development double (163). This effect may be related to the preservation of nuclear morphology, which was observed when polyamines were added to isolated rat liver nuclei during incubation at 37° (241).

C. SUMMARY

It is not possible at this time to state if the above observations on the relationships between polyamines and nucleic acids are related to the function of the polyamines in the cell. It seems clear, however, from the studies with mutants and from the widespread occurrence of the amines in natural materials that the polyamines do have an essential role *in vivo*, but that it will be necessary for future experiments to delineate which of the many observations summarized in this review are related to the physiological functions of the polyamines.

References

1. Abraham, K. A., *Eur. J. Biochem.*, 5, 143–146 (1938).
2. Abrams, A., *J. Biol. Chem.*, 240, 3675–3681 (1965).
3. Adachi, O., Yamada, H., and Ogata, K., *Agr. Biol. Chem.*, 30, 1202–1210 (1966).
4. Adiga, P. R., Hussa, R. O., and Winnick, T., *Biochemistry*, 7, 1808–1817 (1968).
5. Agrell, I. P. S., and Heby, O., *Exp. Cell Res.*, 50, 668–671 (1968).
6. Akopyan, Z. I., Stesina, L. N., and Gorkin, V. Z., *J. Biol. Chem.*, 246, 4610–4618 (1971).
7. Alarcon, R. A., *Arch. Biochem. Biophys.*, 106, 240–242 (1964).
8. Alarcon, R. A., *Arch. Biochem. Biophys.*, 113, 281–287 (1966).
9. Alarcon, R. A., *Fed. Proc.*, 28, 805 (1969).
10. Alarcon, R. A., *Arch. Biochem. Biophys.*, 137, 365–372 (1970).
11. Algranati, I. D., and Lengyel, P., *J. Biol. Chem.*, 241, 1778–1783 (1966).
12. Ames, B. N., and Dubin, D. T., *J. Biol. Chem.*, 235, 769–775 (1960).
13. Ames, B. N., Dubin, D. T., and Rosenthal, S. M., *Science*, 127, 814–816 (1958).
14a. Andersson, G., and Heby, O., *J. Nat. Cancer Inst.*, 45, 165–172 (1972).
14b. Audit, C., Viala, B., and Robin, Y., *Comp. Biochem. Physiol.*, 22, 775–785 (1967).
15. Bachrach, U., *Nature*, 194, 377–378 (1962).
16. Bachrach, U., *J. Biol. Chem.*, 237, 3443–3448 (1962).
17. Bachrach, U., *Ann. N.Y. Acad. Sci.*, 171, 939–956 (1970).
18. Bachrach, U., *Ann. Rev. Microbiol.*, 24, 109–134 (1970).
19. Bachrach, U., Abzug, S., and Bekierkunst, A., *Biochim. Biophys. Acta*, 134, 174–181 (1967).
20. Bachrach, U., Bekierkunst, A., and Abzug, S., *Israel J. Med. Sci.*, 3, 474–477 (1967).
21. Bachrach, U., and Ben-Joseph, M., *FEBS Lett.*, 15, 75–77 (1971).
22. Bachrach, U., and Cohen, I., *J. Gen. Microbiol.*, 26, 1–9 (1961).
23. Bachrach, U., and Don, S., *J. Gen. Virol.*, 11, 1–9 (1971).

24. Bachrach, U., and Eilon, G., *Biochim. Biophys. Acta*, *145*, 418–426 (1967).
25. Bachrach, U., and Eilon, G., *Biochim. Biophys. Acta*, *179*, 473–483 (1969).
26. Bachrach, U., and Eilon, G., *Biochim. Biophys. Acta*, *179*, 494–496 (1969).
27. Bachrach, U., and Leibovici, J., *Israel J. Med. Sci.*, *1*, 541–551 (1965).
28. Bachrach, U., and Leibovici, J., *J. Mol. Biol.*, *19*, 120–132 (1966).
29. Bachrach, U., and Oser, I. S., *J. Biol. Chem.*, *238*, 2098–2101 (1963).
30. Bachrach, U., and Persky, S., *Biochem. Biophys. Res. Commun.*, *24*, 135–140 (1966).
31. Bachrach, U., and Persky, S., *Biochim. Biophys. Acta*, *179*, 484–493 (1969).
32. Bachrach, U., Rabina, S., Loebenstein, G., and Eilon, G., *Nature*, *208*, 1095–1096 (1965).
33. Bachrach, U., and Reches, B., *Anal. Biochem.*, *17*, 38–48 (1966).
34a. Bachrach, U., and Robinson, E., *Israel J. Med. Sci.*, *1*, 247–250 (1965).
34b. Bachrach, U., Tabor, C. W., and Tabor, H., *Biochim. Biophys. Acta*, *78*, 768–770 (1963).
35. Bagni, N., G. *Bot. Ital.*, *102*, 67–72 (1968); Bagni, N., Caldarera, C. M., and Moruzzi, G., *Experientia*, *23*, 139–141 (1967).
36. Ballard, P. L., and Williams-Ashman, H. G., *J. Biol. Chem.*, *241*, 1602–1615 (1966).
37a. Barbiroli, B., Corti, A., and Caldarera, C. M., *FEBS Lett.*, *13*, 169–172 (1971).
37b. Barbiroli, B., Corti, A., and Caldarera, C. M., *Biochem. J.*, *123*, 123–124 (1971).
37c. Bard, D. S., and Ketcham, A. S., *Cancer*, *19*, 1149–1154 (1966).
38. Bardsley, W. G., Hill, C. M., and Lobley, R. W., *Biochem. J.*, *117*, 169–176 (1970).
39. Beer, S. V., and Kosuge, T., *Virology*, *40*, 930–938 (1970).
40. Berlin, C. M., and Schimke, R. T., *Mol. Pharmacol.*, *1*, 149–156 (1965).
41. Bertossi, F., Bagni, N., Moruzzi, G., and Caldarera, C. M., *Experientia*, *21*, 80–83 (1965).
42. Blaschko, H., and Bonney, R., *Proc. Roy. Soc., Ser. B*, *156*, 268–279 (1962).
43. Blethen, S. L., Boeker, E. A., and Snell, E. E., *J. Biol. Chem.*, *243*, 1671–1677 (1968).
44. Bluestein, H. G., Allende, C. C., Allende, J. E., and Cantoni, G. L., *J. Biol. Chem.*, *243*, 4693–4699 (1968).
45. Boeker, E. A., Fischer, E. H., and Snell, E. E., *J. Biol. Chem.*, *244*, 5239–5245 (1969).
46. Boeker, E. A., Fischer, E. H., and Snell, E. E., *J. Biol. Chem.*, *246*, 6776–6781 (1971).
47. Boeker, E. A., and Snell, E. E., *J. Biol. Chem.*, *243*, 1678–1684 (1968).
48. Boeker, E. A., and Snell, E. E., in *Methods in Enzymology*, Vol. 17, Part B, H. Tabor and C. W. Tabor, Eds., Academic Press, New York, 1971, pp. 657–662.
49. Bowman, W. H., Tabor, C. W., and Tabor, H., manuscript in preparation.
50. Boyle, S. M., and Cohen, P. S., *J. Bacteriol.*, *96*, 1266–1272 (1968).
51. Bremer, H. J., Kohne, E., and Endres, W., *Clin. Chim. Acta*, *32*, 407–418 (1971).

52. Brewer, E. N., and Rusch, H. P., *Biochem. Biophys. Res. Commun.*, *25*, 579–584 (1966).
53. Brock, M. L., *Virology*, *26*, 221–227 (1965).
54. Buffoni, F., *Pharmacol. Rev.*, *18*, 1163–1199 (1966).
55. Caldarera, C. M., Barbiroli, B., and Moruzzi, G., *Biochem. J.*, *97*, 84–88 (1965).
56. Caldarera, C. M., Casti, A., De Sanctis, B., and Moruzzi, G., *FEBS Lett.*, *14*, 29–32 (1971).
57. Caldarera, C. M., Cozzani, C., and Moruzzi, M. S., *Experientia*, *22*, 579–580 (1966).
58. Caldarera, C. M., Giorgi, P. P., and Casti, A., *J. Endocrinol.*, *46*, 115–116 (1970).
59. Caldarera, C. M., and Moruzzi, G., *Ann. N.Y. Acad. Sci.*, *171*, 709–722 (1970).
60. Caldarera, C. M., Moruzzi, M. S., Barbiroli, B., and Moruzzi, G., *Biochem. Biophys. Res. Commun.*, *33*, 266–271 (1968).
61. Caldarera, C. M., Moruzzi, M. S., Rossoni, C., and Barbiroli, B., *J. Neurochem.*, *16*, 309–316 (1969).
62. Campello, A. P., Tabor, C. W., and Tabor, H., *Biochem. Biophys. Res. Commun.*, *19*, 6–9 (1965).
63. Cantoni, G. L., in *Methods in Enzymology*, Vol. 2, S. P. Colowick and N. O. Kaplan, Eds., Academic Press, New York, 1955, pp. 254–256.
64. Cantoni, G. L., and Durrell, J., *J. Biol. Chem.*, *225*, 1033–1048 (1957).
65. Carr, C. W., *Fed. Proc.*, *30*, 1146 Abstr. (1971).
66. Castelli, A., and Rossoni, C., *Experientia*, *24*, 1119–1120 (1968).
67. Chang, K. Y., and Carr, C. W., *Biochim. Biophys. Acta*, *157*, 127–139 (1968).
68. Changchien, L., and Aronson, J. N., *J. Bacteriol.*, *103*, 734–740 (1970).
69. Chevaillier, P., *Exp. Cell Res.*, *58*, 213–224 (1969).
70. Choi, Y. S., and Carr, C. W., *J. Mol. Biol.*, *25*, 331–345 (1967).
71. Choi, Y. S., and Carr, C. W., *Nature*, *217*, 556–557 (1968).
72. Chu, T. M., Mallette, M. F., and Mumma, R. O., *Biochemistry*, *4*, 1399–1406 (1968).
73. Cleaves, G. R., and Cohen, P. S., *J. Bacteriol.*, *103*, 697–701 (1970).
74. Cocucci, S., and Bagni, N., *Life Sci. (Oxford)*, *7*, 113–120 (1968).
75. Cohen, S., O'Malley, B. W., and Stastny, M., *Science*, *170*, 336–338 (1970).
76a. Cohen, S. S., *Virus-Induced Enzymes*, Columbia University Press, New York, 1968.
76b. Cohen, S. S., *Ann. N.Y. Acad. Sci.*, *171*, 869–881 (1970).
76c. Cohen, S. S., *Introduction to the Polyamines*, Prentice-Hall, Englewood Cliffs, New Jersey, 1971.
77. Cohen, S. S., Hoffner, N., Jansen, M., Moore, M., and Raina, A., *Proc. Nat. Acad. Sci. U.S.*, *57*, 721–728 (1967).
78. Cohen, S. S., and Lichtenstein, J., *J. Biol. Chem.*, *235*, 2112–2116 (1960).
79a. Cohen, S. S., Morgan, S., and Streibel, E., *Proc. Nat. Acad. Sci. U.S.*, *64*, 669–676 (1969).

79b. Cohen, S. S., and Raina, A., in *Organizational Biosynthesis*, H. J. Vogel, J. O. Lampen, and V. Bryson, Eds., Academic Press, New York, 1967, pp. 157–182.

80a. Costa, M. T., Rotilio, G., Finazzi-Agro, A. F., Vallogini, M. P., and Mondovi, B., *Arch. Biochem. Biophys.*, *147*, 8–13 (1971).

80b. Creveling, C. R., and Daly, J. W., in *Methods in Enzymology*, Vol. 17, Part B, H. Tabor and C. W. Tabor, Eds., Academic Press, New York, 1971, pp. 846–850.

81. Damiani, A., Giglio, E., Puliti, R., and Ripamonti, A., *J. Mol. Biol.*, *11*, 441–442 (1965).

82. Damiani, A., Liquori, A. M., Puliti, R., and Ripamonti, A., *J. Mol. Biol.*, *11*, 438–440 (1965).

83. Darrow, R. A., and Rodstrom, R., *Proc. Nat. Acad. Sci. U.S.*, *55*, 205–212 (1966).

84. Datta, R. K., Sen, S., and Ghosh, J. J., *Biochem. J.*, *114*, 847–854 (1969).

85. Davis, G. R. F., *Comp. Biochem. Physiol.*, *19*, 619–627 (1966).

86. Davis, R. H., *J. Bacteriol.*, *96*, 389–395 (1968).

87. Davis, R. H., Lawless, M. B., and Port, L. A., *J. Bacteriol.*, *102*, 299–305 (1970).

88. De Meis, L., *Amer. J. Physiol.*, *212*, 92–96 (1967).

89. De Meis, L., *J. Biol. Chem.*, *243*, 1174–1179 (1968).

90. Derrick, N., and Laki, K., *Biochem. Biophys. Res. Commun.*, *22*, 82–88 (1966).

91a. Dion, A. S., and Cohen, S. S., *Fed. Proc.*, *30*, 1263 Abstr. (1971).

91b. Dion, A. S., and Cohen, S. S., *Proc. Nat. Acad. Sci. U.S.*, *69*, 213–217 (1972).

92. Dion, A. S., and Herbst, E. J., *Proc. Nat. Acad. Sci. U.S.*, *58*, 2367–2371 (1967).

93. Dion, A. S., and Herbst, E. J., *Ann. N.Y. Acad. Sci.*, *171*, 723–734 (1970).

94. Doctor, B. P., Fournier, M. J., and Thornsvard, C., *Ann. N.Y. Acad. Sci.*, *171*, 863–868 (1970).

95. Downey, K. M., So, A. G., and Davie, E. W., *Biochemistry*, *4*, 1702–1709 (1965).

96. Dubin, D. T., *Biochem. Biophys. Res. Commun.*, *1*, 262–265 (1959).

97. Dubin, D. T., and Rosenthal, S. M., *J. Biol. Chem.*, *235*, 776–782 (1960).

98. Dudley, H. W., Rosenheim, M. C., and Rosenheim, O., *Biochem. J.*, *18*, 1263–1272 (1924).

99. Dudley, H. W., Rosenheim, O., and Starling, W. W., *Biochem. J.*, *20*, 1082–1094 (1926).

100. Duerre, J. A., in *Methods in Enzymology*, Vol. 17, Part B, H. Tabor and C. W. Tabor, Eds., Academic Press, New York, 1971, pp. 411–415.

101. Dvilansky, A., Britten, A. F. H., and Loewy, A. G., *Brit. J. Haematol.*, *18*, 399–410 (1970).

102. Dykstra, W. G., Jr., and Herbst, E. J., *Science*, *149*, 428–429 (1965).

103. Edlin, G., and Broda, P., *Bacteriol. Rev.*, *32*, 206–226 (1968).

104. Eilon, G., and Bachrach, U., *Biochim. Biophys. Acta*, *179*, 464–472 (1969).

105. Elliott, B., and Michaelson, I. A., *Anal. Biochem.*, *19*, 184–187 (1967).

106. Erdmann, V. A., Thomas, G. A., Norton, J. W., and Herbst, E. J., *Biochim. Biophys. Acta*, *157*, 43–51 (1968).

107. Erskine, J. M., *Appl. Microbiol.*, *19*, 638–642 (1970).

108. Ezekiel, D. H., and Brockman, H., *J. Mol. Biol.*, *31*, 541–552 (1968).

109. Fabro, S., Smith, R. L., and Williams, R. T., *Nature*, *208*, 1208–1209 (1965).

110. Fair, W. R., and Wehner, N., *Appl. Microbiol.*, *21*, 6–8 (1971).

111. Falaschi, A., and Kornberg, A., *Proc. Nat. Acad. Sci. U.S.*, *54*, 1713–1720 (1965).

112. Fausto, N., *Biochim. Biophys. Acta*, *190*, 193–201 (1969).

113. Fausto, N., *Cancer Res.*, *30*, 1947–1952 (1970).

114. Fausto, N., *Biochim. Biophys. Acta*, *238*, 116–128 (1971).

115. Feldman, M. J., and Russell, D. H., *Fed. Proc.*, *30*, 1297 Abstr. (1971).

116. Feldman, M. J., Levy, C. C., and Russell, D. H., *Biochem. Biophys. Res. Commun.*, *44*, 675–681 (1971); *Biochemistry*, *11*, 671–677 (1972).

117. FerroLuzzi-Ames, G., and Ames, B. N., *Biochem. Biophys. Res. Commun.*, *18*, 639–647 (1965).

118. Finazzi-Agro, A., Rotilio, G., Costa, M. T., and Mondovi, B., *FEBS Lett.*, *4*, 31–32 (1969).

119. Folk, J. E., and Cole, P. W., *Biochim. Biophys. Acta*, *122*, 244–264 (1966).

120. Fox, C. F., Gumport, R. I., and Weiss, S. B., *J. Biol. Chem.*, *240*, 2101–2109 (1965).

121. Fraser, D., and Mahler, H. R., *J. Amer. Chem. Soc.*, *80*, 6456 (1958).

122. Friedman, H., Lu, P., and Rich, A., *Nature*, *223*, 909–913 (1969).

123. Friedman, M. E., and Bachrach, U., *J. Bacteriol.*, *92*, 49–55 (1966).

124. Friedman, S., *Fed. Proc.*, *16*, 183 (1957).

125a. Friedman, S. J., Bellantone, R. A., and Canellakis, E. S., *Biochim. Biophys. Acta*, *261*, 188–193 (1972).

125b. Friedman, S. J., and Canellakis, E. S., *Fed. Proc.*, *30*, 1298 Abstr. (1971).

125c. Friedman, S. J., Halpern, K. V., and Canellakis, E. S., *Biochim. Biophys. Acta*, *261*, 181–187 (1972).

126. Friedman, S. M., Axel, R., and Weinstein, I. B., *J. Bacteriol.*, *93*, 1521–1526 (1967).

127. Friedman, S. M., and Weinstein, I. B., *Proc. Nat. Acad. Sci. U.S.*, *52*, 988–996 (1964).

128. Fuchs, E., Millette, R. L., Zillig, W., and Walter, G., *Eur. J. Biochem.*, *3*, 183–193 (1967).

129. Fukami, H., Tomida, I., Morino, T., Yamada, H., Oki, T., Kawasaki, H., and Ogata, K., *Biochem. Biophys. Res. Commun.*, *28*, 19–24 (1967).

130. Gabbay, E. J., Glasser, R., and Gaffney, B. L., *Ann. N.Y. Acad. Sci.*, *171*, 810–826 (1970).

131. Gabbay, E. J., and Shimshak, R. R., *Biopolymers*, *6*, 255–268 (1968).

132. Gale, E. F., *Biochem. J.*, *36*, 64–75 (1942).

133. Gale, E. F., in *Advances in Enzymology*, Vol. 6, F. F. Nord, Ed., Interscience, New York, 1946, pp. 1–32.

134. George, D. J., and Phillips, A. T., *Fed. Proc.*, *28*, 668 (1969).

135. George, D. J., and Phillips, A. T., *J. Biol. Chem.*, *245*, 528–537 (1970).

136. Giglio, E., Liquori, A. M., Puliti, R., and Ripamonti, A., *Acta Crystallogr.*, *20*, 652–659 (1966).
137. Giorgi, P. P., *Biochem. J.*, *120*, 643–651 (1970).
138. Givot, I. L., Smith, T. A., and Abeles, R. H., *J. Biol. Chem.*, *244*, 6341–6353 (1969).
139. Glaser, R., and Gabbay, E. J., *Biopolymers*, *6*, 243–254 (1968).
140. Goldstein, J., *Exp. Cell Res.*, *37*, 494–497 (1965).
141. Goldstein, J., *Biochim. Biophys. Acta*, *123*, 620–623 (1966).
142a. Goryachenkova, E. V., Shcherbatiuk, L. I., and Voronina, E. A., *Biokhimiya*, *32*, 398–402 (1967); English ed., *32*, 330–333 (1967).
142b. Gorycahenkova, E. V., Shcherbatiuk, L. I., Zamaraev, K. I., in *Pyridoxal Catalysis: Enzymes and Model Systems*, E. E. Snell, A. E. Braunstein, E. S. Severin, and Y. M. Torchinsky, Eds., Interscience, New York, 1968, pp. 391–402.
142c. Goryachenkova, E. V., Zamaraev, K. I., and Scherbatiuk, L. I., *Dokl. Akad. Nauk Uzb SSR*, *173*, 707–710 (1967); *Dokl. Biol. Sci.*, *173*, 87–89 (1967).
143. Greene, R. C., *J. Amer. Chem. Soc.*, *79*, 3929 (1957).
144. Groman, N. B., and Suzuki, G., *J. Bacteriol.*, *92*, 1735–1740 (1966).
145. Grossowicz, N., and Ariel, M., *J. Bacteriol.*, *85*, 293–300 (1963).
146. Grossowicz, N., and Ariel, M., *Israel J. Med. Sci.*, *1*, 320 (1965).
147. Guirard, B. M., and Snell, E. E., *J. Bacteriol.*, *88*, 72–80 (1964).
148a. Gumport, R. I., *Ann. N.Y. Acad. Sci.*, *171*, 915–938 (1970).
148b. Gumport, R. I., and Weiss, S. B., *Biochemistry*, *8*, 3618–3628 (1969).
149. Ham, R. G., *Biochem. Biophys. Res. Commun.*, *14*, 34–38 (1964).
150. Hamalainen, R., *Acta Soc. Med. "Duodecim," Ser. A*, *23*, 97–165 (1941).
151. Hammond, J. E., and Herbst, E. J., *Anal. Biochem.*, *22*, 474–484 (1968).
152a. Hannonen, P., Jänne, J., and Raina, A., *Biochem. Biophys. Res. Commun.*, *46*, 341–348 (1972).
152b. Hannonen, P., Raina, A., and Khawaja, J. A., *Scand. J. Clin. Lab. Invest.*, *27*, *Suppl. 116*, 4 (1971).
153. Hanson, K. R., and Havir, E. A., *Fed. Proc.*, *28*, 602 (1969).
154. Hanson, K. R., and Havir, E. A., *Arch. Biochem. Biophys.*, *141*, 1–17 (1970).
155. Hardy, S. J. S., and Turnock, G., *Nature New Biol.*, *229*, 17–19 (1971).
156. Hardy, S. J. S., and Turnock, G., *Nature New Biol.*, *232*, 152–153 (1971).
157. Harold, F. M., *J. Bacteriol.*, *88*, 1416–1420 (1964).
158. Hasse, K., Ratych, O. T., and Salnikow, J., *Hoppe-Seyler's Z. Physiol. Chem.*, *348*, 843–851 (1967).
159. Hasse, K., and Schmid, G., *Biochem. Z.*, *337*, 69–79 (1963).
160a. Hatano, H., Sumizu, K., Rokushika, S., and Murakami, F., *Anal. Biochem.*, *35*, 377–383 (1970).
160b. Hayashi, S., Aramaki, Y., and Noguchi, T., *Biochem. Biophys. Res. Commun.*, *46*, 795–800 (1972).
161. Heby, O., and Agrell, I., *Hoppe-Seyler's Z. Physiol. Chem.*, *352*, 29–38 (1971).
162. Heby, O., and Lewan, L., *Virchows Arch. Abt. B Zellpathol.*, *8*, 58–66 (1971).
163. Hennen, S., *Proc. Nat. Acad. Sci. U.S.*, *66*, 630–637 (1970).

164. Herbst, E. J., and Dion, A. S., *Fed. Proc.*, *29*, 1563–1567 (1970).
165. Herbst, E. J., and Dykstra, W. G., Jr., *Int. Biochem. Meetings, New York, Abstr. III-24*, p. 230 (1964).
166. Herbst, E. J., and Snell, E. E., *J. Biol. Chem.*, *176*, 989–990 (1948).
167. Herbst, E. J., and Snell, E. E., *J. Biol. Chem.*, *181*, 47–54 (1949).
168. Herbst, E. J., Weaver, R. H., and Keister, D. L., *Arch. Biochem. Biophys.*, *75*, 171–177 (1958).
169. Hettinger, T. P., and Craig, L. C., *Biochemistry*, *7*, 4147–4153 (1968).
170. Hettinger, T. P., and Craig, L. C., *Biochemistry*, *9*, 1224–1232 (1970).
171. Hettinger, T. P., Kurylo-Borowska, Z., and Craig, L. C., *Biochemistry*, *7*, 4153–4160 (1968).
172. Hettinger, T. P., Kurylo-Borowska, Z., and Craig, L. C., *Ann. N.Y. Acad. Sci.*, *171*, 1002–1009 (1970).
173. Higgins, M. L., Tillman, M. C., Rupp, J. P., and Leach, F. R., *J. Cell Physiol.*, *74*, 149–154 (1969).
174. Hill, J. M., in *Methods in Enzymology*, Vol. 17, Part B, H. Tabor and C. W. Tabor, Eds., Academic Press, New York, 1971, pp. 730–735.
175. Hirashima, A., Asano, K., and Tsugita, A., *Biochim. Biophys. Acta*, *134*, 165–173 (1967).
176. Hirschman, S. Z., Leng, M., and Felsenfeld, G., *Biopolymers*, *5*, 227–233 (1967).
177. Hirshfield, I. N., Rosenfeld, H. J., Leifer, Z., and Maas, W. K., *J. Bacteriol.*, *101*, 725–730 (1970).
178. Hodgins, D. S., and Abeles, R. H., *J. Biol. Chem.*, *242*, 5158–5159 (1967).
179. Hodgins, D. S., and Abeles, R. H., *Arch. Biochem. Biophys.*, *130*, 274–285 (1969).
180. Holder, S., and Bremer, H. J., *J. Chromatogr.*, *25*, 48–57 (1966).
181. Holm, B., and Emanuelsson, H., *Z. Zellforsch. Mikroskop. Anat.*, *115*, 593–598 (1971).
182. Holm, B., and Heby, O., *Z. Naturforsch.*, *B*, *26*, 604–606 (1971).
183. Horacek, P., and Cernohorsky, I. J., *Biochem. Biophys. Res. Commun.*, *32*, 956–962 (1968).
184. Hurwitz, C., and Rosano, C. L., *J. Biol. Chem.*, *242*, 3719–3722 (1967).
185. Igarashi, K., and Takeda, Y., *Biochim. Biophys. Acta*, *213*, 240–243 (1970).
186. Iitaka, Y., and Huse, Y., *Bull. Chem. Soc. Jap.*, *37*, 437–439 (1964).
187. Iitaka, Y., and Huse, Y., *Acta Crystallogr.*, *Sect. B*, *18*, 110–121 (1965).
188. Ikemura, T., *Biochim. Biophys. Acta*, *195*, 389–395 (1969).
189. Inouye, M., and Pardee, A. B., *Ann. N.Y. Acad. Sci.*, *171*, 901–909 (1970).
190. Inouye, M., and Pardee, A. B., *J. Bacteriol.*, *101*, 770–776 (1970).
191. Ishida, T., and Sueoka, N., *J. Biol. Chem.*, *243*, 5329–5336 (1968).
192. Jakoby, W. B., and Fredericks, J., *J. Biol. Chem.*, *234*, 2145–2150 (1959).
193. Janne, J., *Acta Physiol. Scand.*, *Suppl. 300*, 1–71 (1967).
194. Janne, J., and Raina, A., *Acta Chem. Scand.*, *22*, 1349–1351 (1968).
195. Janne, J., and Raina, A., *Biochim. Biophys. Acta*, *174*, 769–772 (1969).
196. Janne, J., Raina, A., and Siimes, M., *Biochim. Biophys. Acta*, *166*, 419–426 (1968).

197. Janne, J., Schenone, A., and Williams-Ashman, H. G., *Biochem. Biophys. Res. Commun.*, *42*, 758–764 (1971).

198. Janne, J., and Williams-Ashman, H. G., *Biochem. J.*, *119*, 595–597 (1970).

199. Janne, J., and Williams-Ashman, H. G., *J. Biol. Chem.*, *246*, 1725–1732 (1971).

200. Janne, J., and Williams-Ashman, H. G., *Biochem. Biophys. Res. Commun.*, *42*, 222–229 (1971).

201. Janne, J., Williams-Ashman, H. G., and Schenone, A., *Biochem. Biophys. Res. Commun.*, *43*, 1362–1368 (1971).

202a. Jellinck, P. H., and Cox, J., *Experientia*, *26*, 1066 (1970).

202b. Jellinck, P. H., and Perry, G., *Biochim. Biophys. Acta*, *137*, 367–374 (1967).

203. Johnson, H. G., and Bach, M. K., *Proc. Nat. Acad. Sci. U.S.*, *55*, 1453–1456 (1966).

204. Johnson, M. W., and Markham, R., *Virology*, *17*, 276–281 (1962).

205. Kaiser, D., Tabor, H., and Tabor, C. W., *J. Mol. Biol.*, *6*, 141–147 (1963).

206. Kakimoto, Y., Nakajima, T., Kumon, A., Matsuoka, Y., Imaoka, N., Sano, I., and Kanazawa, A., *J. Biol. Chem.*, *244*, 6003–6007 (1969).

207. Kapeller-Adler, R., *Amine Oxidases and Methods for Their Study*, Wiley-Interscience, New York, 1970.

208. Katz, E., Goldblum, T., Bachrach, U., and Goldblum, N., *Israel J. Med. Sci.*, *3*, 575–576 (1967).

209. Kay, J. E., and Cooke, A., *FEBS Lett.*, *16*, 9–12 (1971).

210a. Kayne, M. S., and Cohn, M., *Biochem. Biophys. Res. Commun.*, *46*, 1285–1291 (1972).

210b. Kedracki, R., and Szer, W., *Acta Biochim. Pol.*, *14*, 163–168 (1967).

211a. Khawaja, J. A., *Biochim. Biophys. Acta*, *254*, 117–128 (1971).

211b. Khawaja, J. A., and Raina, A., *Biochem. Biophys. Res. Commun.*, *41*, 512–518 (1970).

212. Kim, K., *J. Biol. Chem.*, *239*, 783–786 (1964).

213. Kim, K., and Tchen, T. T., *Biochem. Biophys. Res. Commun.*, *9*, 99–102 (1962).

214. Kim, K., and Tchen, T. T., in *Methods in Enzymology*, Vol. 17, Part B, H. Tabor and C. W. Tabor, Eds., Academic Press, New York, 1971, pp. 812–815.

215. Kim, W. K., *Can. J. Bot.*, *49*, 1119–1122 (1971).

216. Kimes, B. W., Weiss, R. L., and Morris, D. R., *Fed. Proc.*, *30*, 1204 Abstr. (1971).

217. Kimes, B. W., and Morris, D. R., *Biochim. Biophys. Acta*, *228*, 223–234 (1971).

218. Kimes, B. W., and Morris, D. R., *Biochim. Biophys. Acta*, *228*, 235–244 (1971).

219. Kleczkowski, K., and Wielgat, B., *Bull. Acad. Pol. Sci. Ser. Sci. Biol.*, *16*, 521–526 (1968).

220. Kobayashi, Y., Kupelian, J., and Maudsley, D. V., *Science*, *172*, 379–380 (1971).

221. Kohne, E., and Bremer, H. J., *Klin. Wochenschr.*, *47*, 214–220 (1969).

222. Kostyo, J. L., *Biochem. Biophys. Res. Commun.*, *23*, 150–155 (1966).

223. Krakow, J. S., *Biochim. Biophys. Acta*, *72*, 566–571 (1963).
224. Kremzner, L. T., *Anal. Biochem.*, *15*, 270–277 (1966).
225a. Kremzner, L. T., *Fed. Proc.*, *29*, 1583–1588 (1970).
225b. Kremzner, L. T., Barrett, R. E., and Terrano, M. J., *Ann. N.Y. Acad. Sci.*, *171*, 735–748 (1970).
226. Kremzner, L. T., and Harter, D. H., *Biochem. Pharmacol.*, *19*, 2541–2550 (1970).
227. Kullnig, R., Rosano, C. L., and Hurwitz, C., *Biochem. Biophys. Res. Commun.*, *39*, 1145–1148 (1970).
228. Kumagai, H., Nagate, T., Yamada, H., and Fukami, H., *Biochim. Biophys. Acta*, *185*, 242–244 (1969).
229a. Kupchan, S. M., Davies, A. P., Barboutis, S. J., Schnoes, H. K., and Burlingame, A. L., *J. Amer. Chem. Soc.*, *89*, 5718–5719 (1967); *J. Org. Chem.*, *34*, 3888–3893 (1969).
229b. Kurylo-Borowska, Z., and Tatum, E. L., *Biochim. Biophys. Acta*, *114*, 206–209 (1966).
230. Kuttan, R., Radhakrishnan, A. N., Spande, T., and Witkop, B., *Biochemistry*, *10*, 361–365 (1971).
231. Laki, K., Tyler, H. M., and Yancey, S. T., *Biochem. Biophys. Res. Commun.*, *24*, 776–781 (1966).
232. Lampson, G. P., Tytell, A. A., Field, A. K., Nemes, M. M., and Hilleman, M. R., *Proc. Soc. Exp. Biol. Med.*, *132*, 212–218 (1969).
233. Leboy, P. S., *Ann. N.Y. Acad. Sci.*, *171*, 895–900 (1970); *Biochemistry*, *9*, 1577–1584 (1970).
234. Leboy, P. S., *FEBS Lett.*, *16*, 117–120 (1971).
235. Lewenhoeck, A., *Phil. Trans.*, *12*, 1040–1043 (1678). Transl. in Alle de brieven van Antoni van Leeuwenhoek, J. J. Swart, Ed., N. V. Swets and Zeitlinger, Amsterdam, 1941.
236. Linskens, H. F., Kochuyt, A. S. L., and So, A., *Planta*, *82*, 111–122 (1968).
237. Liquori, A. M., Ascoli, F., De Santis Savino, M., *J. Mol. Biol.*, *24*, 123–124 (1967).
238. Liquori, A. M., Costantino, L., Crescenzi, V., Elia, V., Giglio, E., Puliti, R., De Santis Savino, M., and Vitagliano, V., *J. Mol. Biol.*, *24*, 113–122 (1967).
239. Lockwood, D. H., Lipsky, J. J., Meronk, F., Jr., and East, L. E., *Biochem. Biophys. Res. Commun.*, *44*, 600–607 (1971).
240. Maas, W. K., Leifer, Z., and Poindexter, J., *Ann. N.Y. Acad. Sci.*, *171*, 957–967 (1970).
241. MacGregor, R. R., and Mahler, H. R., *Arch. Biochem. Biophys.*, *120*, 136–157 (1967).
242. Maestre, M. F., and Tinoco, I., Jr., *J. Mol. Biol.*, *23*, 323–335 (1967).
243. Mager, J., *Nature*, *176*, 933–934 (1955).
244. Mager, J., *Nature*, *183*, 1827–1828 (1959).
245. Mager, J., *Biochim. Biophys. Acta*, *36*, 529–531 (1959).
246. Mahler, H. R., and Green, G., *Ann. N.Y. Acad. Sci.*, *171*, 783–800 (1970).
247. Mahler, H. R., and Mehrotra, B. D., *Biochim. Biophys. Acta*, *68*, 211–233 (1963).

248. Mahler, H. R., Mehrotra, B. D., and Sharp, C. W., *Biochem. Biophys. Res. Commun.*, *4*, 79–82 (1961).

249. Maretzki, A., Thom, M., and Nickell, L. G., *Phytochemistry*, *8*, 811–818 (1969).

250. Matsuo, K., and Tsuboi, M., *Bull. Chem. Soc. Jap.*, *39*, 347–352 (1966).

251. Mazia, D., in *Glutathione*, S. Colowick, A. Lazarow, E. Racker, D. R. Schwarz, E. Stadtman, and H. Waelsch, Eds., Academic Press, New York, 1954, pp. 209–228.

252. Melnykovych, G., and Snell, E. E., *J. Bacteriol.*, *76*, 518–523 (1958).

253. Michaels, R., and Kim, K., *Biochim. Biophys. Acta*, *115*, 59–64 (1966).

254. Michaels, R., and Tchen, T. T., *J. Bacteriol.*, *95*, 1966–1967 (1968).

255. Michaels, R., and Tchen, T. T., *Bacteriol. Proc.*, 61 (1968).

256. Michaels, R., and Tchen, T. T., *Biochem. Biophys. Res. Commun.*, *42*, 545–549 (1971).

257. Michaelson, I. A., *Eur. J. Pharmacol.*, *1*, 378–382 (1967).

258. Michaelson, I. A., Coffman, P. Z., and Vedral, D. F., *Biochem. Pharmacol.*, *17*, 2435–2441 (1968).

259. Michaelson, I. A., and Smithson, H. R., *Proc. Soc. Exp. Biol. Med.*, *136*, 660–663 (1971).

260. Mills, J., and Dubin, D. T., *Mol. Pharmacol.*, *2*, 311–318 (1966).

261. Mizusaki, S., Tanabe, Y., and Noguchi, M., *Agr. Biol. Chem.*, *34*, 972–973 (1970).

262. Moller, M. L., and Kim, K., *Biochem. Biophys. Res. Commun.*, *20*, 46–52 (1965).

263a. Mondovi, B., Costa, M. T., Finazzi-Agro, A., and Rotilio, G., *Arch. Biochem. Biophys.*, *119*, 373–381 (1967).

263b. Mondovi, B., Costa, M. T., Finazzi-Agro, A., and Rotilio, G., in *Symposium on Pyridoxal Enzymes*, K. Yamada, N. Katunuma, and H. Wada, Eds., Maruzen, Tokyo, 1968, pp. 143–146.

263c. Mondovi, B., Costa, M. T., Finazzi-Agro, A., and Rotilio, G., in *Pyridoxal Catalysis: Enzymes and Model Systems*, E. E. Snell, A. E. Braunstein, E. S. Severin, and Y. M. Torchinsky, Eds., Interscience, New York, 1968, pp. 403–421.

264a. Mondovi, B., Rotilio, G., Costa, M. T., and Finazzi-Agro, A., in *Methods in Enzymology*, Vol. 17, Part B, H. Tabor and C. W. Tabor, Eds., Academic Press, New York, 1971, pp. 735–740.

264b. Mondovi, B., Rotilio, G., Costa, N. T., Finazzi-Agro, A., Chiancone, E., Hansen, R. E., and Beinert, H., *J. Biol. Chem.*, *242*, 1160–1167 (1967).

264c. Mondovi, B., Rotilio, G., Finazzi-Agro, A., and Antonini, E., in *Magnetic Resonances in Biological Research*, C. Franconi, Ed., Gordon and Breach, New York, 1972, in press.

265. Mondovi, B., Rotilio, G., Finazzi-Agro, A., Vallogini, M. P., Malmstrom, B. G., and Antonini, E., *FEBS Lett.*, *2*, 182–184 (1969).

266. Morris, D. R., in *Methods in Enzymology*, Vol. 17, Part B, H. Tabor and C. W. Tabor, Eds., Academic Press, New York, 1971, pp. 850–853.

267. Morris, D. R., and Jorstad, C. M., *J. Bacteriol.*, *101*, 731–737 (1970).

268. Morris, D. R., and Koffron, K. L., *J. Bacteriol.*, *94*, 1516–1519 (1967).

269. Morris, D. R., and Koffron, K. L., *J. Biol. Chem.*, *244*, 6094–6099 (1969).
270. Morris, D. R., Koffron, K. L., and Okstein, C. J., *Anal. Biochem.*, *30*, 449–453 (1969).
271. Morris, D. R., and Pardee, A. B., *Biochem. Biophys. Res. Commun.*, *20*, 697–702 (1965).
272. Morris, D. R., and Pardee, A. B., *J. Biol. Chem.*, *241*, 3129–3135 (1966).
273. Morris, D. R., Wu, W. H., Applebaum, D., and Koffron, K. L., *Ann. N.Y. Acad. Sci.*, *171*, 968–976 (1970).
274. Moruzzi, G., Barbiroli, B., and Caldarera, C. M., *Biochem. J.*, *107*, 609–613 (1968).
275. Moulton, B. C., and Leonard, S. L., *Endocrinology*, *84*, 1461–1465 (1969).
276. Mudd, S. H., *J. Biol. Chem.*, *238*, 2156–2163 (1963).
277. Mudd, S. H., and Cantoni, G. L., *J. Biol. Chem.*, *231*, 481–492 (1958).
278. Mycek, M. J., Clarke, D. D., Neidle, A., and Waelsch, H., *Arch. Biochem. Biophys.*, *84*, 528–540 (1959).
279a. Nakajima, T., and Matsuoka, Y., *J. Neurochem.*, *18*, 2547–2548 (1971).
279b. Nakajima, T., Zack, J. F., Jr., and Wolfgram, F., *Biochim. Biophys. Acta*, *184*, 651–652 (1969).
280. Nakamoto, T., and Hamel, E., *Proc. Nat. Acad. Sci. U.S.*, *59*, 238–245 (1968).
281. Nara, S., Igaue, I., Gomes, B., and Yasunobu, K. T., *Biochem. Biophys. Res. Commun.*, *23*, 324–328 (1966).
282. Neish, W. J. P., and Key, L., *Biochem. Pharmacol.*, *17*, 497–502 (1968).
283. Neish, W. J. P., and Key, L., *Comp. Biochem. Physiol.*, *27*, 709–714 (1968).
284. Norton, J. W., Erdmann, V. A., and Herbst, E. J., *Biochim. Biophys. Acta*, *155*, 293–295 (1968).
285. O'Brien, R. L., Olenick, J. G., and Hahn, F. E., *Proc. Nat. Acad. Sci. U.S.*, *55*, 1511–1517 (1966).
286. Oki, T., Kawasaki, H., Ogata, K., Yamada, H., Tomida, I., Morino, T., and Fukami, H., *Agr. Biol. Chem.*, *32*, 1349–1354 (1968).
287. Oki, T., Kawasaki, H., Ogata, K., Yamada, H., Tomida, I., Morino, T., and Fukami, H., *Agr. Biol. Chem.*, *33*, 994–1000 (1969).
288. Otsuka, H., *J. Cell. Sci.*, *9*, 71–84 (1971).
289. Padmanabhan, R., and Kim, K., *Biochem. Biophys. Res. Commun.*, *19*, 1–5 (1965).
290. Panko, W. B., and Kenney, F. T., *Biochem. Biophys. Res. Commun.*, *43*, 346–350 (1971).
291. Pearce, L. A., and Schanberg, S. M., *Science*, *166*, 1301–1303 (1969).
292. Pegg, A. E., *Biochim. Biophys. Acta*, *177*, 361–364 (1969).
293. Pegg, A. E., *Ann. N.Y. Acad. Sci.*, *171*, 977–987 (1970).
294. Pegg, A. E., *Biochim. Biophys. Acta*, *232*, 630–642 (1971).
295. Pegg, A. E., Lockwood, D. H., and Williams-Ashman, H. G., *Biochem. J.*, *117*, 17–31 (1970).
296. Pegg, A. E., and Williams-Ashman, H. G., *Biochem. Biophys. Res. Commun.*, *30*, 76–82 (1968).
297. Pegg, A. E., and Williams-Ashman, H. G., *Biochem. J.*, *108*, 533–539 (1968).
298. Pegg, A. E., and Williams-Ashman, H. G., *Biochem. J.*, *115*, 241–247 (1969).

299. Pegg, A. E., and Williams-Ashman, H. G., *J. Biol. Chem.*, *244*, 682–693 (1969).

300. Pegg, A. E., and Williams-Ashman, H. G., *Arch. Biochem. Biophys.*, *137*, 156–165 (1970).

301. Perry, T. L., Hansen, S., and MacDougall, L., *J. Neurochem.*, *14*, 775–782 (1967).

302. Petersen, E. E., Kroger, H., and Hagen, U., *Biochim. Biophys. Acta*, *161*, 325–330 (1968).

303. Pett, D. M., and Ginsberg, H. S., *Fed. Proc.*, *27*, 615 (1968).

304*a*. Pohjanpelto, P., and Raina, A., *Nature New Biol.*, *235*, 247–249 (1972).

304*b*. Potier, P., Le Men, J., Janot, M., Bladon, P., Brown, A. G., and Wilson, C. S., *Tetrahedron Lett.*, *5*, 293–300 (1963).

305. Quash, G., and Taylor, D. R., *Clin. Chim. Acta*, *30*, 17–23 (1970).

306. Quigley, J. W., and Cohen, S. S., *J. Biol. Chem.*, *244*, 2450–2458 (1969).

307. Raina, A., *Acta Physiol. Scand.*, *60*, *Suppl. 218*, 1–81 (1963).

308. Raina, A., and Cohen, S. S., *Proc. Nat. Acad. Sci. U.S.*, *55*, 1587–1593 (1966).

309. Raina, A., and Hannonen, P., *Acta Chem. Scand.*, *24*, 3061–3064 (1970).

310. Raina, A., and Hannonen, P., *FEBS Lett.*, *16*, 1–4 (1971).

311. Raina, A., and Janne, J., *Acta Chem. Scand.*, *22*, 2375–2378 (1968).

312. Raina, A., and Janne, J., *Fed. Proc.*, *29*, 1568–1574 (1970).

313. Raina, A., Janne, J., Hannonen, P., and Holtta, E., *Ann. N.Y. Acad. Sci.*, *171*, 697–708 (1970).

314. Raina, A., Janne, J., and Khawaja, J. A., *Scand. J. Clin. Lab. Invest.*, *25*, *Suppl. 113*, p. 98 (1970).

315. Raina, A., Janne, J., and Siimes, M., *Biochim. Biophys. Acta*, *123*, 197–201 (1966).

316. Raina, A., Jansen, M., and Cohen, S. S., *J. Bacteriol.*, *94*, 1684–1696 (1967).

317. Raina, A., and Telaranta, T., *Biochim. Biophys. Acta*, *138*, 200–203 (1967).

318. Ray, P. H., and Brock, T. D., *J. Gen. Microbiol.*, *66*, 133–135 (1971).

319. Razin, S., and Rozansky, R., *Arch. Biochem. Biophys.*, *81*, 36–54 (1959).

320. Riley, W. D., and Snell, E. E., *Biochemistry*, *7*, 3520–3528 (1968).

321. Rinaldini, L. M., *Anal. Biochem.*, *36*, 352–367 (1970).

322. Robin, Y., Audit, C., and Landon, M., *Comp. Biochem. Physiol.*, *22*, 787–797 (1967).

323. Robin, Y., and Van Thoai, N., *C. R. Hebd. Seances Acad. Sci., Paris*, *252*, 1224–1226 (1961).

324. Robison, B., and Zimmerman, T. P., *J. Biol. Chem.*, *246*, 110–117 (1971).

325. Robison, B., and Zimmerman, T. P., *J. Biol. Chem.*, *246*, 4664–4670 (1971)

326. Rogers, G. T., Ulbricht, T. L. V., and Szer, W., *Biochem. Biophys. Res. Commun.*, *27*, 372–377 (1967).

327. Rolle, I., Payer, R., and Soeder, C. J., *Arch. Mikrobiol.*, *77*, 185–195 (1971).

328. Rometsch, R., Marxer, A., and Miescher, K., *Helv. Chim. Acta*, *34*, 1611–1618 (1951).

329. Roon, R. J., and Barker, H. A., *Fed. Proc.*, *30*, 1297 Abstr. (1971); *J. Bacteriol.*, *109*, 44–50 (1972).

330. Rosano, C. L., and Hurwitz, C., *Biochem. Biophys. Res. Commun.*, *37*, 677–683 (1969).
331. Rosenthal, S. M., and Dubin, D. T., *J. Bacteriol.*, *84*, 859–863 (1962).
332. Rosenthal, S. M., Fisher, E. R., and Stohlman, E. F., *Proc. Soc. Exp. Biol. Med.*, *80*, 432–434 (1952).
333. Rosenthal, S. M., and Tabor, C. W., *J. Pharmacol. Exp. Ther.*, *116*, 131–138 (1956).
334. Rotilio, G., Calabrese, L., Finazzi-Agro, A., and Mondovi, B., *Biochim. Biophys. Acta*, *198*, 618–620 (1970).
335. Rucker, R. B., and O'Dell, B. L., *Biochim. Biophys. Acta*, *235*, 32–43 (1971).
336a. Russell, D. H., *Ann. N.Y. Acad. Sci.*, *171*, 772–782 (1970).
336b. Russell, D. H., *Proc. Nat. Acad. Sci. U.S.*, *68*, 523–527 (1971).
336c. Russell, D. H., *Nature New Biol.*, *233*, 144–145 (1971); *Cancer Res.*, *32*, 7–10 (1972).
337. Russell, D. H., and Levy, C. C., *Cancer Res.*, *31*, 248–251 (1971).
338. Russell, D. H., Levy, C. C., Taylor, R. L., Gfeller, E., and Sterns, D. N., *Fed. Proc.*, *30*, 1093 Abstr. (1971).
339a. Russell, D. H., and Lombardini, J. B., *Biochim. Biophys. Acta*, *240*, 273–286 (1971).
339b. Russell, D. H., and McVicker, T. A., *Biochim. Biophys. Acta*, *244*, 85–93 (1971).
340. Russell, D. H., Medina, V. J., and Snyder, S. H., *J. Biol. Chem.*, *245*, 6732–6738 (1970).
341. Russell, D., and Snyder, S. H., *Proc. Nat. Acad. Sci. U.S.*, *60*, 1420–1427 (1968); *Endocrinology*, *84*, 223–228 (1969).
342. Russell, D. H., and Snyder, S. H., *Mol. Pharmacol.*, *5*, 253–262 (1969).
343. Russell, D. H., Snyder, S. H., and Medina, V. J., *Life Sci. (Oxford)*, *8*, Part II, 1247–1254 (1969).
344. Russell, D. H., Snyder, S. H., and Medina, V. J., *Endocrinology*, *86*, 1414–1419 (1970).
345. Russell, D. H., and Taylor, R. L., *Endocrinology*, *88*, 1397–1403 (1971).
346. Salas, M., Hille, M. B., Last, J. A., Wahba, A. J., and Ochoa, S., *Proc. Nat. Acad. Sci. U.S.*, *57*, 387–394 (1967).
347. Satake, K., and Fujita, H., *J. Biochem.*, *40*, 547–556 (1953).
348. Schindler, J., *Experientia*, *21*, 697–698 (1965).
349. Schlenk, F., and Dainko, J. L., *Arch. Biochem. Biophys.*, *113*, 127–133 (1966).
350a. Schlenk, F., and Zydek, C. R., *Biochem. Biophys. Res. Commun.*, *31*, 427–432 (1968).
350b. Schlenk, F., Zydek-Cwick, C. R., and Hutson, H. K., *Arch. Biochem. Biophys.*, *142*, 144–149 (1971).
351. Schrock, T. R., Oakman, N. J., and Bucher, N. L. R., *Biochim. Biophys. Acta*, *204*, 564–577 (1970).
352. Schwimmer, S., *Biochim. Biophys. Acta*, *166*, 251–254 (1968).
353a. Seiler, N., *J. Chromatog.*, *63*, 97–112 (1971).
353b. Seiler, N., and Askar, A., *J. Chromatog.*, *62*, 121–127 (1971).
353c. Seiler, N., and Schneider, H. H., *J. Chromatogr.*, *59*, 367–371 (1971).

264   HERBERT TABOR AND CELIA WHITE TABOR

353d. Seiler, N., and Schroder, J. M., *Brain Res.*, *22*, 81–103 (1970).
354. Seiler, N., Werner, G., Fischer, H. A., Knotgen, B., and Hinz, H., *Hoppe-Seyler's Z. Physiol. Chem.*, *350*, 676–682 (1969).
355. Seiler, N., and Wiechmann, M., *Experientia*, *21*, 203–204 (1965).
356. Seiler, N., and Wiechmann, M., *Hoppe-Seyler's Z. Physiol. Chem.*, *348*, 1285–1290 (1967).
357. Sercarz, E. E., and Gorini, L., *J. Mol. Biol.*, *8*, 254–262 (1964).
358. Shalitin, C., *J. Virol.*, *1*, 569–575 (1967).
359. Shalitin, C., and Sarid, S., *J. Virol.*, *1*, 559–568 (1967).
360. Shapiro, J. T., Stannard, B. S., and Felsenfeld, G., *Biochemistry*, *8*, 3233–3241 (1969).
361. Shapiro, S. K., and Schlenk, F., Eds., *Transmethylation and Methionine Biosynthesis*, University of Chicago Press, 1965.
362. Shaw, M. K., *J. Bacteriol.*, *95*, 221–230 (1968).
363. Sher, I. H., and Mallette, M. F., *Arch. Biochem. Biophys.*, *53*, 370–380 (1954).
364a. Shiba, T., Kubota, I., and Kaneko, T., *Tetrahedron*, *26*, 4307–4311 (1970).
364b. Shimizu, H., Kakimoto, Y., and Sano, I., *Arch. Biochem. Biophys.*, *110*, 368–372 (1965).
364c. Shimizu, H., Kakimoto, Y., and Sano, I., *Nature*, *207*, 1196–1197 (1965).
365. Siimes, M., *Acta Physiol. Scand., Suppl. 298*, 1–66 (1967).
366. Siimes, M., *Scand. J. Clin. Lab. Invest.*, *21, Suppl. 101*, 10 (1968).
367. Siimes, M., and Janne, J., *Acta Chem. Scand.*, *21*, 815–817 (1967).
368. Silman, N., Artman, M., and Engelberg, H., *Biochim. Biophys. Acta*, *103*, 231–240 (1965).
369. Silver, S., Wendt, L., Bhattacharyya, P., and Beauchamp, R. S., *Ann. N.Y. Acad. Sci.*, *171*, 838–862 (1970).
370. Simon, E. J., Cohen, S. S., and Raina, A., *Biochem. Biophys. Res. Commun.*, *24*, 482–488 (1966).
371. Simon, E. J., Schapira, L., and Wurster, N., *Mol. Pharmacol.*, *6*, 577–587 (1970).
372. Smith, J. K., *Biochem. J.*, *103*, 110–119 (1967).
373. Smith, T. A., *Phytochemistry*, *2*, 241–252 (1963).
374. Smith, T. A., *Phytochemistry*, *4*, 599–607 (1965).
375. Smith, T. A., *Phytochemistry*, *8*, 2111–2117 (1969).
376. Smith, T. A., *Phytochemistry*, *9*, 1479–1486 (1970).
377. Smith, T. A., *Anal. Biochem.*, *33*, 10–15 (1970).
378. Smith, T. A., *Biochem. Biophys. Res. Commun.*, *41*, 1452–1456 (1970).
379. Smith, T. A., *Ann. N.Y. Acad. Sci.*, *171*, 988–1001 (1970).
380. Smith, T. A., *Biol. Rev.*, *46*, 201–241 (1971); Smith, T. A., *Endeavour*, *31*, 22–28 (1972).
381. Smith, T. A., and Garraway, J. L., *Phytochemistry*, *3*, 23–26 (1964).
382. Smith, T. A., and Sinclair, C., *Ann. Bot.*, *31*, 103–111 (1967).
383. Sneath, P. H. A., *Nature*, *175*, 818 (1955).
384. Snyder, S. H., Kreuz, D. S., Medina, V. J., and Russell, D. H., *Ann. N.Y. Acad. Sci.*, *171*, 749–771 (1970).
385. Snyder, S. H., and Russell, D. H., *Fed. Proc.*, *29*, 1575–1582 (1970).

386. So, A. G., Davie, E. W., Epstein, R., and Tissieres, A., *Proc. Nat. Acad. Sci. U.S.*, *58*, 1739–1746 (1967).
387. Soda, K., and Moriguchi, M., in *Methods in Enzymology*, Vol. 17, Part B, H. Tabor and C. W. Tabor, Eds., Academic Press, New York, 1971, pp. 677–681.
388. Southren, A. L., Kobayashi, Y., Brenner, P., and Weingold, A. B., *J. Appl. Physiol.*, *20*, 1048–1051 (1965).
389. Stastny, M., and Cohen, S., *Biochem. Biophys. Acta*, *204*, 578–589 (1970); *261*, 177–180 (1972).
390. Stevens, L., *Biochem. J.*, *103*, 811–815 (1967).
391. Stevens, L., *Biochem. J.*, *113*, 117–121 (1969).
392. Stevens, L., *Biol. Rev.*, *45*, 1–27 (1970).
393. Stevens, L., and McCann, L. M., *Ann. N.Y. Acad. Sci.*, *171*, 827–837 (1970).
394. Stevens, L., and Morrison, M. R., *Biochem. J.*, *108*, 633–640 (1968).
395. Stirpe, F., and Novello, F., *Eur. J. Biochem.*, *15*, 505–512 (1970).
396a. Strausbach, P. H., and Fischer, E. H., *Biochemistry*, *9*, 233–238 (1970).
396b. Suwalsky, M., Traub, W., Shmueli, U., and Subirana, J. A., *J. Mol. Biol.*, *42*, 363–373 (1969).
397. Suzuki, H., Ogura, Y., and Yamada, H., *J. Biochem.*, *69*, 1065–1074 (1971).
398. Szer, W., *J. Mol. Biol.*, *16*, 585–587 (1966).
399. Szer, W., *Biochem. Biophys. Res. Commun.*, *22*, 559–564 (1966).
400. Szer, W., *Ann. N.Y. Acad. Sci.*, *171*, 801–809 (1970).
401. Tabor, C. W., *Biochem. Biophys. Res. Commun.*, *2*, 117–120 (1960).
402. Tabor, C. W., *J. Bacteriol.*, *83*, 1101–1111 (1962).
403. Tabor, C. W., in *Methods in Enzymology*, Vol. 5, S. P. Colowick and N. O. Kaplan, Eds., Academic Press, New York, 1962, pp. 756–760.
404. Tabor, C. W., in *Methods in Enzymology*, Vol. 5, S. P. Colowick and N. O. Kaplan, Eds., Academic Press, New York, 1962, pp. 761–765.
405. Tabor, C. W., *Biochem. Biophys. Res. Commun.*, *30*, 339–342 (1968).
406. Tabor, C. W., and Dobbs, L. G., *J. Biol. Chem.*, *245*, 2086–2091 (1970).
407. Tabor, C. W., and Kellogg, P. D., *J. Biol. Chem.*, *242*, 1044–1052 (1967).
408. Tabor, C. W., and Kellogg, P. D., *J. Biol. Chem.*, *245*, 5424–5433 (1970); in *Methods in Enzymology*, Vol. 17, Part B, H. Tabor and C. W. Tabor, Eds., Academic Press, New York, 1971, pp. 746–753.
409. Tabor, C. W., and Rosenthal, S. M., *J. Pharmacol. Exp. Ther.*, *116*, 139–155 (1956).
410. Tabor, C. W., and Tabor, H., *J. Biol. Chem.*, *241*, 3714–3723 (1966).
411. Tabor, C. W., and Tabor, H., *Biochem. Biophys. Res. Commun.*, *41*, 232–238 (1970).
412. Tabor, C. W., and Tabor, H., *Fed. Proc.*, *30*, 1068 Abstr. (1971).
413. Tabor, C. W., Tabor, H., and Bachrach, U., *J. Biol. Chem.*, *239*, 2194–2203 (1964).
414. Tabor, C. W., Tabor, H., McEwen, C. M., Jr., and Kellogg, P. D., *Fed. Proc.*, *23*, 385 (1964).
415. Tabor, H., *Pharmacol. Rev.*, *6*, 299–343 (1954).
416. Tabor, H., *Biochem. Biophys. Res. Commun.*, *3*, 382–385 (1960).
417. Tabor, H., *Biochem. Biophys. Res. Commun.*, *4*, 228–231 (1961).

418. Tabor, H., *Biochemistry*, *1*, 496–501 (1962).
419. Tabor, H., Rosenthal, S. M., and Tabor, C. W., *J. Amer. Chem. Soc.*, *79*, 2978–2979 (1957).
420. Tabor, H., Rosenthal, S. M., and Tabor, C. W., *J. Biol. Chem.*, *233*, 907–914 (1958).
421. Tabor, H., and Tabor, C. W., *Pharmacol. Rev.*, *16*, 245–300 (1964).
422. Tabor, H., and Tabor, C. W., *Fed. Proc.*, *25*, 879–880 (1966).
423. Tabor, H., and Tabor, C. W., *J. Biol. Chem.*, *244*, 2286–2292 (1969).
424. Tabor, H., and Tabor, C. W., *J. Biol. Chem.*, *244*, 6383–6387 (1969).
425. Tabor, H., and Tabor, C. W., in *Methods in Enzymology*, Vol. 17, Part B, H. Tabor and C. W. Tabor, Eds., Academic Press, New York, 1971, pp. 393–397.
426. Tabor, H., and Tabor, C. W., in *Methods in Enzymology*, Vol. 17, Part B, H. Tabor and C. W. Tabor, Eds., Academic Press, New York, 1971, pp. 815–833.
427. Tabor, H., Tabor, C. W., and DeMeis, L., in *Methods in Enzymology*, Vol. 17, Part B, H. Tabor and C. W. Tabor, Eds., Academic Press, New York, 1971, pp. 829–833.
428. Tabor, H., Tabor, C. W., and Rosenthal, S. M., *Ann. Rev. Biochem.*, *30*, 579–604 (1961).
429. Takeda, Y., *J. Biochem.*, *66*, 345–349 (1969).
430. Takeda, Y., *Biochim. Biophys. Acta*, *182*, 258–261 (1969).
431. Takeda, Y., and Igarashi, K., *Biochem. Biophys. Res. Commun.*, *37*, 917–924 (1969).
432. Takeda, Y., and Igarashi, K., *J. Biochem.*, *68*, 937–940 (1970).
433. Tamura, C., Sim, G. A., Jeffreys, J. A. D., Bladon, P., and Ferguson, G., *Chem. Commun.*, 485–486 (1965).
434. Tanner, M. J. A., *Biochemistry*, *6*, 2686–2693 (1967).
435. Thomas, T. D., Lyttleton, P., Williamson, K. I., and Batt, R. D., *J. Gen. Microbiol.*, *58*, 381–390 (1969).
436. Tobari, J., and Tchen, T. T., *J. Biol. Chem.*, *246*, 1262–1265 (1971).
437. Ueyama, E., Davis, C. L., and Wolin, M. J., *J. Dairy Sci.*, *48*, 73–76 (1965).
438. Unemoto, T., Ikeda, K., Hayashi, M., and Miyaki, K., *Chem. Pharm. Bull. (Tokyo)*, *11*, 148–151 (1963).
439. Van Dijk-Salkinoja, M. S., and Planta, R. J., *FEBS Lett.*, *12*, 9–13 (1970).
440. Wall, R. A., *J. Chromatogr.*, *37*, 549–551 (1968).
441. Wang, J. H., Humniski, P. M., and Black, W. J., *Biochemistry*, *7*, 2037–2044 (1968).
442. Waring, M., *J. Mol. Biol.*, *54*, 247–279 (1970).
443. Weaver, R. H., and Herbst, E. J., *J. Biol. Chem.*, *231*, 637–646 (1958).
444. Weinstein, I. B., Ochoa, M., Jr., and Friedman, S. M., *Biochemistry*, *5*, 3332–3339 (1966).
445. Weiss, R. L., and Morris, D. R., *Biochim. Biophys. Acta*, *204*, 502–511 (1970).
446a. Wheaton, T. A., and Stewart, I., *Nature*, *206*, 620–621 (1965).
446b. White, W. F., Cohen, A. I., Rippel, R. H., Story, J. C., and Schally, A. V., *Endocrinology*, *82*, 742–752 (1968).

447. Wickner, R. B., *J. Biol. Chem.*, *244*, 6550–6552 (1969).
448. Wickner, R. B., Tabor, C. W., and Tabor, H., *J. Biol. Chem.*, *245*, 2132–2139 (1970).
449. Wickner, R. B., Tabor, C. W., and Tabor, H., in *Methods in Enzymology*, Vol. 17, Part B, H. Tabor and C. W. Tabor, Eds., Academic Press, New York, 1971, pp. 647–651.
450. Williams-Ashman, H. G., and Lockwood, D. H., *Ann. N.Y. Acad. Sci.*, *171*, 882–894 (1970).
451. Williams-Ashman, H. G., Pegg, A. E., and Lockwood, D. H., in *Advances in Enzyme Regulations*, Vol. 7, G. Weber, Ed., Pergamon Press, Oxford, 1969, pp. 291–323.
452. Wilson, O. H., and Holden, J. T., *J. Biol. Chem.*, *244*, 2737–2742 (1969).
453. Witkin, S. S., and Rosenberg, E., *J. Bacteriol.*, *103*, 641–649 (1970).
454. Wu, W. H., and Morris, D. R., *Bacteriol. Proc.*, 133–134 (1970).
455. Yamada, H., in *Methods in Enzymology*, Vol. 17, Part B, H. Tabor and C. W. Tabor, Eds., Academic Press, New York, 1971, pp. 726–730.
456. Yamada, H., and Adachi, O., in *Methods in Enzymology*, Vol. 17, Part B, H. Tabor and C. W. Tabor, Eds., Academic Press, New York, 1971, pp. 705–709.
457. Yamada, H., Adachi, O., Kumagai, H., and Ogata, K., *Mem. Res. Inst. Food Sci. (Kyoto Univ.)*, *26*, 21–23 (1965).
458. Yamada, H., Adachi, O., and Ogata, K., *Agr. Biol. Chem.*, *29*, 864–869 (1965).
459. Yamada, H., Adachi, O., and Ogata, K., *Agr. Biol. Chem.*, *29*, 912–917 (1965).
460. Yamada, H., Adachi, O., and Ogata, K., *Agr. Biol. Chem.*, *29*, 1148–1149 (1965).
461. Yamada, H., Kawasaki, H., Oki, T., Tomida, I., Fukami, H., and Ogata, K., *Mem. Res. Inst. Food Sci. (Kyoto Univ.)*, *29*, 11–17 (1968).
462. Yamada, H., Kumagai, H., Kawasaki, H., Matsui, H., and Ogata, K., *Biochem. Biophys. Res. Commun.*, *29*, 723–727 (1967).
463. Yamada, H., Tanaka, A., and Ogata, K., *Mem. Res. Inst. Food Sci. (Kyoto Univ.)*, *26*, 1–9 (1965).
464. Yamada, H., Tanaka, A., and Ogata, K., *Agr. Biol. Chem.*, *29*, 260–261 (1965).
465. Yamasaki, E. F., Swindell, R., and Reed, D. J., *Biochemistry*, *9*, 1206–1210 (1970).
466a. Yasunobu, K. T., Achee, F., Chevvenka, C., and Wang, T. M., in *Symposium on Pyridoxal Enzymes*, K. Yamada, N. Katunuma, and H. Wada, Eds., Maruzen, Tokyo, 1968, pp. 139–142.
466b. Yasunobu, K. T., and Smith, R. A., in *Methods in Enzymology*, Vol. 17, Part B, H. Tabor and C. W. Tabor, Eds., Academic Press, New York, 1971, pp. 698–704.
467. Yoshida, D., *Plant Cell Physiol.*, *10*, 393–397 (1969).
468. Young, D. V., and Srinivasan, P. R., *Biochim. Biophys. Acta*, *238*, 447–463 (1971).

469. Zappia, V., Cortese, R., Zydek-Cwick, C. R., and Schlenk, F., *Rend. Class. Sci. Fisch.*, *46*, 191–195 (1969).
470. Zeller, E. A., in *The Enzymes*, Vol. 8, P. D. Boyer, H. Lardy, and K. Myrback, Eds., Academic Press, New York, 1963, pp. 313–335.
471. Zillig, W., Krone, W., and Albers, M., *Z. Physiol. Chem.*, *317*, 131–143 (1959).

ACYL CARRIER PROTEIN

By DAVID J. PRESCOTT, *Bryn Mawr, Pennsylvania* and
P. ROY VAGELOS, *St. Louis, Missouri*

CONTENTS

Much of our information on the nature of the proteins involved in the biosynthesis of fatty acids has resulted from studies on bacterial systems. The fatty acid synthetase systems of *E. coli* and *Clostridium kluyveri* are found in the soluble fraction when cells are ruptured by various procedures, and, more important, the individual protein components are not associated in a multienzyme complex (4,33–35,56). Thus many of the proteins have been purified and obtained in a homogeneous state by standard fractionation techniques. In contrast, the fatty acid synthetase of yeast (59), avian liver (13), rat liver (18), rat adipose tissue (70), and lactating mammary gland (20) exist as tight multienzyme complexes that have not yet been dissociated into the active component proteins. However, the evidence is clear that in all biological systems, the critical thiol, to which is attached the growing fatty acyl chain as a thioester, is contributed by 4'-phosphopantetheine linked through a serine hydroxyl to a unique protein. This protein, called acyl carrier protein (ACP), was first isolated from E. coli (33–35, 63,118), where it has been studied most extensively.

The observation of Lynen (59) that acetoacetate appeared to be bound to protein in the yeast fatty acid synthetase led to the postulate that all fatty acyl intermediates in long chain fatty acid synthesis were protein-bound. Extracts of E. coli, which catalyze fatty acid biosynthesis, were initially fractionated into two components, originally designated as Enzyme I and Enzyme II (4,33,34). Although Enzyme I was shown to be extremely sensitive to sulfhydryl inhibitors, inhibition of Enzyme II by sulfhydryl poisons could only be demonstrated when the protein was reduced immediately before introduction of the inhibitor. Enzyme II exhibited properties expected for a cofactor that was protein in nature (4); it was subsequently designated ACP (63). This protein cofactor functions in the biosynthesis of fatty acids in a manner somewhat analogous to the function of coenzyme A in the oxidation of fatty acids. Protein-bound intermediates were isolated, identified, and shown to be thioesters metabolically active in the E. coli fatty acid synthetase system. This article will serve as a review of the current knowledge of ACP with special emphasis on the properties, function, and interaction of this protein with enzymes of fatty acid biosynthesis. Papers appearing since 1967 will be stressed, and the reader is referred to a prior review article (68) for thorough coverage before 1967. The field of fatty acid biosynthesis will be reviewed only in so far as it applies to these objectives.

I. Purification and Physical Properties

Acyl carrier protein can be prepared easily from extracts of E. coli (63,67,68). The steps include removal of nucleic acids from the cell extract and precipitation of protein by ammonium sulfate to 80% of saturation. The 80% ammonium sulfate supernatant, containing the ACP, is acidified to pH 1; the ACP is recovered in the precipitate, which is then chromatographed on DEAE cellulose, and finally on

DEAE Sephadex. The protein appears to be homogeneous on DEAE Sephadex chromatography, polyacrylamide disk gel electrophoresis, and ultracentrifugal analysis. The steps are summarized in Table I. Modifications of this procedure have yielded homogeneous preparations of ACP from *Arthrobacter*, avocado, spinach (96), *Mycobacterium phlei* (71), and *Clostridium butyricum* (3).

TABLE I
Purification of ACP from *E. coli*

Step	Purity $\left(\dfrac{\mu g \text{ ACP}}{\text{mg protein}}\right)$	Recovery %
Cell extract	2.5	100
Streptomycin sulfate, ammonium sulfate, and acid precipitation	20	50–60
DEAE cellulose chromatography	200–400	30–40
DEAE sephadex chromatography	900–1000	25–35

Acyl carrier protein from *E. coli* exhibits properties expected of a globular protein (102). Thus the ratio of the frictional coefficient observed to the minimum frictional coefficient in aqueous solution was determined to be 1.12. The optical rotatory dispersion curve for the protein in the ultraviolet region showed a negative trough at 232 nm, a shoulder near 213 nm, and a positive peak at 198 mμ (102), all typical of α-helical polypeptides (26). The mean residue rotation at 233 nm has been reported as approximately $-4500°$ (102) or approximately $-8400°$ (86). The reason for this discrepancy is not understood. The sedimentation constant for completely reduced ACP was found to be 1.34 S (102). An earlier report of 1.44 S (63) probably reflected the presence of dimers of ACP resulting from disulfide bonds through the prosthetic group. Guanidine hydrochloride caused reversible denaturation. At high concentrations of the denaturatant (6 M), the protein exhibited optical rotatory dispersion properties similar to those of a random coil, but the ORD spectrum returned to that of the native protein upon dilution of the guanidine hydrochloride to 0.6 M (102). The molecular weight determined from the primary structure is 8847 (112).

The physical properties of ACP isolated from *M. phlei* resemble those of the *E. coli* protein. The ratio of frictional coefficients is 1.12, indicating that ACP (*M. phlei*) is also a globular protein in solution

(71). The sedimentation coefficient is 1.49 S, which when used with the diffusion constant of 13.1 × 10⁻⁷ cm²/sec, yielded a molecular weight of 10,450. The optical rotatory dispersion curve of this protein also displayed characteristics typical of α-helical proteins. The mean residue rotation at 233 nm, is −7800° (71). Acyl carrier protein (*M. phlei*) also exhibited optical rotatory dispersion curves characteristic of a random coil at high concentrations of guanidine HCl. Thus the size and physical properties of *M. phlei* ACP suggest that it closely resembles *E. coli* ACP.

II. Structural Studies of *E. coli* ACP

A. PROSTHETIC GROUP AND PRIMARY SEQUENCE

Sulfhydryl titration of *E. coli* ACP revealed the presence of 1 mole of sulfhydryl group/mole protein (63), which was identified as 2-mercaptoethylamine (90). Other components of CoA were identified in the protein. Thus 1 mole of ACP was found to contain 1 mole of β-alanine (63,65,90), organic phosphate and pantoic acid (65). The complete structure of the prosthetic group was established enzymatically (65). Mild alkaline treatment resulted in loss of the prosthetic group from ACP. The alkaline cleavage product could be converted to CoA in the presence of dephospho CoA pyrophosphorylase, dephospho CoA kinase, and ATP. The structure of the prosthetic group was thus established as 4′-phosphopantetheine. Acyl derivatives of ACP, which are intermediates in fatty acid synthesis, are thioesters in which the sulfhydryl group of the 4′-phosphopantetheine is esterified.

Since the attachment of the prosthetic group exhibited acid stability and alkali lability, the possibility of a phosphodiester link through a serine hydroxyl was proposed. This was established through studies of peptic peptides prepared from enzymatically synthesized 2-¹⁴C-malonyl-ACP. Two radioactive peptides were purified and analyzed with the following results (65,66). Peptide PA-3 contained, on a mole per mole basis, 2-¹⁴C-malonate, phosphate, pantolactone, β-alanine, taurine (the oxidation product of 2-mercaptoethylamine), and the amino acids, serine, aspartic acid, and leucine, while PA-1 contained an additional residue each of glycine and alanine. Thus these peptides both contained a serine residue. The chemical characteristics of the covalent bond joining the prosthetic group to the protein identified this linkage as a phosphate ester through the serine hydroxyl group (65,

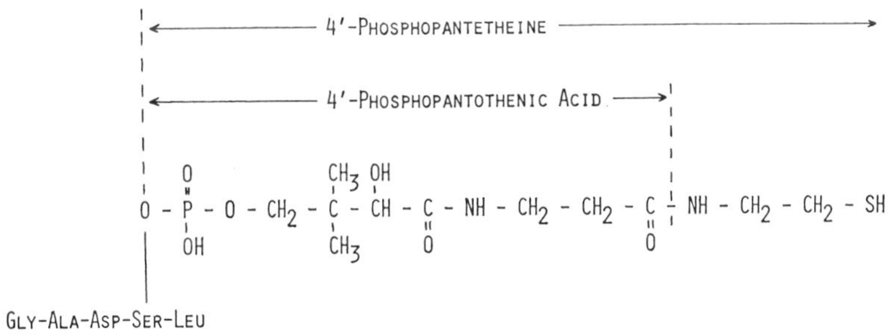

$$4'\text{-PHOSPHOPANTETHEINE}$$

$$4'\text{-PHOSPHOPANTOTHENIC ACID}$$

$$\begin{array}{c} O \\ \| \\ O - P - O - CH_2 - C - CH - C - NH - CH_2 - CH_2 - C - NH - CH_2 - CH_2 - SH \\ | \quad\quad\quad | \quad | \quad \| \quad\quad\quad\quad\quad\quad\quad \| \\ OH \quad\quad CH_3 \ OH \ O \quad\quad\quad\quad\quad\quad\quad\quad O \end{array}$$

GLY-ALA-ASP-SER-LEU

Fig. 1. The structure of peptide PA-1 isolated from a peptic digest of *E. coli* ÂCP.

66). The relative alkali stability of the phosphate ester linkage to pantetheine indicated that the phosphate is in the 4', rather than the 2' position, of the pantothenic acid (10).

The primary sequence of the isolated radioactive peptic peptide was determined by several methods. Edman degradations, dinitrophenylation, hydrazinolysis, and nitrous acid treatment showed the structure for the peptide PA1 found in Figure 1 (66). When this peptide was treated under mild alkaline conditions, 4'-phosphopantetheine was released. Since the phosphate of the prosthetic group is a good leaving group, a β-elimination mechanism was proposed (Fig. 2). The observation that serine disappearance was accompanied by an equivalent appearance of pyruvate supports the proposed mechanism (66).

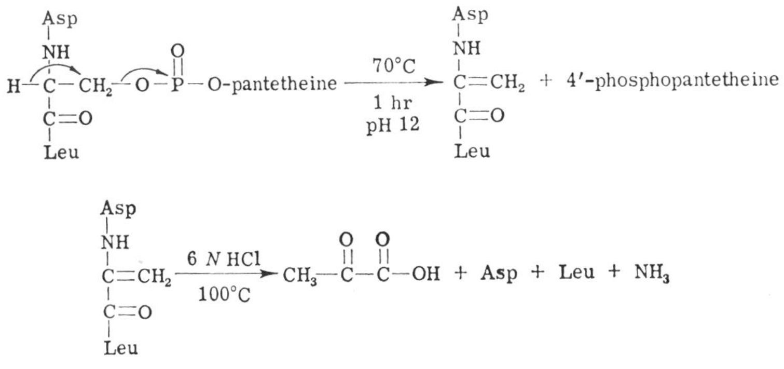

Fig. 2. The β-elimination mechanism for removal of 4'-phosphopantetheine and acid conversion of dehydroalanine to pyruvate.

The complete amino acid sequence of *E. coli* ACP has been determined (112). Several features of the amino acid composition and sequence are noteworthy. The molecule consists of 14 residues of glutamic acid and 8 residues of aspartic acid out of a total of 77 residues. There are only 4 lysine residues and 1 each of arginine and histidine. Therefore, ACP is quite acidic in nature with an isoelectric point of about pH 4.2, and at this pH the protein is probably least soluble (112). As is apparent in Figure 3, the acidic residues occur throughout the sequence, while the basic residues appear clustered at both amino and carboxyl ends. Residues 47 to 49 are in the sequence of Glu-Glu-Glu and 56 to 58 are Asp-Glu-Glu. In fact, 9 out of 14 residues between residues 47 and 60 are acidic, while there are no extended sequences rich in hydrophobic side chains. The possible exception is the sequence from residues 62 to 69. The prosthetic group is clearly attached to Serine 36 and not to Serine 27 as was originally postulated (61), and thus lies midway along the sequence. In view of the preponderance of charged residues and the fact that ACP behaves as a typical globular protein (102), the three-dimensional structure of the molecule may show a preponderance of charged residues on the surface (112). This

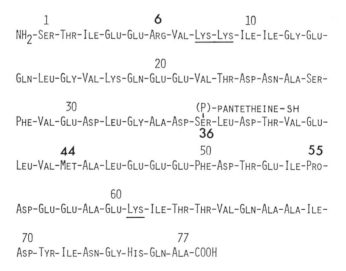

Fig. 3. The complete amino acid sequence of *E. coli* ACP. From Vanaman, Wakil, and Hill (112).

may have important implications for the interactions of acyl ACPs with the various enzymes of the fatty acid synthetase system.

B. STRUCTURE–ACTIVITY RELATIONSHIPS

At least 10 enzymes are known to react with ACP substrates. With the elucidation of the primary sequence of ACP, investigations were initiated to attempt to understand what parts of the protein structure are important for the activity of these different enzymes. These studies have indicated that few alterations of the polypeptide chain of ACP are tolerated by the enzymes of fatty acid biosynthesis or by the enzyme, holo-ACP synthetase, which catalyzes the synthesis of holo-ACP from apo-ACP and CoA, reaction 2 (Section III.A). Preparations of modified ACP were assayed in the overall fatty acid synthesis reaction or in the more sensitive malonyl CoA–CO_2 exchange reaction, which is dependent upon the first 3 enzymes of fatty acid synthesis (4). Preparations of modified apo-ACP were tested enzymatically in the holo-ACP synthetase reaction. As seen in Table II, treatment of ACP with carboxypeptidase A led to the removal of the three carboxyl terminal residues of the polypeptide chain, yielding peptide 1 through 74. This peptide was fully active in fatty acid synthesis and in the malonyl CoA–CO_2 exchange reaction (61,86). In addition, the apopeptide functioned as well as apo-ACP in accepting the prosthetic group from CoA in the presence of holo-ACP synthetase (86). However, the apopeptide did not exhibit inhibition in the concentration

TABLE II
Structure–Activity Relationships of Several Peptides of ACP[a]

Peptide	Method of preparation	Activity of holopeptide in CO_2 exchange reaction (μM CO_2/(min)/ (mg protein)	Conversion of apopeptide to holopeptide in presence of ACP synthetase, %
ACP (1 → 77)	—	6.9	100
1 → 74	Carboxypeptidase A	7.0	100
19 → 61	Trypsin	0	<5
1 → 44	CNBr	0	—
7 → 77	Trypsin treatment of acetylated ACP	0	<5

[a] From Majerus (61,62) and Prescott, Elovson, and Vagelos (86).

range where apo-ACP exhibited strong substrate inhibition. The K_m values for apo-ACP and apopeptide 1 through 74 are $5.5 \times 10^{-7} M$ and $2.0 \times 10^{-6} M$, respectively. Therefore, although the carboxyl terminal tripeptide did not appear to be essential for the function of ACP in fatty acid biosynthesis, it may play a role in the regulation of intracellular levels of ACP. It should be mentioned that the fatty acids produced with holopeptide 1 through 74 in the *in vitro* fatty acid synthetase system should be examined, since subtle effects on some of the biosynthetic enzymes might be revealed in such studies.

Trypsin treatment of ACP yielded peptide 19 through 61, which is inactive in fatty acid synthesis and in the malonyl CoA–CO_2 exchange reaction (61); the apopeptide 19 through 61 is inactive in the holo-ACP synthetase reaction. Malonyl peptide 19 through 61 was synthesized by chemical acylation of the peptide in order to determine whether it was active in the condensation reaction of fatty acid synthesis, catalyzed by β-ketoacyl ACP synthetase (reactions 4 and 6). This enzyme was found to be completely inactive with the peptide substrate, thereby explaining the inactivity of peptide 19 through 61 in fatty acid synthesis. Acetoacetyl peptide 19 through 61 was also tested with the enzyme, β-ketoacyl ACP reductase (reaction 7) and found to be only slightly active compared to the native substrate. The peptide substrate had a K_m of $1.7 \times 10^{-3} M$ and a V_{max} of 0.21 μM/(min) (mg protein), compared to $6.6 \times 10^{-5} M$ and 1.36 μM/(min) (mg protein) for acetoacetyl ACP. This difference suggested that the site which confers the high reactivity of the ACP substrate was missing in the tryptic peptide. The existence of such a site was indicated by the fact that free ACP competitively inhibited ($K_i = 2.2 \times 10^{-4} M$) the reduction of acetoacetyl ACP by β-ketoacyl ACP reductase, whereas the tryptic peptide did not inhibit reduction of acetoacetyl ACP.

These experiments suggested that ACP interacts with β-ketoacyl ACP reductase at a site different from the prosthetic group-active site interaction, and that this site was altered or lost in the tryptic peptide (61). Recent investigations of β-ketoacyl ACP reductase with model compounds, as well as ACP substrates, have led the authors (92) to the same conclusion. Work of Simoni, Criddle, and Stumpf (96) also indicated that there may be more than one specific site on the ACP molecule. These authors purified ACP from several sources. While the ACP from *E. coli*, *Arthrobacter*, avocado, and spinach exhibited strikingly similar amino acid compositions, subtle structural differences were indicated. Although the ACP from these various sources functioned interchangeably in the *E. coli* or plant fatty acid synthetase systems, the products of fatty acid synthesis varied according to the ACP used. This suggested that slight structural variations in ACP can alter the specificity of ACP substrates in some, but not all, reactions of fatty acid synthesis. In addition, it has been reported that ACP of *M. phlei* does not function as well as *E. coli* ACP in the *E. coli* malonyl CoA–CO_2 exchange reaction (71). In fact, ACP (*M. phlei*) is an inhibitor of ACP (*E. coli*) in the fatty-acid-synthesizing system of *E. coli*.

Cyanogen bromide treatment of ACP yielded peptide 1 through 44, which is inactive. On the other hand, acetylation of the four lysine residues and the amino group of the terminal serine residue had no effect on the ability of ACP to function in fatty acid synthesis (62). Trypsin treatment of fully acetylated ACP yielded peptide 7 through 77, since cleavage occurred only at the single arginine residue. This peptide, which lacked the amino terminal hexapeptide, was completely inactive in fatty acid synthesis and the malonyl CoA–CO_2 exchange reaction (62); moreover, the apopeptide was inactive in the holo-ACP synthetase reaction (86). Thus it is obvious that the peptide structure is very important for the activity of ACP both with the enzymes of fatty acid synthesis and with holo-ACP synthetase.

Chemical modifications accompanied by optical rotatory dispersion measurements have attempted to further the understanding of the structure and function of ACP (1). Nitration of tyrosine led to a decrease in melting temperature, determined by optical activity measurements. No effect on the malonyl CoA–CO_2 exchange reaction or fatty acid synthesis *in vitro* was observed, while acylation of glycerol-3-phosphate by the palmitate derivative of the modified ACP and a purified *E. coli* membrane fraction was decreased. Alkylation of methionine had only small effects on structure and activity measured in the malonyl CoA–CO_2 exchange reaction. However, *in vitro* fatty acid synthesis and acylation of glycerol-3-phosphate were decreased. The modification of surface carboxyl groups with glycine ethyl ester and dicyclohexylcarbodiimide led to slight structural changes, accompanied by a complete loss of biological activity. Finally, acetylation of the α-amino group strongly affected the structure, as measured by the melting temperature (1), but did not affect biological activity (62). The removal of the *N*-terminal hexapeptide led to the loss of all organized structure as determined by optical measurements. This last finding led the authors to propose that the terminal hexapeptide stabilizes the structure, possibly through a salt bridge with the sequence Ala-Asp-Ser-Leu of the substrate binding site (1).

C. CHEMICAL SYNTHESIS OF ACP

It is apparent that important information was derived from studies of ACP peptides and the specific amino acid modifications. However, not all parts of a protein structure can be explored through these procedures; therefore, a more general approach was undertaken: the chemical synthesis of analogues of ACP. The first step in this plan was to synthesize active ACP. Although ACP contains 77 residues, peptide 1 through 74 was selected for synthesis, since it had been demonstrated that the tripeptide at the C-terminus was not required for biological activity. This was convenient also in that peptide 1 through 74 did not contain histidine, tryptophan, or cysteine—the amino acids that are generally difficult to use in a solid phase synthesis. The plan was to prepare the apopeptide 1 through 74 and, after deprotection of the peptide, to add the prosthetic group enzymatically to form the holopeptide.

The synthetic scheme employed was in essence the solid phase procedure of Merrifield (40,73), although several minor modifications were used (41,69). The completed peptide was cleaved from the support with hydrobromic acid and trifluoroacetic acid. The product was extracted with 50% acetic acid, sized on Sephadex G-25, and the nitroarginine was deprotected by hydrogenation. The prosthetic group labeled with tritium was transferred by holo-ACP synthetase (reaction 2) to the peptide from [^{3}H] pantetheine–CoA to form [^{3}H] holo-ACP 1 through 74. The product from this reaction was purified by ion exchange chromatography on DEAE cellulose. The specific radioactivity (dpm/mg protein) of tritiated holo-ACP indicated that approximately 40% of the peptide had accepted the tritiated prosthetic group. The yield of protein at this stage was 1.5%, based on the first amino acid in the synthesis. The synthetic ACP was tested in the malonyl CoA–CO$_2$ exchange reaction, where it was found to be 30% as active as native ACP 1 through 74. Figure 4 shows that the purified synthetic product (solid circles) and native ACP (open circles) cochromatographed on DEAE-cellulose when a shallow lithium chloride gradient was used. When tritiated synthetic ACP was chromatographed alone, it was noted that the tritium counts (solid circles) and

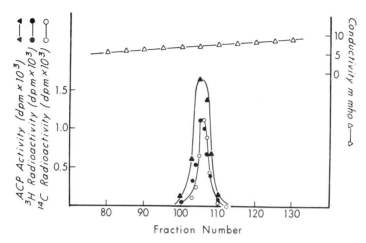

Fig. 4. DEAE-cellulose chromatography of synthetic ACP 1–74 (●–●) and native ACP 1–74 (○–○). Synthetic ACP contained ^{3}H and native ACP contained ^{14}C in the prosthetic group, 4′-phosphopantetheine. ACP activity was assayed in the malonyl CoA–CO$_2$ exchange reaction (▲–▲). The eluting solvent was lithium chloride (△–△).

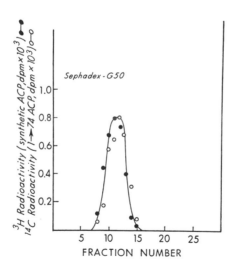

Fig. 5. Sephadex G-50 chromatography of synthetic ACP 1–74 (●–●) and native ACP 1–74 (○–○). Radioactive labels were the same as in Fig. 4.

279

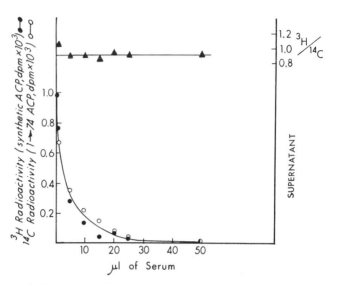

Fig. 6. Antibody precipitation of synthetic ACP 1–74 (●–●) and native ACP 1–74 (○–○). Radioactive labels were the same as in Fig. 4. Increasing aliquots of antibody made against native ACP were added to a mixture of radioactive synthetic and native ACP. After the precipitates were removed by centrifugation, the ^{3}H and ^{14}C in the supernatant and in the precipitates were measured. The radioactivity in the supernatant and the ratio of ^{3}H:^{14}C in the precipitates are presented.

the activity in the malonyl CoA–CO$_2$ exchange reaction cochromatographed. Thus synthetic and native ACP had similar charge properties. Figure 5 shows that synthetic (solid circles) and native (open circles) ACP cochromatographed on Sephadex G-50, indicating that the two were of similar size. Figure 6 indicates that synthetic and native ACP were coprecipitated by antibodies prepared against holo-ACP. As anticipated, the counts left in the supernatant decreased as the amount of added serum antibody was increased. The ^{3}H:^{14}C ratio of the precipitating material remained essentially constant.

Since ACP does not contain cysteine, only the peptide material in the preparation that was active with holo-ACP synthetase would contain a sulfhydryl group, that contributed by the prosthetic group, 4'-phosphopantetheine. Therefore, affinity chromatography was used to purify the product by the binding of the sulfhydryl group to a p-chloromercuribenzoyl agarose column. Although 25% of the material washed through the column with the original buffer (Fig. 7), the protein

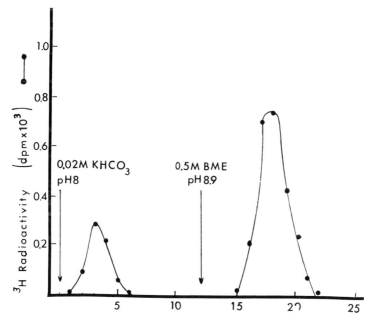

Fig. 7. Affinity chromatography of synthetic ^{3}H-ACP 1–74 on PCMB Agarose.
The column was eluted initially with 0.02 M KHCO$_3$ (pH 8.0) and then with
0.5 M 2-mercaptoethanol (BME). Fractions were assayed for ^{3}H radioactivity.

eluted by 0.5 M 2-mercaptoethanol was 75% active in the malonyl
CoA–CO$_2$ exchange reaction. Thus affinity chromatography has
yielded the most active synthetic ACP thus far.

Table III illustrates the activities of native ACP, synthetic ACP,
and the analogues that have been prepared thus far in the two assays
that have been utilized. The left column presents the activity of the
apo-ACP preparations with holo-ACP synthetase, and the right column
presents the activity of the holo-ACP preparations in the malonyl
CoA–CO$_2$ exchange reaction. In the first line are shown the activities
of 1 through 77 ACP in both assays as a reference. The second line
indicates that 1 through 74 ACP, prepared by carboxypeptidase
treatment of native ACP, was fully active, having 100% activity in
both assays. When native apo-ACP was subjected to the conditions
of deprotection required in the chemical synthesis, line 3, only 45%
of the activity was retained in both assays. In line 4 are shown the
activities of the deprotected synthetic product, which was 35 and

TABLE III
Activity of Synthetic and Native Peptides of ACP

	Activity	
	Apo with ACP synthetase (^{3}H dpm/μg protein)	ACP activity in malonyl CoA–CO$_2$ exchange (^{14}CO$_2$ dpm/20 $\mu\mu$M)
1. 1→ 77 Native ACP	7900	2500
2. 1→ 74 Native ACP	7900 (100%)	2460 (100%)
3. Deprotected 1→ 74 native ACP	3560 (45%)	1120 (45%)
4. 1→ 74 Synthetic ACP	2365 (35%)	750 (30%)
5. 7→ 77 Native ACP	0 (0%)	0 (0%)
6. Arg NO$_2$[6] 1→ 74 synthetic ACP	2405 (31%)	390 (16%)
7. 3→ 74 Synthetic ACP	3160 (40%)	1000 (40%)
8. Arg NO$_2$[6] norleu[8,9,18] 1→ 74 synthetic ACP	2370 (30%)	

30% active in the two assays, thus indicating that the synthetic product had activities very similar to native ACP, which had been similarly treated (as in line 3). Line 5 shows that removal of the N-terminal hexapeptide of native ACP by trypsin yielded a product, peptide 7 through 77, which was completely inactive in both assays. Thus an attempt was made to delineate the critical amino acid residue in this hexapeptide, which contained serine, threonine, isoleucine, two glutamic acids, and the single arginine residue of the protein. The importance of the arginine residue was discounted (see line 6) when it was discovered that synthetic apo-ACP containing nitroarginine in position 6 was 31% active in the holo-ACP synthetase reaction. This nitroarginine derivative also had significant activity in the malonyl CoA–CO$_2$ exchange reaction. The possible importance of the N-terminus itself was ruled out by the observations that both acetylation of the N-terminal serine and the omission of either ser-1 or ser-1 thr-2 in the synthesis gave products that were active in both assays. For example, line 7 indicates that synthetic 3 through 74 ACP was active. These results suggest that the sequence ile[3]-glu[4]-glu[5] is the part of the hexapeptide that is required for biological activity, and derivatives lacking these residues are being prepared (W. S. Hancock, unpublished results). The observation that acetylation did not affect the activity of ACP suggested that the ε-amino groups of lysine are not essential. Therefore the peptide (line 8) was synthesized in which norleucine was substituted for lysine at positions 8, 9, and 18. This peptide also contained nitroarginine at position 6. This product was equally active with holo-ACP synthetase as unsubstituted synthetic 1 through 74. It is therefore apparent that ACP can be synthesized by the Merrifield solid phase method and that some interesting analogues have already been prepared. It is hoped that the production of a large number of analogues will help elucidate the mechanism of interaction of this protein with the enzymes involved in fatty acid biosynthesis.

III. Synthesis and Turnover of the Prosthetic Group of ACP

A. ENZYMES

Since the prosthetic group of ACP is 4'-phosphopantetheine (65), the metabolism of CoA and ACP could be expected to be closely related. Several enzymes that are involved with the metabolism of the prosthetic group of ACP have been identified and partially characterized. ACP hydrolase catalyzes the removal of 4'-phosphopantetheine from ACP according to reaction 1.

$$\text{Holo-ACP} + \text{H}_2\text{O} \xrightarrow{\text{Mn}^{+2}} \text{4'-phosphopantetheine} + \text{apo-ACP} \qquad (1)$$

The enzyme is cytoplasmic and has been partially purified from extracts of *E. coli* (110). Nonspecific proteolysis by this enzyme was excluded by identification of intact apo-ACP as a reaction product. The analysis of amino acid composition of the isolated apo-ACP demonstrated the specific loss of β-alanine and 2-mercaptoethylamine. In addition, 4'-phosphopantetheine was identified as the other reaction product by coupling the hydrolase reaction with the enzymes, dephospho CoA pyrophosphorylase and dephospho CoA kinase, which convert 4'-phosphopantetheine to CoA. The CoA thus formed was assayed with phosphotransacetylase (99). ACP hydrolase is completely dependent on divalent metal cations for activity, with manganous ion giving maximal activity at $2.5 \times 10^{-5}\ M$. Other divalent metal cations, such as magnesium, cobalt, iron, and zinc were effective activators at higher concentrations. However, CuSO_4, CdCl_2, and

$CrCl_2$ did not stimulate the reaction at any concentration tested nor were trivalent metal cations, such as $FeCl_3$ or $AlCl_3$, effective.

Perhaps the most striking characteristic of ACP hydrolase is its specificity for hydrolysis of the phosphodiester of intact ACP. The enzyme is inactive with large peptides of ACP, including the tryptic peptide 19 through 61 described above (Section II.B). Coenzyme A and glycerylphosphorylserine were also tested and found to be inactive as substrates. However, ACP from $C.$ $butyricum$ did serve as the substrate for ACP hydrolase from $E.$ $coli$, although the rate of hydrolysis was approximately one-third that observed with ACP from $E.$ $coli$ (110). Thus the enzyme is a phosphodiesterase with strict specificity for ACP.

The enzyme, holo-ACP synthetase, catalyzes the synthesis of holo-ACP from apo-ACP and CoA according to reaction 2:

$$\text{apo-ACP} + \text{CoA} \xrightarrow{\text{Mg}^{2+}} \text{holo-ACP} + 3',5'\text{-adenosine diphosphate} \qquad (2)$$

Initial attempts at purifying this enzyme from extracts of $E.$ $coli$ were unsuccessful because the enzyme is very unstable. However, after it was discovered that ACP synthetase was markedly protected from inactivation by the presence of half-saturating concentrations of reduced CoA, a 780-fold purification was achieved by carrying out the fractionation in the presence of 10^{-4} M CoA (26). This protective effect was specific for CoA, since CoA analogues, dephospho-CoA or oxidized CoA did not serve as protective agents. The purified enzyme was not homogeneous, but the preparation was free of apo-ACP, holo-ACP, and ACP hydrolase. The enzyme has apparent K_m values of 4×10^{-7} M for apo-ACP and 1.5×10^{-4} M for CoA. Dephospho-CoA and oxidized CoA were inactive as substrates. Although the enzyme requires either Mg^{2+} or Mn^{2+} for activity, both cations gave complex saturation curves, with Mg^{2+} showing substrate activation between 10^{-2} M and 10^{-1} M.

Because quantitative conversion of apo-ACP to holo-ACP can readily be achieved with an excess of synthetase, this enzyme can be utilized in an assay for apo-ACP in biological material. As noted above (Section II.C) the synthetase has been utilized in the conversion of chemically synthesized apo-ACP to holo-ACP, and the high degree of specificity that the enzyme exhibits toward the peptide structure has been useful in the purification of biologically active synthetic holo-ACP (41). The specificity of the synthetase toward apopeptides is demonstrated in

Table II, where it is noted that apopeptide 1 through 74, which is active as the holopeptide in the malonyl CoA–CO_2 reaction, is also active in the holo-ACP synthetase reaction. On the other hand, apopeptides 19 through 61 and 7 through 77 are completely inactive with the synthetase, and this parallels the inactivity of the analogous holopeptides in fatty acid synthesis.

While holo-ACP synthetase represents a very small proportion of the soluble protein of *E. coli*, in the order of 0.01 %, the enzymatic activity is sufficient to convert 10 mμM of apo-ACP to holo-ACP per minute per gram of cells, and *E. coli* contains only 50 mμM of ACP per gram of cells. The intracellular concentration of CoA has been shown to vary from 20 to 200 μM depending on growth conditions (17), and while these concentrations would not be saturating for the enzyme (K_m for CoA approximately 0.15 mM), the intracellular enzyme activity would be sufficient to account for all synthesis of ACP in exponentially growing cells (26).

B. TURNOVER STUDIES

The demonstration that the prosthetic group of ACP is 4′-phosphopantetheine was soon followed by the discovery of holo-ACP synthetase, which catalyzes the transfer of 4′-phosphopantetheine from CoA to apo-ACP to form holo-ACP. Pantothenate is a component of 4′-phosphopantetheine (Fig. 1); in the majority of organisms most of the intracellular pantothenate is found in the 4′-phosphopantetheine of ACP and CoA, with CoA containing most of this vitamin. An early study demonstrated that an *E. coli* pantothenate auxotroph utilized the pantothenate of endogenous CoA for the synthesis of ACP when the organism was transferred from pantothenate-supplemented to pantothenate-deficient medium. These observations suggested a precursor–product relationship between CoA and ACP (7). In order to obtain more direct evidence for this relationship, kinetic isotope experiments were carried out with radioactive pantothenate in the *E. coli* pantothenate auxotroph (84).

Two models were devised for considering the interrelationships among pantothenate-containing compounds *in vivo*. In Model *a* [Fig. 8(*a*)] which incorporates the enzymatic activity of holo-ACP synthetase, CoA is the direct precursor for the prosthetic group of ACP and mass flows in the direction of

4′-phosphopantetheine → dephospho CoA → CoA →

ACP → 4′-phosphopantetheine

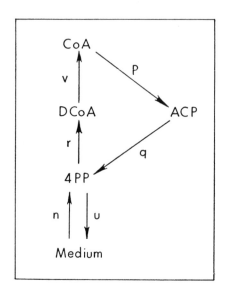

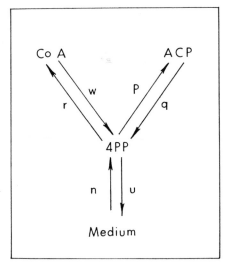

Fig. 8. Possible models of ACP synthesis and prosthetic group turnover. Model a, 4PP, 4'-phosphopantetheine; DCoA, dephospho CoA. Model b, flow rates represented by n, p, q, r, u, v, and w.

Model b [Fig. 8(b)] depicts 4'-phosphopantetheine, or a compound kinetically indistinguishable from it, as the direct precursor of the prosthetic group, and mass can flow reversibly between both 4'-phosphopantetheine and CoA or between 4'-phosphopantetheine and ACP. Model b, therefore, tested the exclusiveness of Model a. Both models provided for growth-independent turnover of the prosthetic group of ACP by means of ACP hydrolase, and both models had provisions for the exit of pantothenate-containing material from the cell 4'-phosphopantetheine compartment into the medium.

Model a [Fig. 8(a)] was consistent with what is known of the sequence of CoA biosynthesis from 4'-phosphopantetheine (17). In addition, it was consistent with the demonstrated enzymatic pathways of ACP synthesis (26) and ACP hydrolysis (110). The excretion of panto-thenate-containing compounds, which had been observed previously (23,74), was accommodated. Most important was the fact that a unique choice of rate constants for flows of radioactivity of pantothe-nate through the various pools was satisfactory for pulse, chase, and recovery from pantothenate starvation experiments (84).

In contrast to the consistent fit of all experimental data in Figure 8a, the data in Figure 8b could not be made to fit all the experiments with a unique choice of rate constants. Model b could be made to fit each type of experiment, but the rate constants had to be modified for each type of experiment. Moreover, some of these choices were mutually exclusive (84). Thus Model a was to be preferred. This implied that CoA is indeed the precursor of the prosthetic group of ACP *in vivo*.

The turnover of the prosthetic group of ACP was observed directly in experiments in which the pantothenate auxotroph recovered from pantothenate starvation. More than 90% of the radioactivity was found in ACP in cells that had stopped growing because they had exhausted the radioactive pantothenate of the medium (7). When these cells were transferred to medium containing a high concentration of unlabeled pantothenate, CoA synthesis quickly recovered, and label could be directly observed flowing from ACP into the 4'-phosphopante-theine (and CoA) compartment. In exponentially growing cells, the fractional rate constant for the loss of ACP radioactivity was 0.04/min, or 4% of the ACP in the compartment was hydrolyzed to 4'-phospho-pantetheine and apo-ACP per minute. For comparison, in exponentially growing cells with a generation time of about 70 min, the increase in mass of the ACP compartment is 1%/min. Thus, in exponentially growing cells, the rate of ACP synthesis was about five times what it would be if turnover did not occur (84). The turnover of the 4′-phosphopantetheine prosthetic group of ACP and CoA may be a special consequence of the role these coenzymes play in metabolism. However, the specific role of this turnover is not understood.

The average proteins of *E. coli* do not undergo turnover during exponential cell growth (45). However, in view of the rapid turnover of the prosthetic group of ACP, it was important to ascertain whether the protein component undergoes simultaneous turnover; recent experiments (G. L. Powell and A. R. Larrabee, personal communication) indicate that the protein component of *E. coli* ACP does not undergo such turnover. Investigations of the rat-liver fatty acid synthetase complex have indicated that protein turnover of the enzyme complex is much slower than the turnover of the 4′-phosphopantetheine prosthetic group (107). In steady state conditions, rat-liver fatty acid synthetase exhibited a half-life of 71 to 108 hr. The exchange rate of covalently bound 4′-phosphopantetheine in the complex was at least an order of magnitude greater. It should be noted that ACP hydrolase has not yet been identified in the animal system. A model was proposed in which the prosthetic group or a small portion of the complex containing the prosthetic group was periodically removed and replaced in the larger fatty acid synthetase complex. This action must take place many times before the individual complex is catabolized. Such prosthetic group turnover may be a means of control of overall fatty acid synthetase activity (107).

IV. ACP from Other Sources

While the majority of the studies attempting to delineate the structure and function of ACP have been carried out with *E. coli* ACP, homogeneous preparations of ACP have also been obtained from *Arthrobacter*, avocado, spinach (96), *M. phlei* (71), and *C. butyricum* (3). The amino acid compositions of these proteins are shown in Table IV. The compositions of *E. coli* (112) and yeast ACP (121) are listed for comparison. The most striking feature of these compositions is the presence of 1 mole each of taurine (the oxidation product of 2-mercapto-ethylamine) and β-alanine, representing the 4′-phosphopantetheine prosthetic group. Five of the seven ACPs have only a single sulfhydryl

TABLE IV

Amino Acid Composition of ACP from Several Sources[a]

Amino acid	E. coli[b]	arthrobacter[c]	Avocado[c]	Spinach[c]	M. phlei[d]	C. butyricum[e]	Yeast[f]
Cysteic acid	0	0	1	0	0	0	1
Taurine	1	1	1	1	1	1	1
β-Alanine	1	1	1	1	1	1	1
Aspartic	9	14	12	12	11	13	14
Threonine	6	2	7	6	5	1	8
Serine	3	6	10	5	5	3	12
Glutamic	18	10	22	16	20	14	19
Proline	1	1	3	2	4	1	7
Glycine	4	5	7	4	6	1	13
Alanine	7	12	11	9	13	6	14
Valine	7	6	10	7	8	7	8
Methionine	1	1	1	1	1	4	2–3
Isoleucine	7	6	5	5	6	8	7
Leucine	5	6	9	7	8	7	13
Tyrosine	1	0	1	0	2	1	4–5
Phenylalanine	2	3	3	2	2	3	6
Lysine	4	5	10	9	5	4	10
Histidine	1	1	1	1	0	2	2–3
Arginine	1	1	1	0	3	0	5

Values for amino acid composition of yeast ACP were calculated assuming 1 mole of β-alanine/mole protein, and rounding to the nearest integer.

[a] Nearest integer values.
[b] Vanaman, Wakil, and Hill (112).
[c] Simoni, Criddle, and Stumpf (96).
[d] Matsumura, Brindley, and Bloch (72).
[e] Ailhaud, Vagelos, and Goldfine (3).
[f] Willecke, Ritter, and Lynen (121).

group, that contributed by the prosthetic group. Avocado and yeast ACP both have an additional sulfhydryl group contributed by a cysteine residue; however, Simoni, Criddle, and Stumpf (96) have shown conclusively that the cysteine of avocado ACP does not function as an acyl carrier. All the proteins, except that derived from yeast, are rich in glutamic and aspartic acid residues, thus accounting for the acidic nature of the proteins. Molecular weights of these proteins range from approximately 8600 for *C. butyricum* ACP (3) to 11,900 for avocado ACP (96).

The report of the isolation of yeast ACP represents the first isolation of ACP from one of the multienzyme complexes. Very small amounts of material were available for the amino acid composition studies given in Table I, and the value of 16,000 for the molecular weight, based on amino acid composition, must be considered only tentative (121). Future work on ACP from multienzyme complexes may reveal the reasons for the extremely tight association exhibited by these systems. Examination of the preliminary amino acid composition of the yeast ACP reveals some important differences from the ACPs isolated from the easily dissociated fatty acid synthetase systems. The proportion of acidic residues is lower than that for ACP of the other sources; the proportion of aromatic residues and proline is higher. Whether these differences are significant in the association of yeast ACP with the enzymes of yeast fatty acid synthetase must await further investigation.

Mycobacterium phlei contains two types of fatty acid synthetase systems (72). One system exists as a Type 1 multienzyme complex of molecular weight 1.7×10^6 (14), which is similar to the complex isolated from yeast (59). The other system (Type 2) consists of individual enzymes that can be purified separately, and it resembles the *E. coli* system in the types of reactions that are catalyzed, with one notable exception. While the *E. coli* system utilizes acetyl groups effectively as a primer in fatty acid synthesis, the *Mycobacterium* Type 2 system utilized palmityl or stearyl groups exclusively, while short chain precursors were inactive (72). Therefore, the Type 2 fatty acid synthetase in this organism functions in chain-elongation only. The characteristics of the ACP from *Mycobacterium* discussed above are those of the ACP from the Type 2 system. Whether this ACP might be derived from dissociation of the multienzyme complex has not been established (72). Therefore, comparative studies of ACP from the

Types 1 and 2 systems have not yet been carried out. ACP from
E. coli was active in the *Mycobacterium* Type 2 system, although some
differences in rate of elongation were noted compared to ACP from
Mycobacterium. These experiments were performed with stearyl CoA
as a primer (72).

It has been noted that all the preparations of ACP studied thus far,
except yeast, have been obtained from organisms which contain fatty
acid synthetase systems that are found dissociated when the cell
membranes are ruptured. Preliminary studies of yeast ACP have been
reported (121), but ACP has not been isolated from other organisms that
have a fatty acid synthetase multienzyme complex. However, the presence of protein-bound 4'-phosphopantetheine has been shown in the
multienzymes complexes isolated from rat adipose tissue (53), pigeon
liver (19), rat liver (18), lactating rat mammary gland (98), and *M. phlei*
(14). In addition, the presence of ACP in the multienzyme complex
isolated from etiolated cells of *Euglena gracilis* is indicated by the fact
that this complex catalyzes *de novo* fatty acid synthesis from acetyl
CoA and malonyl CoA in the absence of added ACP (24).

V. Enzymes of Fatty Acid Biosynthesis

The role of ACP in the biosynthesis of saturated fatty acids is
illustrated in reactions 3 through 9.

$$CH_3CO—S—CoA + HS—ACP \rightleftharpoons CH_3CO—S—ACP + CoA—SH \tag{3}$$

$$CH_3CO—S—ACP + HS—E_{cond} \rightleftharpoons CH_3CO—S—E_{cond} + ACP—SH \tag{4}$$

$$HOOCCH_2CO—S—CoA + HS—ACP \rightleftharpoons$$
$$HOOCCH_2CO—S—ACP + CoA—SH \tag{5}$$

$$HOOCCH_2CO—S—ACP + CH_3CO—S—E_{cond} \rightleftharpoons$$
$$CO_2 + HS—E_{cond} + CH_3COCH_2CO—S—ACP \tag{6}$$

$$CH_3COCH_2CO—S—ACP + NADPH + H^+ \rightleftharpoons$$
$$NADP^+ + CH_3CHOHCH_2CO—S—ACP \tag{7}$$

$$CH_3CHOHCH_2CO—S—ACP \rightleftharpoons CH_3CH{=}CHCO—S—ACP + H_2O \tag{8}$$

$$CH_3CH{=}CHCO—S—ACP + NADPH + H^+ \rightleftharpoons$$
$$NADP^+ + CH_3CH_2CH_2CO—S—ACP \tag{9}$$

In the initial reaction (reaction 3), the acetyl group of acetyl CoA is
transferred to the sulfhydryl group of ACP by acetyl CoA–ACP
transacylase, forming acetyl ACP. The acetyl group is then transferred
to the sulfhydryl group of the condensing enzyme (E_{cond}) to form an

292   DAVID J. PRESCOTT AND P. ROY VAGELOS

acetyl enzyme intermediate and liberate ACP (reaction 4). Malonyl CoA–ACP transacylase then catalyzes the transfer of a malonyl group from CoA to ACP in reaction 5; this is followed by the condensation reaction (reaction 6), which takes place between malonyl ACP and acetyl enzyme to produce acetoacetyl ACP, CO_2, and the free condensing enzyme. Acetoacetyl ACP is then reduced by NADPH to form specifically D-(−)-β-hydroxybutyryl-ACP (reaction 7); the latter is dehydrated to form the *trans* unsaturated thioester, crotonyl ACP (reaction 8); and crotonyl ACP is reduced by NADPH to form the saturated thioester, butyryl ACP (reaction 9).

In the normal biosynthetic sequence, butyryl ACP reacts with the condensing enzyme to form butyryl enzyme (reaction 4), thereby liberating ACP, which can accept a malonyl group and thus initiate another elongation, reduction, dehydration, and reduction sequence. After appropriate repetitions of this series of reactions, the normal product, palmityl ACP is formed. Small amounts of myristate and stearate are also produced, but the major saturated fatty acid produced *in vivo* contains 16 carbon atoms. The *in vitro* products of the *E. coli* fatty acid synthetase are free fatty acids; this is probably a result of thioester hydrolysis catalyzed by a specific palmityl thioesterase that has been recently characterized (11). *In vivo* palmityl ACP and other long chain acyl-ACPs produced by the synthetase probably react directly with *sn*-glycero-3-phosphate and a specific membranous acyltransferase to form lysophosphatidic acid, the first intermediate in the pathway to phospholipid synthesis (see Section VI). In the discussion of specific reactions of fatty acid biosynthesis, emphasis will be put on enzymes isolated from *E. coli*, since these will serve to illustrate best the interactions of specific enzymes with ACP. For a discussion of fatty acid biosynthesis in multienzyme complexes and the regulation of fatty acid synthesis, the reader is referred to a recent review (109).

A. ACETYL CoA–ACP TRANSACYLASE

E. coli acetyl CoA–ACP transacylase catalyzes reaction 3 (5). The transfer of the acetyl groups from CoA to ACP is accompanied by an equivalent disappearance of ACP sulfhydryl groups. Moreover treatment of the acetyl ACP with neutral hydroxylamine released the acetyl group as acetyl hydroxamic acid. Thus acetyl ACP was shown to be a thioester (63). The enzyme is relatively specific for acetyl groups, since propionyl-, butyryl-, hexanoyl-, and octanoyl CoA were transacylated very slowly (122). Malonyl CoA is inactive with this enzyme. Pantetheine can replace ACP, and acetylpantetheine readily substitutes for acetyl CoA. The enzyme is severely inhibited by both N-ethylmaleimide and iodoacetamide, and this inhibition is prevented by prior incubation with acetyl CoA (122). Incubation of the enzyme with ^{14}C-acetyl CoA led to the formation of ^{14}C-acetyl enzyme, which was separated from the reaction mixture by filtration through Sephadex G-25. The ^{14}C-acetyl enzyme was able to transfer the ^{14}C-acetyl group to either CoA or ACP. Although the nature of the ^{14}C-acetyl enzyme was not further characterized, Williamson and Wakil (122) have proposed that the enzyme intermediate is a thioester that functions as follows:

$$\text{acetyl-S-CoA} + \text{HS-enzyme} \rightarrow \text{acetyl-S-enzyme} + \text{CoA-SH} \qquad (10)$$

$$\text{acetyl-S-enzyme} + \text{HS-ACP} \rightarrow \text{acetyl-S-ACP} + \text{HS-enzyme} \qquad (11)$$

The sum of these two reactions is reaction 3.

The acetyl transacylase activity of pigeon-liver fatty acid synthetase multienzyme complex has been studied (49,83). Incubation of pigeon-liver fatty acid synthetase with acetyl CoA indicated that acetyl groups may bind to two different thiol groups plus a nonthiol in the complex. One thiol has been identified as the central thiol of 4′-phosphopantetheine and the other thiol as a cysteine residue, presumably on the condensing enzyme of the complex (49). From the results of differential inhibition studies using N-ethylmaleimide and iodoacetamide, these authors have suggested that the order of acetyl transfer is first to a nonthiol, tentatively identified as a serine hydroxyl, then to the central thiol (4′-phosphopantetheine), and finally to the peripheral sulfhydryl of the condensing enzyme. Prior studies had also shown that both acetyl and malonly groups bind to the central thiol (48). These suggestions are similar to those proposed earlier by Lynen et al. (60) for yeast fatty acid synthetase.

The pigeon-liver fatty acid synthetase has been shown to catalyze transacylation of acetyl groups from acetyl CoA to $E.\ coli$ ACP. Iodoacetic acid, while inhibiting fatty acid synthesis, did not inhibit this transacylase activity (83). In addition, CoA is a competitive inhibitor of acetyl transacylase with respect to $E.\ coli$ ACP. Transacylase activities have been reported to dissociate from the synthetase complex in low yield in the presence of 0.5 M guanidine HCl (83).

B. MALONYL CoA–ACP TRANSACYLASE

E. coli malonyl CoA–ACP transacylase catalyzes the analogous transfer of a malonyl group to ACP, reaction 5 (5,122). The transfer of the malonyl group to ACP results in an equivalent disappearance of the sulfhydryl groups of ACP. When the malonyl ACP was treated with neutral hydroxylamine, malonyl hydroxamic acid was identified as the product (63). Therefore malonyl groups are bound to ACP through thioester links. This enzyme is very specific for the malonyl group, since there is no activity with acetyl CoA as a substrate. Pantetheine can substitute efficiently for ACP in this reaction, and malonyl pantetheine substitutes for malonyl CoA. Incubation of the enzyme with ^{14}C-malonyl CoA leads to the formation of ^{14}C-malonyl enzyme which can transfer the ^{14}C-malonyl group to either CoA or ACP (50,109; Ruch and Vagelos, unpublished results). Therefore the mechanism of malonyl transacylation involves a malonyl-enzyme intermediate as follows:

$$\text{Malonyl-S-CoA} + \text{enzyme} \rightarrow \text{malonyl enzyme} + \text{CoA-SH} \qquad (12)$$

$$\text{Malonyl enzyme} + \text{HS-ACP} \rightarrow \text{malonyl-S-ACP} + \text{enzyme} \qquad (13)$$

The sum of these two reactions is reaction 5. The details of this transacylase mechanism are not clear at this time, since not all of the data concerning the enzyme are in agreement. Ruch and Vagelos (unpublished results) have found that the enzyme is stimulated by sulfhydryl compounds such as dithiothreitol, and that it is severely inhibited by low concentrations of *p*-chloromercuribenzoate. This inhibition was readily demonstrated in the overall reaction 5 and in the partial reaction 12. Thus the enzyme appears to contain a critical sulfhydryl residue. On the other hand, Joshi and Wakil (50) report that the enzyme is insensitive to sulfhydryl inhibitors and is strongly inactivated by phenylmethane-sulfonylfluoride, suggesting that it possesses an active serine residue. They further report that the malonyl-enzyme intermediate is stable to performic acid oxidation, and that hydrolysis of the intermediate with pepsin yields peptic peptides which are stable to performic acid. These observations led these authors to propose that a nonthiol group of the protein is involved in the enzyme intermediate, and that the hydroxyl group of a serine residue appears to be most likely. Further experiments are warranted to fully characterize the malonyl-enzyme intermediate.

Malonyl transacylase activity of pigeon-liver fatty acid synthetase has been studied (49,83) and found to catalyze the transfer of malonyl groups from CoA to *E. coli* ACP. Protection of the transacylase activity against inhibition by *N*-ethylmaleimide could be achieved by preincubation with either acetyl CoA or malonyl CoA. Iodoacetamide, while inhibiting fatty acid biosynthesis, did not inhibit the transacylase activity. Acetyl CoA and CoA were competitive inhibitors of the malonyl transacylase activity (49). Small amounts of malonyl transacylase activity could be dissociated from the complex by treatment with .5 M guanidine HCl.

Incubation of pigeon-liver synthetase with ^{14}C-malonyl CoA was shown to label the protein (15,49). This labeling was attributed to involvement of the central thiol (4′-phosphopantetheine) and a non-thiol site tentatively identified as a serine hydroxyl (49). No labeling of the peripheral sulfhydryl, which was labeled when acetyl CoA was used, was observed with malonyl groups. It is not known whether the serine hydroxyls in the nonthiol binding sites are identical in the case of both acetyl and malonyl transacylation (49). Earlier studies by Lynen and his co-workers (60,94) with the yeast fatty acid synthetase complex had led to a similar proposal that nonthiol sites, probably the hydroxyl groups of serine residues, are involved in both acetyl and malonyl transacylation reactions.

C. β-KETOACYL ACP SYNTHETASE

This enzyme, which is involved in reactions 4 and 6, has been isolated from extracts of *E. coli* and characterized (5,6,37,85,106). The enzyme is extremely specific for thioesters of ACP, since CoA thioesters do not serve as substrates (6). Since the K_m for either acetyl ACP or malonyl ACP was not affected by increasing concentrations of the other substrate, the binding of the two condensing units is nonconsecutive and at independent binding sites (6). The enzyme is stimulated by 2-mercaptoethanol and inhibited by iodoacetamide and *N*-ethylmaleimide (6,37,106). In addition, the enzyme is inhibited by iodoacetamide when 1 mole of cysteine/mole of enzyme is alkylated, and acetyl ACP protects the enzyme against alkylation (37). Evidence for the existence of an acetyl-enzyme intermediate has been presented; this intermediate has been shown to have the properties of a thioester. The acetyl enzyme was shown to be metabolically active in transferring the acetyl group to ACP (reaction 4) or in condensing with malonyl ACP to form acetoacetyl ACP (reaction 6).

β-Ketoacyl ACP synthetase has been shown to be composed of subunits. The molecular weight of the active species is 66,000 (37). In the presence of a reducing agent and guanidine hydrochloride or sodium dodecyl sulfate, the molecular weight of the protein becomes 34,000 $\pm$ 3000 (85). The enzymatic activity can be partially recovered after dilution of the concentrated guanidine hydrochloride solution. Maximal enzymatic activity is observed after incubating purified enzyme in the presence of high concentrations (0.2 M) of phosphate; this activation may be caused by a conformational change (85). The enzyme crystallizes from phosphate-buffered ammonium sulfate solution. The activity is stable in the crystalline state. Both the yeast and pigeon liver fatty acid synthetase complexes have been shown to be unstable (46,78), and this instability is thought to be caused by the β-ketoacyl ACP synthetase, which is easily denatured. Maximal activity of these complexes is also observed in high phosphate concentrations. Peptide mapping data of reduced and alkylated $E.$ $coli$ β-ketoacyl-ACP synthetase are consistent with the subunits being identical (85).

1. Enzyme Specificity in Regulation of Fatty Acid Chain Length

The predominating saturated fatty acid in $E.$ $coli$ is palmitate (hexadecanoate), whereas the predominating unsaturated fatty acids are palmitoleate (cis-9-hexadecenoate) and cis-vaccenate(cis-11-octadecenoate) (52,56). The fatty acid synthetase of this organism synthesizes both saturated and unsaturated fatty acids in $vitro$, and, as shown by Bloch (12), the critical reaction in the biosynthetic pathway that leads to unsaturated fatty acids is catalyzed by β-hydroxydecanoyl thioester dehydrase. Figure 9 demonstrates the fact that β-hydroxydecanoyl ACP is the intermediate at the branch point between the pathways to saturated and unsaturated fatty acids. Dehydration of this compound in pathway A gives rise to cis-3-decenoyl ACP. This thioester presumably condenses with malonyl ACP to initiate chain elongation of the cis-unsaturated acyl ACP intermediates. Other hypothetical intermediates expected to undergo condensation with malonyl ACP in this pathway include cis-5-dodecenoyl ACP, cis-7-tetradecenoyl ACP, and cis-9-hexadecenoyl ACP

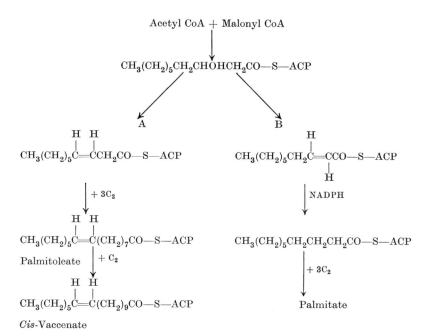

Fig. 9. Fatty acid biosynthesis in *E. coli*. Reaction A initiates the pathway to unsaturated acids and reaction B, the pathway to saturated acids.

(palmitoleate). On the other hand, pathway B is initiated by a dehydration that forms *trans*-2-decenoyl ACP, the normal intermediate in saturated fatty acid synthesis. The thioester intermediates expected to undergo chain elongation in the condensation reactions of pathway B include decanoyl ACP, dodecanoyl ACP, and tetradecanoyl ACP.

The availability of the pure condensing enzyme and many of the postulated ACP intermediates has permitted investigations to determine whether this enzyme catalyzes all the condensations in the biosynthesis of both saturated and unsaturated fatty acids, and whether the specificity of this enzyme can explain the accumulation of fatty acids of particular chain lengths in the cell (38). As shown in Table V, reactions with acetyl ACP, decanoyl ACP, and dodecanoyl ACP gave approximately similar results for both K_m and V_{max}. The enzyme is slightly less active with tetradecanoyl ACP. However, it is completely inactive with hexadecanoyl ACP, the C_{16} saturated fatty acid that accumulates in the cell. Assay of the enzyme with *cis*-3-decenoyl ACP and *cis*-5-dodecenoyl ACP, two early intermediates in the synthesis of unsaturated fatty acids, indicated that they are both as active as acetyl ACP or decanoyl ACP and have similar K_m's. The activity of *cis*-9-hexadecenoyl ACP is decreased to approximately one-fifth of the rate of the *cis*-5-dodecenoyl ACP. The *cis*-11-octadecenoyl ACP is completely inactive. These results indicate that this condensing enzyme can function in both saturated and unsaturated fatty acid synthesis in *E. coli*. Furthermore, the specificity of this enzyme explains how chain elongation is terminated specifically at C_{16} in the saturated pathway and at C_{16} and C_{18} in the unsaturated pathway.

TABLE V

Activity of β-Ketoacyl ACP Synthetase with Various
Intermediates in Fatty Acid Synthesis in *E. coli*[a]

Intermediate	K_m (μM)	V_{max} (μM product/min/mg)
Acetyl ACP	0.52	2.8
Decanoyl ACP	0.33	2.8
Dodecanoyl ACP	0.27	0.97
Tetradecanoyl ACP	0.28	0.31
Hexadecanoyl ACP	—	N.A.[b]
cis-3-Decenoyl ACP	0.71	1.9
cis-5-Dodecenoyl ACP	0.20	1.7
cis-9-Hexadecenoyl ACP	0.37	0.37
cis-11-Octadecenoyl ACP	—	N.A.

[a] Assays were carried out according to Greenspan et al. (38). Kinetic constants were determined from Lineweaver-Burk plots of the data obtained.

[b] N. A., no activity.

D. β-KETOACYL ACP REDUCTASE

The initial reduction in *de novo* fatty acid biosynthesis is catalyzed by β-ketoacyl ACP reductase according to reaction 7. The reduction product of acetoacetyl ACP has the D ($-$) configuration (5). This is in contrast to the L ($+$) configuration of β-hydroxyacyl CoA intermediates in β-oxidation of fatty acids (54). The *E. coli* enzyme can utilize CoA thioesters, although kinetic measurements indicated that both the K_m and V_{max} values for acetoacetyl CoA make this a much poorer substrate than the ACP derivative. Thioesters of pantetheine exhibited approximately the same affinity and maximal velocity with the enzyme as the CoA derivative (6). Acetoacetyl ACP exhibited substrate inhibition above 0.05 M concentrations (6), and the reaction was absolutely specific for NADPH; no reaction was observed when NADH was substituted as the cofactor. The reversibility of the reaction at high pH values (9.0) allowed the stereochemistry of the β-hydroxyacyl substrate to be determined (6). At lower pH values, however, the reaction strongly favored reduction of the β-keto derivative with an equilibrium constant of 3.9 $\times$ 10^{-7} M (105).

The broad specificity of the reductase regarding the thiol moiety of the thioester substrate has allowed this enzymatic activity to be demonstrated in the fatty acid synthetase complex of yeast (59) and brain (89). The $E.$ $coli$ reductase also has a broad specificity for the chain length of the β-ketoacyl ACP substrate (105). Recent experiments (Birge and Vagelos, unpublished results) have indicated that this enzyme is active with the β-ketoacyl ACP derivatives which are intermediates in the synthesis of both saturated and unsaturated fatty acids. Salts have been reported to stimulate fatty acid synthesis in $E.$ $coli$ extracts (93), and several enzymatic activites are affected. β-Ketoacyl ACP reductase is stimulated 20- to 40-fold by a number of mono- and divalent cations. Activation of the enzyme by salts was not observed when acetoacetyl-N-acetyl-cysteamine was utilized as substrate. The apparent activation observed with ACP substrates was attributed to cations complexing with the protein moiety of the ACP substrates, thus facilitating binding of the substrates to the enzyme. The cations may increase the reaction rate by reducing the repulsion between negatively charged carboxyl groups of the ACP substrate and the enzyme (93). Kinetic studies with model compounds have indicated that the substrate, acetoacetyl ACP, is bound to the reductase through its protein moiety, the pantoyl residue of the 4′-phosphopantetheine prosthetic group, and the β-ketoester group (92).

E. β-HYDROXYACYL ACP DEHYDRASES

β-Hydroxyacyl ACP dehydrase catalyzes the reversible dehydration of D $(-)$-β-hydroxyacyl ACP substrates according to reaction 8. The $E.$ $coli$ enzyme catalyzing the dehydration of β-hydroxybutyryl ACP was first described by Majerus, Alberts, and Vagelos (64), who partially characterized the reaction. The dehydration is readily reversible, and one assay for this enzyme is based on the disappearance of absorbance at 263 nm, due to the *trans*-2 conjugated double bond of the thioester product (5,64). The enzyme exhibits absolute specificity for the thiol moiety in a manner similar to β-ketoacyl ACP synthetase. Model substrates of CoA, pantetheine or *N*-acetylcysteamine are not metabolized. The reversibility of the enzymatic reaction has allowed the stereospecificity of the substrate metabolized to be determined conveniently. Thus the formation of β-hydroxyacyl ACP was coupled to either β-ketoacyl ACP reductase, specific for the D (−) isomer, or β-hydroxyacyl CoA dehydrogenase from pig heart, which is specific for the L (+) isomer. In this manner, the hydration of crotonyl ACP was shown to yield only the D (−) isomer of β-hydroxy butyryl ACP (64). In contrast, only the L (+) isomer is formed by the crotonase enzyme of fatty acid degradation.

β-Hydroxyacyl ACP dehydrase was highly purified from *E. coli* extracts in order to investigate whether a single enzyme catalyzes the dehydration of β-hydroxyacyl ACP intermediates in the pathways of both saturated and unsaturated fatty acid synthesis (Birge and Vagelos, unpublished results). The 500-fold purified enzyme preparation was found to be active with all the intermediates (C_4 through C_{16}) in the saturated pathway with highest activity at C_4 and lowest activity at C_{10}. In addition, the enzyme was active with the *cis*-unsaturated β-hydroxyacyl ACP intermediates of the unsaturated fatty acid biosynthetic pathway. Moreover, the ratio of activities with saturated and unsaturated pathway intermediates remained essentially constant throughout the 500-fold purification; inactivation of the purified enzyme with sulfhydryl inhibitors and by heating caused a parallel drop in activity with both saturated and unsaturated pathway intermediates. Thus it appears that a single β-hydroxyacyl ACP dehydrase can catalyze the *trans* dehydration of all the intermediates in fatty acid synthesis in *E. coli*. On the other hand, Wakil and his co-workers have reported the presence in *E. coli* extracts of three different β-hydroxyacyl ACP dehydrases active in the saturated pathway that vary with respect to chain length specificity, and the evidence for this has been recently reviewed (116). The relationship between these various dehydrase activities is not clear at this time.

One other *E. coli* dehydrase has been studied extensively by Bloch (12) and his co-workers, the β-hydroxydecanoyl thioester dehydrase. This enzyme catalyzes the conversion of β-hydroxydecanoyl ACP to *cis*-3-decenoyl ACP, the first intermediate in the pathway of unsaturated fatty acid synthesis (Fig. 9, reaction A). The enzyme, which has been obtained as a homogeneous protein, is active with N-acetylcysteamine, pantetheine, and CoA derivatives, as opposed to the dehydrase discussed above. The length of the carbon chain is crucial for the rate of the dehydrase-catalyzed reaction. Peak activity was found with 10-carbon substrates, the activities with C_9 and C_{11} compounds were decreased, while activities with the C_8 and C_{12} homologues were extremely low. The stringent chain length specificity permits this enzyme to function in the introduction of *cis* double bonds at only the 10-carbon-chain length during fatty acid synthesis. The mechanism of this interesting reaction has been investigated in great detail, and this work has been recently reviewed (12).

F. ENOYL ACP REDUCTASE

An enoyl ACP reductase preparation has been purified 250-fold from crude extracts of *E. coli* (119). Available evidence indicates that there are two distinct enzymes in the purified preparation that catalyze reduction of enoyl ACP derivatives according to reaction 9. An NADPH specific enzyme is inactive at pH values greater than 8.0 and has absolute specificity for acyl ACP substrates. This enzyme also exhibits greatest activity with crotonyl ACP as substrate, compared to longer chain acyl ACP substrates. The presence of an essential sulfhydryl group is indicated, since the enzyme is readily inhibited by para-hydroxymercuribenzoate, iodoacetate, and N-ethylmaleimide.

In contrast to the NADPH specific enzyme, an NADH specific enzyme exhibits activity over a wide range of pH values. This enzyme exhibits higher activity with decenoyl ACP as substrate compared to crotonyl ACP. Moreover, enoyl CoA thioesters are active substrates with the NADH specific reductase, and a different pattern of inhibition by sulfhydryl reagents was observed (119). The two enoyl reductase activities have not yet been separated.

VI. Other Reactions Involving ACP

A. PHOSPHOLIPID BIOSYNTHESIS

Fatty acyl groups in *E. coli* are found almost exclusively in the esterified form in phospholipids, which comprise about 10% of the dry weight of the cell. The fatty acids are synthesized by a series of reactions in which they are bound covalently to ACP as thioesters (Section V). This raised the question as to whether the acyl residues from the immediate end products of fatty acid biosynthesis, mainly palmityl ACP, palmitoleyl ACP, and *cis*-vaccenyl-ACP, were transferred directly to *sn*-glycero-3-phosphate, or were transferred via the corresponding CoA esters. Attempts to answer this question were made initially with membranous enzyme preparations from both *E. coli* (2), and *C. butyricum* (30,31). Palmityl ACP incubated with *sn*-glycero-3-phosphate and enzyme gave rise primarily to monoglyceride in the *E. coli* system and lysophosphatidic acid in the *C. butyricum* system. The observation that *sn*-glycero-3-phosphate was an obligatory acyl acceptor, which could not be replaced by glycerol, strongly suggested an initial formation of lysophosphatidic acid in both systems according to reaction 14:

sn-glycero-3-phosphate + palmityl-S-ACP →

lysophosphatidic acid + ACP—SH   (14)

Further investigations of the *E. coli* system have demonstrated that lysophosphatidate was formed stoichiometrically from *sn*-glycero-3-phosphate and palmityl ACP, and that it was a precursor of the monoglyceride which accumulated under certain conditions (113). *sn*-Glycero-3-phosphate acylation occurred with a large number of acyl CoA and acyl ACP thioesters, but conversion of *sn*-glycero-3-phosphate into lipid was most effective in the presence of palmityl CoA. A decrease in chain length or the introduction of a double bond in the acyl chain reduced the rate of *sn*-glycero-3-phosphate acylation. A similar selectivity was found for the acyl derivatives of ACP. It was demonstrated that with palmityl CoA or palmityl ACP, over 90% of the acylation took place at C-1 of *sn*-glycero-3-phosphate, giving rise to 1-acyllysophosphatidate. Further acylation of this intermediate to phosphatidate was catalyzed by a separate acyltransferase (87,113), which exhibited marked specificity for unsaturated acyl CoA and acyl ACP esters (reaction 15):

1-acyllysophosphatidic acid + palmitoleyl-S-ACP →

ACP—SH + phosphatidic acid   (15)

Although the rates of acylation were higher with acyl CoA esters, it was shown that this lysophosphatidate acyltransferase was even more specific for the acyl moiety of acyl ACP than acyl CoA esters. Thus phosphatidate synthesis from sn-glycero-3-phosphate and acyl ACP esters was only observed with unsaturated acyl ACP esters and not with palmityl ACP. In the latter case, essentially only lysophosphatidate formed because of the strict specificity for unsaturated thioesters of the lysophosphatidate acyltransferase enzyme. The accumulation of monoglyceride was shown to be due to a lysophosphatidic acid phosphatase (113), which had optimal activity at pH 7.0. Thus acyltransferase reactions carried out at pH 8.5 yielded almost entirely the phosphorylated products, lysophosphatidic, and phosphatidic acids. ·

These experiments have conclusively demonstrated that long chain acyl ACP and acyl CoA derivatives are active in the acylation of both sn-glycero-3-phosphate and lysophosphatidic acid. Since both of these acyl donors are active in the synthesis of phosphatidate in vitro, it is impossible to conclude from these studies which cofactor functions under physiological conditions. It is possible that both types of thioesters function under different conditions. For instance, thioesters of ACP might be the substrates when phosphatidate is synthesized from newly synthesized fatty acids, whereas thioesters of CoA could be substrates during renewal of phospholipid chains with preexisting fatty acids or for the incorporation of exogenous fatty acids taken up from the medium. In support of this proposal are the experiments of Overath, Pauli, and Schairer (80), which have indicated that the first enzyme of the β-oxidation pathway, acyl CoA synthetase, is required for the incorporation of exogenous fatty acids into phospholipids in vivo. Mutants deficient in this enzyme incorporate endogenously produced fatty acids into phospholipids normally but are unable to incorporate exogenous fatty acids.

Additional experiments testing the specificity of the sn-glycero-3-phosphate acyltransferase specificity have indicated that, although palmityl groups are found almost entirely in the 1-position of lysophosphatidate, when the acyl donor was an unsaturated thioester, the fatty acid of the lysophosphatidate produced was mainly in the 2-position (87,97,113). Studies of temperature-sensitive sn-glycero-3-phosphate acyltransferase mutants (87) have shown that a single enzyme catalyzes the transfer of unsaturated acyl groups to position 2 or palmitate to position 1 of sn-glycero-3-phosphate. In view of the stringent specificity that is also exhibited by lysophosphatidate acyltransferase, it is apparent that these in vitro reactions result in products that are remarkably consistent with the fatty acid composition of phospholipids found in intact cells (82,95,115).

Taylor and Heath (104) have shown a very selective transfer of long chain β-hydroxy fatty acids from ACP thioesters to the 1-position of sn-phosphatidyl-ethanolamine. However, the significance of this reaction is not clear, as β-hydroxy fatty acids are not found in phospholipids *in vivo* (55,58,79).

The fatty acid synthetase systems of both higher and lower plants exhibit properties very similar to that of *E. coli*, and they are dependent upon ACP (75,76,81). Extracts prepared from photoauxotrophic cells of *Euglena gracilis* synthesize fatty acyl ACP esters as products of the chloroplast fatty acid synthetase (27), they elongate medium chain fatty acyl ACP derivatives, and they catalyze the desaturation of stearyl ACP to oleate (75,76). Thus experiments were performed to determine if acyl ACP esters were acyl donors for complex lipid synthesis in this organism (88). In contrast to the bacterial systems discussed above, fatty acyl ACPs did not serve as substrates for phospholipid synthesis, whereas acyl groups were transferred from CoA thioesters to phospholipids. On the other hand, these extracts did catalyze the transfer of stearyl and oleyl groups from thioesters of *E. coli* ACP to monogalactosyl diglycerides. Since this incorporation was stimulated by sn-glycero-3-phosphate, it was felt to represent net synthesis of galactolipid. As in the case of phosphatidate synthesis in bacterial extracts, CoA esters, as well as ACP esters, were active acyl donors in monogalactosyl diglyceride synthesis.

B. ENZYME-BOUND 4'-PHOSPHOPANTETHEINE IN CYCLIC PEPTIDE BIOSYNTHESIS

The presence of 4'-phosphopantetheine has been demonstrated in the multienzyme complexes that synthesize gramicidin S and tyrocidine (29,51). The growth of *Bacillus Brevis* in the presence of ^{14}C-pantothenate led to the incorporation of radioactivity into a protein of 460,000 molecular weight. Microbiological assay revealed the presence of 1 mole pantothenate/mole protein. This protein has binding capacity for D- and L-phenylalanine, L-asparagine, L-glutamine, L-tyrosine, L-valine, L-ornithine, and L-leucine, the amino acids found in tyrocidine. Two other proteins, a light fraction (mol. wt. = 100,000) and an intermediate fraction (mol. wt. = 230,000) were necessary to observe tyrocidine synthesis. The light fraction bound D- or L-phenylalanine;

the intermediate fraction, proline. The initial binding of these eight amino acids occurred through sulfhydryl groups that probably were contributed by cysteine. Transfer of the dipeptide Phe-Pro to the charged heavy fraction initiated polymerization, with the 4'-phosphopantetheine, presumably acting in a manner somewhat analogous to the 4'-phosphopantetheine of yeast fatty acid synthetase (51). Similar considerations apply to the case of gramicidin S biosynthesis (29). Thus 4'-phosphopantetheine is thought to serve as a flexible arm that is utilized to transfer sequentially the growing peptide chain from one site to another.

VII. Intracellular Localization of ACP in E. coli

As discussed above, the fatty acid synthetases of bacteria and plants are exemplified by the system studied in detail in E. coli; they are nonassociated in that the component proteins fail to show any tendency to associate in vitro after the cell membranes of the organism are disrupted. The individual soluble enzymes can be isolated by conventional means; ACP is present in the free state, associated with neither the cell membrane nor the enzymes of the fatty acid synthetase system. Yet in these fatty acid synthetases, as in the case of the fatty acid synthetase multienzyme complexes, ACP must interact specifically and consecutively with all the biosynthetic enzymes. In spite of this, the process of fatty acid biosynthesis in E. coli is apparently very efficient, since no acyl ACP intermediates are found in extracts made from cells grown in normal conditions. The possibility was considered that the fatty acid synthetase of E. coli might exhibit some kind of structural organization in vivo. Experiments were designed to localize ACP, a component of the fatty acid synthetase with a unique prosthetic group that facilitates detection, in the bacterial cell. For the localization of ACP, E. coli auxotrophs, requiring either pantothenate or β-alanine, were utilized. When these mutants were grown on limiting concentrations of radioactive pantothenate or β-alanine and allowed to remain in stationary phase for some time, the level of CoA dropped dramatically and the pantothenate or β-alanine was present almost exclusively in ACP (7,84). Initial experiments were conducted to determine if the labeled ACP was present in the periplasmic space (114). Cells, subjected to the osmotic shock treatment described by Heppel (44), retained 92% of the radioactivity inside the shocked cells; while

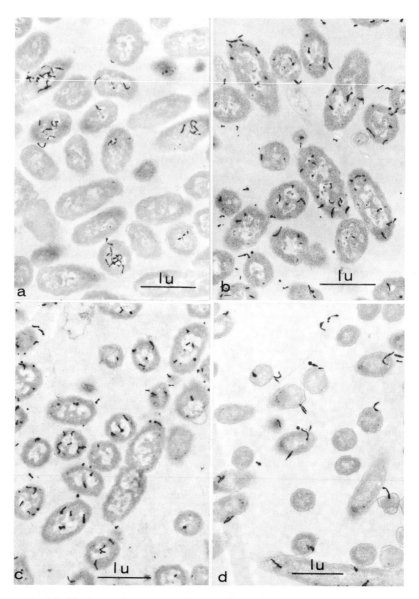

Fig. 10. Electron microscopy and autoradiography of tritiated cells. Washed cells were fixed in OsO_4 and processed for autoradiography, exposure time 6 months. Cells were grown in the presence of the following label: (a) [methyl-^{3}H]-thymidine (6.7 Ci/mM); (b) L-^{3}H-tryptophan (2.5 Ci/mM); (c) ^{3}H-β-alanine (5.2 Ci/mM) with cells containing 11% of the ^{3}H in ACP and 89% in CoA; and (d) ^{3}H-β-alanine (5.2 Ci/mM) with cells containing 89% of the ^{3}H in ACP and 11% in CoA. From van den Bosch, Williamson, and Vagelos (114).

306

81 % of the 5'-nucleotidase, a typical periplasmic enzyme, was released into the osmotic shock fluid. Thus ACP is not in the periplasmic space. In order to localize ACP within the cell by electron microscopy and autoradiography, the cells were grown in the presence of β-alanine of very high specific radioactivity (5.2 Ci/mM) in such a way that 11, 68, or 89 % of the radioactivity was in ACP, the remainder being in CoA in each instance. The distribution of grains, representing β-particle tracks, was determined by electron microscopy and autoradiography. The distance from the grains to the cell surface was measured, and the data were treated statistically. In order to ascertain that the methods utilized would adequately delineate specific cellular areas, a number of controls were added. Figure 10(a) demonstrates that cells in which the DNA was labeled with [methyl-^{3}H]-thymidine had grains predominating over the nuclear areas. Figure 10(b) shows that cells grown on L-^{3}H-tryptophan, in order to label all proteins in general, contained grains over the cytoplasm. In Figure 10(c) are shown the cells grown on ^{3}H-β-alanine with 11 % of the ^{3}H in ACP and 89 % in CoA; in Figure 10(d) are shown the cells grown on ^{3}H-β-alanine with 89 % of the ^{3}H in ACP and 11 % in CoA. It is apparent that the grain distribution of the cells with 89 % of the ^{3}H-β-alanine in CoA was very similar to the distribution of grains in ^{3}H-tryptophan cells. In other words, CoA is distributed in the cytoplasm as are the general proteins of the cell. The grain distribution in the cells with 89 % of the ^{3}H-β-alanine in ACP [Fig. 10(d)] is different from all the others as the majority of the grains are noted to be close to the surface of the cell.

Statistical analyses of all preparations verified the fact that ACP is located on or near the inside surface of the plasma membrane. This location of ACP is consistent with its role in the synthesis of fatty acids and phospholipids that are present almost exclusively in the envelope of E. coli. In addition, the fact that ACP is not randomly distributed in the cell suggests a degree of organization for a critical component of the fatty acid synthetase which had not been appreciated before. This finding suggests that the E. coli fatty acid synthetase may be organized in vivo, perhaps in a typical multienzyme complex that includes ACP.

References

1. Abita, J. P., Lazdunski, M., and Ailhaud, G. P., *Eur. J. Biochem*, *23*, 412 (1971).

2. Ailhaud, G. P., and Vagelos, P. R., *J. Biol. Chem.*, *241*, 3866 (1966).
3. Ailhaud, G. P., Vagelos, P. R., and Goldfine, H., *J. Biol. Chem.*, *242*, 4459 (1967).
4. Alberts, W. A., Goldman, P., and Vagelos, P. R., *J. Biol. Chem.*, *238*, 557 (1963).
5. Alberts, A. W., Majerus, P. W., Talamo, B., and Vagelos, P. R., *Biochemistry*, *3*, 1563 (1964).
6. Alberts, A. W., Majerus, P. W., and Vagelos, P. R., *Biochemistry*, *4*, 2265 (1965).
7. Alberts, A. W., and Vagelos, P. R., *J. Biol. Chem.*, *241*, 5201 (1966).
8. Alberts, A. W., and Vagelos, P. R., *Proc. Nat. Acad. Sci. U.S.*, *59*, 561 (1968).
9. Allmann, D. W., Hubbard, D. D., and Gibson, D. M., *J. Lipid Res.*, *6*, 63 (1965).
10. Baddiley, J., and Thain, E. M., *J. Chem. Soc.*, *246*, 2253 (1951).
11. Barnes, E. M., Jr., and Wakil, S. J., *J. Biol. Chem.*, *243*, (1969).
12. Bloch, K., *Accounts Chem. Res.*, *2*, 193 (1969).
13. Bressler, R., and Wakil, S. J., *J. Biol. Chem.*, *236*, 1643 (1961).
14. Brindley, C. N., Matsumura, S., and Bloch, K., *Nature*, *224*, 666 (1969).
15. Brodie, J. D., Wasson, G., and Porter, J. W., *J. Biol. Chem.*, *239*, 1346 (1964).
16. Brock, D. J., and Bloch, K., *Biochem. Biophys. Res. Commun.*, *23*, 775 (1966).
17. Brown, G. M., *J. Biol. Chem.*, *234*, 379 (1959).
18. Burton, D. N., Haavik, A. G., and Porter, J. W., *Arch. Biochem. Biophys.*, *126*, 141 (1968).
19. Butterworth, P. H., Yang, P. C., Bock, R. M., and Porter, J. W., *J. Biol. Chem.*, *242*, 3508 (1967).
20. Carey, E. M., and Dils, R., *Biochim. Biophys. Acta.*, *210*, 371 (1970).
21. Chesterton, C. J., Butterworth, P. H. W., and Porter, J. W., *Arch. Biochem. Biophys.*, *126*, 864 (1968).
22. Cuatrecasas, P., *J. Biol. Chem.*, *245*, 3059 (1970).
23. Davis, B. D., *Experientia*, *6*, 41 (1950).
24. Delo, J., Ernst-Fonberg, M. L., and Bloch, K., *Arch. Biochem. Biophys.*, *143*, 385 (1971).
25. Donaldson, W. E., Wit-Peeters, E. M., and Scholte, H. R., *Biochem. Biophys. Acta*, *202*, 35 (1970).
26. Elovson, J., and Vagelos, P. R., *J. Biol. Chem.* *243*, 3603 (1968).
27. Ernst-Fonberg, M. L., and Bloch, K., *Arch. Biochem. Biophys.*, *143*, 392 (1971).
28. Gelhorn, A., and Benjamin, W., *Science*, *146*, 1166 (1964).
29. Gilhuus-Moe, C. G., Kristensen, T., Bredesen, J. R., Zimmer, F. L., and Laland, S. G., *Fed. Eur. Biochem. Soc. Lett.*, *7*, 287 (1970).
30. Goldfine, H., *J. Biol. Chem.*, *241*, 3864 (1966).
31. Goldfine, H., Ailhaud, G. P., and Vagelos, P. R., *J. Biol. Chem.*, *242*, 4466 (1967).
32. Goldman, P., *J. Biol. Chem.*, *239*, 3663 (1964).

33. Goldman, P., Alberts, A. W., and Vagelos, P. R., *J. Biol. Chem.*, *238*, 1255 (1963).
34. Goldman, P., Alberts, A. W., and Vagelos, P. R., *J. Biol. Chem.*, *238*, 3579 (1963).
35. Goldman, P., and Vagelos, P. R., *Biochem. Biophys. Res. Commun.*, *5*, 414 (1962).
36. Greenfield, N., Davidson, B., and Fasman, G. D., *Biochemistry*, *6*, 1630 (1967).
37. Greenspan, M. D., Alberts, A. W., and Vagelos, P. R., *J. Biol. Chem.*, *244*, 6477 (1969).
38. Greenspan, M. D., Birge, C. H., Powell, G., Hancock, W. S., and Vagelos, P. R., *Science*, *170*, 1203 (1970).
39. Guchhait, R. B., Putz, G. R., and Porter, J. W., *Arch. Biochem. Biophys.*, *117*, 541 (1966).
40. Gutte, B., and Merrifield, R. B., *J. Biol. Chem.*, *246*, 1922 (1971).
41. Hancock, W. S., Prescott, D. J., Nulty, W. L., Weintraub, J., Vagelos, P. R., and Marshall, G. R., *J. Amer. Chem. Soc.*, *93*, 1799 (1971).
42. Hansen, H. J. M., Carey, W. M., and Dils, R., *Biochem. Biophys. Acta*, *210*, 400 (1970).
43. Harlan, W. R., and Wakil, S. J., *J. Biol. Chem.*, *238*, 2316 (1963).
44. Heppel, L. A., *Science*, *156*, 1451 (1967).
45. Hogness, D. S., Cohen, M., and Monod, J., *Biochim. Biophys. Acta*, *16*, 99 (1955).
46. Hsu, R. V., Wasson, G., and Porter, J. W., *J. Biol. Chem.*, *240*, 3736 (1965).
47. Ilton, M., Jevans, A. W., McCarthy, E. D., Vance, D., White, H. B., III, and Bloch, K., *Proc. Nat. Acad. Sci. U.S.*, *68*, 87 (1971).
48. Jacob, E. J., Butterworth, P. H. W., and Porter, J. W., *Arch. Biochem. Biophys.*, *124*, 392 (1968).
49. Joshi, V. C., Plate, C. A., and Wakil, S. J., *J. Biol. Chem.*, *245*, 2857 (1970).
50. Joshi, V. C., and Wakil, S. J., *Arch. Biochem. Biophys.*, *143*, 493 (1971).
51. Kleinkauf, H., Gevers, W., Roskoski, Jr., R., and Lipmann, F., *Biochem. Biophys. Res. Commun.*, *41*, 1218 (1970).
52. Knivett, V. A., and Cullen, J., *Biochem. J.*, *103*, 299 (1967).
53. Larrabee, A. R., McDaniel, E. G., Bakerman, H. A., and Vagelos, P. R., *Proc. Nat. Acad. Sci. U.S.*, *54*, 267 (1965).
54. Lehninger, A. L., and Greville, G. D., *Biochim. Biophys. Acta*, *12*, 188 (1953).
55. Law, J. H., *Bacteriol. Proc.*, 129 (1961).
56. Lennarz, W. J., Light, R. L., and Bloch, K., *Proc. Nat. Acad. Sci. U.S.*, *48*, 840 (1962).
57. Lorch, E., Abraham, S., and Chaikoff, I. L., *Biochim. Biophys. Acta*, *70*, 627 (1963).
58. Luderitz, O., Jann, K., and Wheat, R., in M. Florkin and E. H. Stotz, Eds., *Comprehensive Biochemistry*, Vol. 26A, Elsevier, Amsterdam, 1968, p. 105.
59. Lynen, F., *Fed. Proc.*, *20*, 941 (1961).
60. Lynen, F., Oesterhelt, D., Schweizer, E., and Willecke, K., *Cellular Compartmentalization and Control of Fatty Acid Metabolism*, Academic Press, New York, 1968, p. 1.

61. Majerus, P. W., J. Biol. Chem., 242, 2325 (1967).
62. Majerus, P. W., Science, 159, 428 (1968).
63. Majerus, P. W., Alberts, A. W., and Vagelos, P. R., Proc. Nat. Acad. Sci. U.S., 51, 1231 (1964).
64. Majerus, P. W., Alberts, A. W., and Vagelos, P. R., J. Biol. Chem., 240, 618 (1965).
65. Majerus, P. W., Alberts, A. W., and Vagelos, P. R., Proc. Nat. Acad. Sci. U.S., 53, 410 (1965).
66. Majerus, P. W., Alberts, A. W., and Vagelos, P. R., J. Biol. Chem., 240, 4723 (1965).
67. Majerus, P. W., Alberts, A. W., and Vagelos, P. R., Biochem. Prep., 12, 56 (1968).
68. Majerus, P. W., and Vagelos, P. R., Advan. Liquid Res., 5, 1 (1967).
69. Marshall, G. R., Hancock, W. S., Prescott, D. J., Nulty, W. L., Weintraub, J., and Vagelos, P. R., Int. Peptide Cong., Vienna, Austria, (1971).
70. Martin, D. B., Horning, M. G., and Vagelos, P. R., J. Biol. Chem., 236, 663 (1961).
71. Matsumura, S., Biochem. Biophys. Res. Commun., 38, 238 (1970).
72. Matsumura, S., Brindley, D. N., and Bloch, K., Biochem. Biophys. Res. Commun., 38, 369 (1970).
73. Merrifield, R. B., J. Amer. Chem. Soc., 85, 2149 (1963).
74. McIlwain, H., Biochem. J., 40, 269 (1946).
75. Nagai, J., and Bloch, K., J. Biol. Chem., 241, 1925 (1966).
76. Nagai, J., and Bloch, K., J. Biol. Chem., 242, 357 (1967).
77. Numa, S., Bortz, W. M., and Lynen, F., Adv. Enzyme Regulation, 3, 407 (1965).
78. Oesterhelt, D., Bauer, H., and Lynen, F., Proc. Nat. Acad. Sci. U.S., 63, 1377 (1969).
79. Osborn, M. J., Ann. Rev. Biochem., 38, 501 (1969).
80. Overath, P. O., Pauli, G., and Schairer, H. V., Eur. J. Biochem., 7, 559 (1969).
81. Overath, P. O., and Stumpf, P. K., J. Biol. Chem., 239, 4103 (1964).
82. Peypoux, F., and Michel, G., Biochim. Biophys. Acta, 218, 453 (1970).
83. Plate, C. A., Joshi, V. C., and Wakil, S. J., J. Biol. Chem., 245, 2868 (1970).
84. Powell, G. L., Elovson, J., and Vagelos, P. R., J. Biol. Chem., 244, 5616 (1969).
85. Prescott, D. J., and Vagelos, P. R., J. Biol. Chem., 245, 5484 (1970).
86. Prescott, D. J., Elovson, J., and Vagelos, P. R., J. Biol. Chem., 244, 4517 (1969).
87. Ray, T. K., Cronan, J. E., Mavis, R. D., and Vagelos, P. R., J. Biol. Chem., 245, 6441 (1970).
88. Renkonen, O., and Bloch, K., J. Biol. Chem., 244, 4899 (1969).
89. Robinson, J. R., Bradley, R. M., and Brady, R. O., J. Biol. Chem., 238, 528 (1963).
90. Sauer, F., Pugh, E. L., Wakil, S. J., Delaney, R., and Hill, R. L., Proc. Nat. Acad. Sci. U.S., 52, 1360 (1964).
91. Scheuerbrandt, G. H., Goldfine, H., Baronowsky, P. E., and Bloch, K., J. Biol. Chem., 236, PC70 (1961).

92. Schulz, H., and Wakil, S. J., *J. Biol. Chem.*, *246*, 1895 (1971).
93. Schulz, H., Weeks, G., Toomey, R. E., Shapiro, M., and Wakil, S. J., *J. Biol. Chem.*, *244*, 6577 (1969).
94. Schweizer, E., Piccinini, F., Duba, C., Gunther, S., Ritter, E., and Lynen, F., *Eur. J. Biochem.*, *15*, 4836 (1970).
95. Silbert, D. F., *Biochemistry*, *9*, 3631 (1970).
96. Simoni, R. D., Criddle, R. S., and Stumpf, P. K., *J. Biol. Chem.*, *242*, 573 (1967).
97. Sinensky, M., *J. Bacteriol.*, *106*, 449 (1971).
98. Smith, S., and Dils, R., *Biochim. Biophys. Acta*, *116*, 23 (1966).
99. Stadtman, E. R., in *Methods in Enzymology*, Vol. 4, S. P. Colowick and N. O. Kaplan, Eds., Academic Press, New York, 1957, p. 228.
100. Stoll, E., Ryder, E., Edwards, J. B., and Lane, M. D., *Proc. Nat. Acad. Sci. U.S.*, *60*, 986 (1968).
101. Sumper, M., Oesterhelt, D., Riepertinger, C., and Lynen, F., *Eur. J. Biochem.*, *10*, 377 (1969).
102. Takagi, T., and Tanford, C., *J. Biol. Chem.*, *243*, 6432 (1968).
103. Taketa, K., and Pogell, B. M., *J. Biol. Chem.*, *241*, 720 (1966).
104. Taylor, S. S., and Heath, E. C., *J. Biol. Chem.*, *244*, 6605 (1969).
105. Toomey, R. E., and Wakil, S. J., *Biochim. Biophys. Acta*, *116*, 189 (1965)
106. Toomey, R. E., and Wakil, S. J., *J. Biol. Chem.*, *241*, 1159 (1966).
107. Tweto, J., Liberti, M., and Larrabee, A. R., *J. Biol. Chem.*, *246*, 2468 (1971).
108. Vagelos, P. R., *Ann. Rev. Biochem.*, *33*, 139 (1964).
109. Vagelos, P. R., *Current Topics in Cellular Regulation*, *3*, 119 (1971).
110. Vagelos, P. R., and Larrabee, A. R., *J. Biol. Chem.*, *242*, 1776 (1967).
111. Vagelos, P. R., Majerus, P. W., Alberts, A. W., Larrabee, A. R., and Ailhaud, G. P., *Fed. Proc.*, *25*, 1485 (1966).
112. Vanaman, T. C., Wakil, S. J., and Hill, R. L., *J. Biol. Chem.*, *243*, 6420 (1968).
113. van den Bosch, H., and Vagelos, P. R., *Biochim. Biophys. Acta*, *218*, 233 (1970).
114. van den Bosch, H., Williamson, J. R., and Vagelos, P. R., *Nature*, *228*, 338 (1970).
115. van Golde, L. M. G., and van Deenen, L. L. M., *Chem. Phys. Lipids*, *1*, 157 (1967).
116. Wakil, S. J., in *Lipid Metabolism*, S. J. Wakil, Ed., Vol. 48, Academic Press, New York, 1970.
117. Wakil, S. J., McClain, L. W., Jr., and Warshaw, J. B., *J. Biol. Chem.*, *235* PC231 (1960).
118. Wakil, S. J., Pugh, E. L., and Sauer, F., *Proc. Nat. Acad. Sci. U.S.*, *52*, 106 (1964).
119. Weeks, G., and Wakil, S. J., *J. Biol. Chem.*, *243*, 1180 (1968).
120. Wells, W. W., Schultz, J., and Lynen, F., *Proc. Nat. Acad. Sci. U.S.*, *56* 633 (1966).
121. Willecke, K., Ritter, E., and Lynen, F., *Eur. J. Biochem.*, *8*, 503 (1969).
122. Williamson, I. P., and Wakil, S. J., *J. Biol. Chem.*, *241*, 2326 (1966).
123. Wit-Peeters, E. M., *Biochim. Biophys. Acta*, *176*, 453 (1969).

AUTHOR INDEX

Numbers in parentheses are reference numbers and show that an author's work is referred to although his name is not mentioned in the text. Numbers in *italics* indicate the pages on which the author is mentioned in the text or the full references appear.

Abeles, R. H., 214(178,179), *256, 257*
Abita, J. P., 277(1), 278(1), *307*
Abraham, K. A., 243(1), *251*
Abraham, S., *309*
Abrams, A., 241(2), *251*
Abzug, S., 227(20), 236(19), *251*
Achee, F., 235(466a), *267*
Acs, G., 18(56), *26*
Adachi, O., 224(3,456,457,459,460), *251, 267*
Adams, M., 93(50), 101(50), 113(50), *124*
Adiga, P. R., 244(4), *251*
Agarwal, K. L., 3(10), *25*
Agrawal, K. M. L., 93(20,94), 96(20,94), 97(20), 104(20,94), 111(20), 112(94), 113(20,94), *123, 125*
Agrell, I., 243, *256*
Agrell, I. P. S., 238(5), 243(5), *251*
Ahmed, M. U., 93(70), 96(70), 113(70), *125*
Ailhaud, G. P., 271(3), 277(1), 278(1), 288(3), 289, 290(3), 302(2,31), *307, 308, 311*
Akanuma, Y., 51(72), 81(72), *86*
Akeson, W., 43(38), *86*
Akopyan, Z. I., 233(6), *251*
Alarcon, R. A., 236(7-10), *251*
Albers, M., 243(471), *268*
Albersheim, P., 93(26), *124*
Alberts, A. W., 269(4,33,34), 270(4,33, 34,63,67), 271(63), 272(63,65,66), 273(66), 275(4), 283(65), 285(7), 287(7), 292(5,63), 294(5,63), 295(5, 6,37), 296(37), 298(5), 299, 300(5,

64), 305(7), *308-311*
Alberty, R. A., 119(241), *129*
Albon, N., 105, *128*
Alder, A. J., 82(152), *89*
Alexander, H. C., 116(217), *129*
Algranati, I. D., 244(11), *251*
Aliapoulios, M. A., 81(148), *89*
Alivisatos, S. G. A., 184(2), 186(2), 188-(2), 189(2), *200*
Allende, C. C., 245(44), *252*
Allende, J. E., 245(44), *252*
Allman, D. W., *308*
Alps, H., 93(57), *124*
Ames, B. N., 208(12,13), 237(12,117), 242, *251, 255*
Aminoff, D., 136(22), *148*
Anagnostopoulos, C., 93(73-75), 105-(73,74), 106(75), *125*
Anders, M., 20(63), 21(63), *26*
Andersson, B., 41(31), 84, *85*
Andersson, G., 227(14a), *251*
Andrews, P., 99(183), *128*
Anfinsen, C. B., 30(1,3,5), 34(1,3,5), 38(1), 42(1), 43(1,5), 45(1), 46(1,5), 47(62), 50(5), 51(5,69), 52(5), 53(5), 57(84-86), 58(1,84-87,89), 62(1), 63, 72(1), 73(1), 75(1,62,144), 77, 80-(144,145), 81(5,69), 83(3), *85-87, 89*
Antonini, E., 233(264c,265), 234(264c, 265), *260*
Appel, H., 101(185), *128*
Applebaum, D., 211(273), 216(273), 217(273), 218(273), *261*
Aramaki, Y., 247(160b), *256*
Archambault, A., 104(195), *128*

313

SUBJECT INDEX

Advances in Enzymology

CUMULATIVE INDEXES, VOLUMES 1–36

A. Author Index

VOL. PAGE

VOL. PAGE

VOL.   PAGE

VOL. PAGE

VOL.   PAGE

B. Subject Index

VOL.   PAGE

VOL.   PAGE